SCIENTIFIC AND TECHNICAL INFORMATION RESOURCES

BOOKS IN LIBRARY AND INFORMATION SCIENCE

A Series of Monographs and Textbooks

EDITOR

ALLEN KENT

Director, Office of Communications Programs
University of Pittsburgh
Pittsburgh, Pennsylvania

Vol. 1 Classified Library of Congress Subject Headings, Volume 1—Classified List, *edited by James G. Williams, Martha L. Manheimer, and Jay E. Daily (out of print)*
Vol. 2 Classified Library of Congress Subject Headings, Volume 2—Alphabetic List, *edited by James G. Williams, Martha L. Manheimer, and Jay E. Daily*
Vol. 3 Organizing Nonprint Materials, *by Jay E. Daily*
Vol. 4 Computer-Based Chemical Information, *edited by Edward McC. Arnett and Allen Kent*
Vol. 5 Style Manual: A Guide for the Preparation of Reports and Dissertations, *by Martha L. Manheimer*
Vol. 6 The Anatomy of Censorship, *by Jay E. Daily*
Vol. 7 Information Science: Search for Identity, *edited by Anthony Debons (out of print)*
Vol. 8 Resource Sharing in Libraries: Why • How • When • Next Action Steps, *edited by Allen Kent (out of print)*
Vol. 9 Reading the Russian Language: A Guide for Librarians and Other Professionals, *by Rosalind Kent*
Vol. 10 Statewide Computing Systems: Coordinating Academic Computer Planning, *edited by Charles Mosmann (out of print)*
Vol. 11 Using the Chemical Literature: A Practical Guide, *by Henry M. Woodburn*
Vol. 12 Cataloging and Classification: A Workbook, *by Martha L. Manheimer (out of print; see Vol. 30)*
Vol. 13 Multi-media Indexes, Lists, and Review Sources: A Bibliographic Guide, *by Thomas L. Hart, Mary Alice Hunt, and Blanche Woolls*
Vol. 14 Document Retrieval Systems: Factors Affecting Search Time, *by K. Leon Montgomery*
Vol. 15 Library Automation Systems, *by Stephen R. Salmon*
Vol. 16 Black Literature Resources: Analysis and Organization, *by Doris H. Clack*
Vol. 17 Copyright—Information Technology—Public Policy: Part I—Copyright—Public Policies; Part II—Public Policies—Information Technology, *by Nicholas Henry*
Vol. 18 Crisis in Copyright, *by William Z. Nasri*
Vol. 19 Mental Health Information Systems: Design and Implementation, *by David J. Kupfer, Michael S. Levine, and John A. Nelson*
Vol. 20 Handbook of Library Regulations, *by Marcy Murphy and Claude J. Johns, Jr.*
Vol. 21 Library Resource Sharing, *by Allen Kent and Thomas J. Galvin*
Vol. 22 Computers in Newspaper Publishing: User-Oriented Systems, *by Dineh Moghdam*
Vol. 23 The On-Line Revolution in Libraries, *edited by Allen Kent and Thomas J. Galvin*
Vol. 24 The Library as a Learning Service Center, *by Patrick R. Penland and Aleyamma Mathai*

Vol. 25 Using the Mathematical Literature: A Practical Guide, *by Barbara Kirsch Schaefer*
Vol. 26 Use of Library Materials: The University of Pittsburgh Study, *by Allen Kent et al.*
Vol. 27 The Structure and Governance of Library Networks, *edited by Allen Kent and Thomas J. Galvin*
Vol. 28 The Development of Library Collections of Sound Recordings, *by Frank W. Hoffmann*
Vol. 29 Furnishing the Library Interior, *by William S. Pierce*
Vol. 30 Cataloging and Classification: A Workbook Second Edition, Revised and Expanded, *by Martha L. Manheimer*
Vol. 31 Handbook of Computer-Aided Composition, *by Arthur H. Phillips*
Vol. 32 OCLC: Its Governance, Function, Financing, and Technology, *by Albert F. Maruskin*
Vol. 33 Scientific and Technical Information Resources, *by Krishna Subramanyam*

Additional Volumes in Preparation

SCIENTIFIC AND TECHNICAL INFORMATION RESOURCES

Krishna Subramanyam
School of Library and Information Science
Drexel University
Philadelphia, Pennsylvania

MARCEL DEKKER, INC.   New York and Basel

Library of Congress Cataloging in Publication Data

Subramanyam, K
   Scientific and technical information resources.

   (Books in library and information science ; v. 33)
   Bibliography: p.
   Includes indexes.
   1. Technical literature. 2. Scientific literature.
3. Reference books--Technology--Bibliography.
4. Reference books--Science--Bibliography. I. Title.
II. Series.
T10.7.S93   507   80-28531
ISBN 0-8247-1356-7

Copyright © 1981 by Marcel Dekker, Inc. All Rights Reserved

Neither this book nor any part may be reproduced or transmitted in any form or by any means, electronic or mechanical, including photocopying, microfilming, and recording, or by any information storage and retrieval system, without permission in writing from the publisher.

Marcel Dekker, Inc.
270 Madison Avenue, New York, New York 10016

Current printing (last digit):
10 9 8 7 6 5 4 3 2 1

Printed in the United States of America

PREFACE

The material presented in this book has its origins in the lectures given since 1973 to graduate students taking the Resources in Science and Technology course at the University of Pittsburgh and the Drexel University. The book is addressed mainly to two groups of users: (1) students and teachers of librarianship and information science, and (2) practicing librarians and technical information officers in science and engineering libraries and information centers.

Guides to the literature of science and technology can either be inventory guides, listing numerous publications (e.g., C. C. Chen's *Scientific and Technical Information Sources*), or expository guides containing narrative descriptions of sources (e.g., D. J. Grogan's *Science and Technology: An Introduction to the Literature*). Inventory guides are more useful as reference sources for practicing librarians than as textbooks for students and teachers of technical information. In view of the diversity and growing volume of reference works and other publications in science and technology, it is impossible for technical librarians and students of technical information to become familiar with large numbers of individual sources of information. Instead, a more useful approach would be to try to understand the total process of scientific and technical communication and the relationship between information needs and information sources.

A major feature of this book is the integration of the inventory approach and the expository approach, and the presentation of a didactic model for scientific and technical communication. The model includes a consideration of the various phases of scientific information—including its generation through R & D, and its recording, surrogation, synthesis, and dissemination. Information is seen as both the source material and the product of R & D activity. The emphasis in this integrative approach to technical information has been on: (1) an understanding of the information needs and information seeking (and dissemination) modes of scientists and engineers, and (2) an overview of the structure and characteristics of the totality of scientific and technical literature and other sources of information.

The emphasis throughout the book has been on current practices in scientific and technical communication, historical aspects, and characteristics and bibliographic control of various forms of scientific and technical literature. Science and engineering librarians and students of technical information cannot afford to remain oblivious to the rapid growth of computerized information dissemination systems and services. Accordingly, recent developments and current trends in the computerized bibliographic control and dissemination of scientific and technical information are also discussed.

Scientific and technical information is a global entity; its bibliographic control and dissemination are supranational concerns transcending linguistic and geographical barriers. Several examples of multinational bibliographic control systems and services are discussed. The products and services of numerous professional societies and national and international agencies including national patent offices and standards organizations are described.

Over 1500 sources have been listed under broad subject categories. These include tertiary sources (guides to literature, bibliographies of bibliographies), secondary sources (bibliographies and catalogs, abstracting and indexing services, databases, reference works of various types), and primary sources (journals, including translated journals). Almost all of the sources listed are English language sources produced in the United States or overseas. All the major branches of the physical and natural sciences and engineering and technology are covered, but health sciences and social sciences are excluded. No attempt has been made to evaluate the sources listed. The listings are illustrative, and not comprehensive.

I am grateful to Professor Allen Kent, Director, Office of Communications Programs, University of Pittsburgh, for his valuable support and encouragement in preparing this book. I wish to thank Alexis Swyderski, head librarian, Yeadon Public Library, Yeadon, Pa., and Vibiana Bowman, systems analyst, Planning Research Corporation, Philadelphia, Pa., for their expert assistance in proofreading.

K. Subramanyam

CONTENTS

Preface   iii

1/ Scientific and Technical Communication   1

    1.1 Introduction   1
    1.2 Characteristics of Scientific Literature   2
    1.3 The Structure of Scientific Literature   4
    1.4 Scientific vs. Technical Literature   10
    1.5 Nonformal Communication   13
    1.6 Information Exchange Groups   15
    1.7 Use of Scientific Literature   16
    References   18

2/ Scientific Societies   21

    2.1 Introduction   21
    2.2 The Royal Society   22
    2.3 Scientific Societies in the United States   24
    2.4 Publications of Scientific Societies   26
    References   28

3/ The Primary Journal   30

    3.1 Historical Overview   30
    3.2 Functions of the Primary Journal   32
    3.3 Problems of the Primary Journal   33
    3.4 Current Trends in Journal Publishing   43
    3.5 Alternatives to the Scientific Journal   47
    3.6 The Future of the Scientific Journal   51
    3.7 Bibliographic Control of Journals   55
    References   61

4/ Conference Literature — 66

- 4.1 Scientific Conferences — 66
- 4.2 Preconference Literature — 68
- 4.3 Literature Generated During the Conference — 69
- 4.4 Postconference Literature — 69
- 4.5 Bibliographic Control of Conference Literature — 69
- 4.6 UNESCO-FID Recommendations — 74
- References — 75

5/ Dissertations, Theses, and Research in Progress — 77

- 5.1 Dissertations — 77
- 5.2 Master's Theses — 79
- 5.3 Foreign Dissertations and Theses — 80
- 5.4 Bibliographic Control of Dissertations and Theses — 82
- 5.5 Research in Progress: SSIE — 85
- References — 87

6/ Patents — 88

- 6.1 Introduction — 88
- 6.2 The United States Patent and Trademark Office (PTO) — 89
- 6.3 Foreign Patents — 91
- 6.4 Patents as a Source of Technological Information — 92
- 6.5 Bibliographic Control of Patents — 93
- Publications of the Patent and Trademark Office — 97
- List of Patent Depositories in the United States — 98
- References — 99

7/ Technical Reports — 100

- 7.1 Introduction — 100
- 7.2 History of Report Literature — 100
- 7.3 Characteristics of Technical Reports — 104
- 7.4 Security Classification — 109
- 7.5 Technical Report Numbers — 110
- 7.6 Bibliographic Control of Technical Reports — 116
- References — 129

8/ Standards and Specifications — 132

- 8.1 Introduction — 132
- 8.2 Specifications — 134
- 8.3 Types of Standards and Specifications — 135

Contents vii

    8.4   Sources of Standards and Specifications   137
    References   148

9/ House Journals   149

    9.1   Introduction   149
    9.2   Internal House Journals   150
    9.3   External House Journals   150
    9.4   Bibliographic Control of House Journals   153
    References   154

10/ Trade Catalogs   155

    10.1   Introduction   155
    10.2   Characteristics and Types of Trade Catalogs   157
    10.3   Acquisition and Control of Trade Catalogs   164
    References   165

11/ Bibliographical Literature   166

    11.1   Introduction   166
    11.2   General Biographical Works   166
    11.3   Specialized Biographical Works   167
    11.4   Biographical Serials   168
    11.5   Collective Biographies   170
    11.6   Biographical Monographs and Autobiographies   171
    11.7   Other Sources of Biographical Information   172
    11.8   Bibliographic Control of Biographical Literature   172
    Selected List of Biographical Works   173
    References   178

12/ Dictionaries and Thesauri   179

    12.1   Dictionaries   179
    12.2   Thesauri   181
    12.3   Bibliographies of Dictionaries   181
    Selected List of Dictionaries   183

13/ Directories and Yearbooks   193

    13.1   Directories   193
    13.2   Yearbooks   195
    13.3   Bibliographies of Directories   196
    Selected List of Directories   196

14/	**Handbooks and Tables**	**208**
	14.1 Handbooks	208
	14.2 Tables	210
	14.3 National Standard Reference Data System (NSRDS)	212
	Selected List of Handbooks and Tables	214
	References	229
15/	**Encyclopedias**	**231**
	15.1 Specialized Encyclopedias	231
	15.2 Single-Volume Encyclopedias	232
	15.3 Multivolume Encyclopedias	233
	15.4 Encyclopedic Dictionaries	234
	15.5 Updating Encyclopedias	235
	Selected List of Encyclopedias	236
16/	**Review Literature**	**242**
	16.1 Introduction	242
	16.2 Review Authors and Review Preparation	244
	16.3 Functions of Reviews	247
	16.4 Characteristics of Reviews	250
	16.5 Types and Sources of Reviews	251
	16.6 Bibliographic Control of Reviews	255
	Selected List of Review Serials	257
	References	263
17/	**Translations**	**266**
	17.1 Introduction	266
	17.2 Translated Journals	268
	17.3 Bibliographic Control of Translations	270
	Indexes of Translations	273
	Selected List of Translated Journals	274
	References	282
18/	**Bibliographic Control of Scientific and Technical Literature**	**283**
	18.1 Proliferation of Literature	283
	18.2 Bibliographies	285
	18.3 Abstracting Services	293
	18.4 Indexing Services	295
	18.5 Characteristics of Abstracting and Indexing Services	297
	18.6 Guides to Literature	304

Contents

18.7 Information Analysis Centers		307
18.8 Decentralized Bibliographic Control		308
Selected List of Guides to Literature, Bibliographies, Catalogs, and Abstracting and Indexing Services		310
References		338

19/ Current Trends and Prospects — 341

19.1 Introduction		341
19.2 Integrated Primary and Secondary Publishing		342
19.3 Computer-Based Bibliographic Control Systems		343
19.4 Online Access to Scientific and Technical Literature		346
19.5 Science Information: A Global Concern		350
Selected List of Bibliographic Databases Available for Online Searching		353
References		360

Bibliography	361
Appendix: Abbreviations	383
Author Index	387
Subject Index	401

SCIENTIFIC AND
TECHNICAL INFORMATION
RESOURCES

1

SCIENTIFIC AND TECHNICAL COMMUNICATION

1.1 Introduction

The date of the first scientific writing is not known precisely. Contributions to science were made by the early civilizations of Assyria, Babylonia, China, Egypt, and India. In these early civilizations, knowledge was transmitted largely through oral communication, and the fragments of papyri and cuneiform clay tablets that are extant from these periods do not give us a precise picture of the pattern of scientific communication during these early periods. The invention of the moveable type by Gutenberg in 1455 was a landmark event in the history of written communication. The printing press made it possible to prepare and disseminate multiple copies of manuscripts.

During the sixteenth and seventeenth centuries, great advances were made in intellectual, economic, technological, and social spheres by natural philosophers such as Francis Bacon and René Descartes, who placed great emphasis on the scientific method of inquiry. During this period, written communication was largely through books and gazettes. The book was not particularly suited to the rapid dissemination of new ideas, since the author had to work for several years and accumulate enough results to warrant publication of a book. Accounts of single observations and discoveries began to be disseminated through booklets or pamphlets. For example, William Harvey's work on the circulation of blood was published as a 72-page booklet in 1628 [1].

Though many changes have taken place in recent times in the modes of dissemination of scientific information, the basic function of scientific literature, namely, to serve as a foundation for advances in science, has remained unchanged. In his opening address to the Royal Society Scientific Information Conference, 1948, Sir Robert Robinson, President of the Royal Society, said [2]:

The sciences have deep human interest and are not devoid of spiritual value. The object of our Founders was declared to be the improvement of natural knowledge. By that they meant, and we still do mean, improvement and spread of knowledge of nature. Neither they could, nor we can, condone the scientific miser who investigates for his own satisfaction, or profit, and keeps the results to himself for selfish reasons, whether they be aesthetic or economic.

Faraday expressed it very well (he always did) when he described the three necessary stages of useful research—the first to begin it, the second to end it, and the third to publish it.

Sir Robinson was reiterating the principle that the march of science rests on its published record, and that ready access to scientific and technical information is a fundamental need of scientists everywhere. More recently, Elmer Hutchisson, director of the American Institute of Physics, asserted his "conviction that the written record of the accomplishments of scientific research constitutes one of civilized man's most important intellectual resources" [3]. Scientists constantly draw upon this growing volume of records, and also strive to contribute their individual share, however small, to the total body of recorded knowledge.

1.2 Characteristics of Scientific Literature

Scientific knowledge is the objective knowledge of the universe and its phenomena, generated by the scientific method of inquiry and validated to conform with empirical observations of natural phenomena. Every new addition to the store of objective knowledge is an extension of the existing body of knowledge as recorded in the primary literature of science. The new knowledge so developed is recorded on tangible media and thus adds to the stockpile of scientific literature. Therefore, scientific literature, which embodies the existing store of objective knowledge, is at once the foundation on which the incremental progress of science rests and also a product of such advances in scientific knowledge. In the humanities, new developments do not necessarily replace past achievements: Bernard Shaw's plays do not make Shakespeare's plays obsolete, and Picasso's paintings do not replace those of Rembrandt. But the nature of the objective knowledge of science is quite different; each incremental advancement in scientific knowledge in some way adds to, modifies, refines, or sometimes totally refutes the prior knowledge on which the advancement was based to begin with. Einstein's general theory of relativity is an extension and a generalization of Newton's classical mechanics; the heliocentric theory of Copernicus rejected and replaced Ptolemy's geocentric theory, then prevailing. This noncumulative quality of science is shared by the literature

1.2 Characteristics of Scientific Literature

of science; hence the clamor of scientists and other users of scientific information for the most recent literature.

The second important attribute of science, which is shared to a large extent by the literature of science, is its universality. Scientific truth is "supranational," and transcends political, sociological, cultural, and linguistic limitations, although these factors influence the organizational dynamics of scientific research in any given society. For example, the organization of scientific research activity in the United States is different from that in the Soviet Union because of the vast differences in the political ideologies and socioeconomic infrastructures of these two countries. However, Soviet physics could not be different from American physics inasmuch as the laws of physics, regardless of the nationality of the physicists who discover them, or of the language in which they are expressed, are as immutable as the natural phenomena they depict. Any aberration that may be deliberately or inadvertently superimposed upon scientific truth by political demagoguery or other ideological considerations, as exemplified by the Lysenko affair in the USSR, is bound to be discovered and rejected sooner or later [4].

Likewise, scientific literature, which is a record of the objective knowledge generated by science, is quintessentially universal, although there may be vast differences in its language, bibliographic format, and physical medium. These differences can be resolved by appropriate transformation (e.g., translation and reformatting), and then the scientific literature produced in one country can be used by the scientists of another country. The abstracts in *Referativnyi Zhurnal*, a Russian abstract journal, are translated and incorporated into *Applied Mechanics Reviews* of the American Society of Mechanical Engineers. Many physics journals produced in the USSR are translated from cover to cover by the American Institute of Physics, Plenum Publishing Corporation, and other agencies, and are made available to English-speaking physicists throughout the world. This could not have been done if scientific literature, like science itself, were not essentially universal. The same cannot be said of the literature of other branches of knowledge, however. Some branches of the social sciences and the humanities are more or less culture specific, and are not transplantable across cultural-geographical interfaces. Islamic law, for example, cannot be practised in the United States, even if books on Islamic law can be translated into English, because of the culture dependency of law. Such translations are useful, though, for academic pursuits.

Scientific literature is the validated record of the achievements of science. Traditionally, scientists have been zealous in guarding the high standards of scholarship and quality of the work reported in scientific literature. Research articles submitted by scientists for publication in scholarly journals are refereed by a panel of experts to ensure accuracy and quality. In order to obtain impartial assessment of the manuscripts, the refereeing process is usually done anonymously. The author is not aware of the identity of the referees, and in some cases, the referees also do not know the identity of the author. Scientific societies play a dominant and useful role in maintaining this tradition of validation of scientific literature.

Scientific literature is also a "public" record of scientific knowledge. The channels of communication (e.g., the primary journal and the conference platform) are accessible to anyone who satisfies the requirements of quality as set forth by scientists themselves. Also, the literature of science is "public" in another sense: With the exception of documents containing proprietary matter or information pertaining to national safety, the literature of science is accessible by anyone for use. Very elaborate bibliographic control mechanisms have been set up to promote easy and rapid access to scientific literature. Since science is sustained by its own literature, the accessibility of scientific literature is crucial for the unimpeded growth of science.

1.3 The Structure of Scientific Literature

The structure of scientific literature can best be understood by tracing the progression of scientific information from its generation as a result of research and development (R & D) endeavors through its dissemination in primary literature, its surrogation in secondary services, and its eventual integration and compaction in reviews, textbooks, and encyclopedias. Figure 1.1 is a schematic diagram of a bibliographic chain showing the progression of scientific information from the idea stage until the new information generated is disseminated through various channels and eventually becomes an integral part of prior scientific knowledge. The numbers within the small inner circles indicate the time frame in years, with research starting from the idea stage at time zero. The products or bibliographic packages emanating from each of the activities are shown in boxes connected to the activity circles in the diagram.

1.3.1 Primary Sources

Unpublished Documents: Primary information derived from R & D activity can be communicated by a variety of channels. When the investigation is still in progress, there is a continual interaction among the members of the research team, and between the research team and members of the larger scientific community who may be interested in the research. At this stage, information flows in both directions: as input to the research team in the form of data and ideas from the scientific community, and as output from the research team in the form of experimental data and preliminary findings. Such interaction almost always takes place through nonformal channels (e.g., oral

1.3 The Structure of Scientific Literature

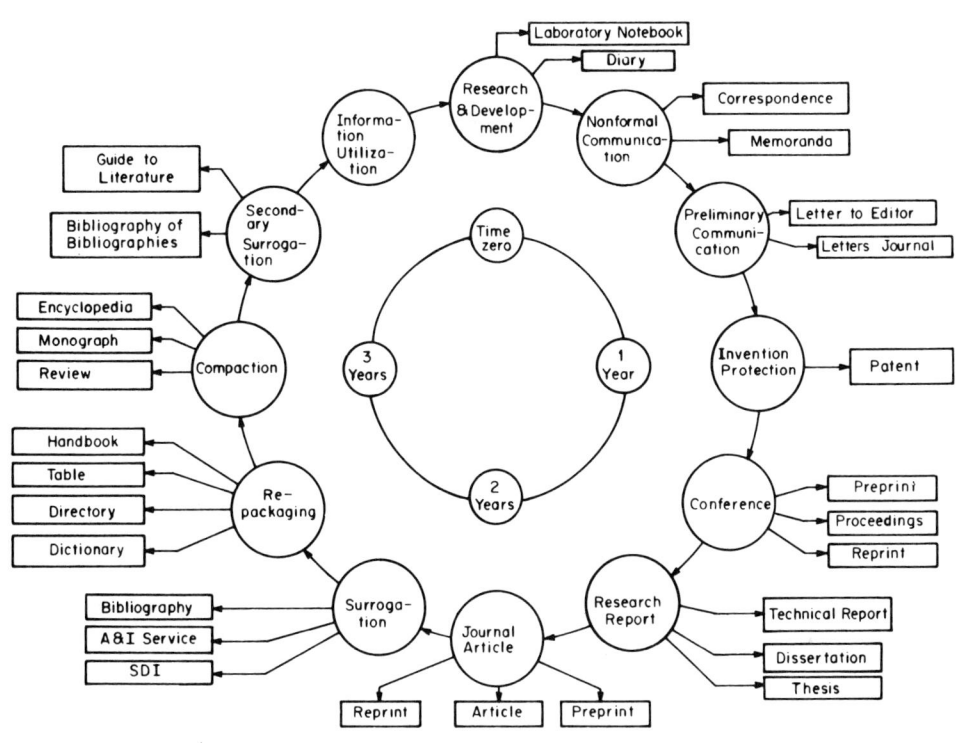

Figure 1.1 The evolution of scientific information. (From *Encyclopedia of Library and Information Science*, vol. 26.)

communication, informal notes and memoranda, and correspondence), and no formal public records are generated. The laboratory notebooks, diaries, or journals maintained by scientists are not published as such, but form the basis for formal primary publication in the form of conference papers, technical reports, dissertations, or journal articles.

Preliminary Communication: When the research investigation has reached a sufficiently advanced stage, scientists prefer to communicate the preliminary findings through the formal channel, in the form of a letter or short communication in a primary journal (such as *Science* or *Nature*), or of a short article in one of the "letters" journals. This communication forms the earliest formal contribution to the existing body of scientific literature. The primary aim of this preliminary communication, which is sometimes made even before the conclusion of the investigation, is to establish priority for an invention or breakthrough that may result from the investigation. A secondary purpose of such preliminary communication is to disseminate nascent information on current research to enable other members of the scientific community to remain abreast of current developments in the wavefront of knowledge. It is generally supposed that a short communication will eventually be followed by a full-length paper, reporting the complete and final findings of the investigation in a standard primary journal. But in practice, it has been found that full-length research papers are not always written as a followup, and the short communication may remain the only printed account of the investigation [5-7].

Patent Specification: Not all scientists, however, communicate the preliminary findings of their research endeavor through the formal channel. Where proprietary interests are involved, no information about the invention is published in the public domain, at least until the invention is protected by applying for a patent. The invention is then described in a patent specification, as required by the patent-granting authority (the Commissioner of Patents and Trademarks in the United States). When the patent is finally granted, the patent specification, printed and distributed by the Patent Office, becomes a part of the primary scientific literature.

Conference Literature: The next possible phase in the dissemination of scientific information is the presentation of a paper in a conference, which may give rise to three types of document: a preprint of the paper distributed by the author or the agency sponsoring the conference, the published proceedings of the conference including the edited version of the paper along with summaries of discussion and related material, and a reprint of the paper distributed during or after the conference. All these documents constitute further additions to the primary literature of science.

Research Report: If the research is done as a part of the requirements for an advanced academic degree, the results are presented in the form of a master's thesis or a doctoral dissertation. In the case of sponsored research, most sponsoring agencies (e.g., the National Science Foundation, the Environmental Protection Agency) require that the results of research be submitted in the form of a report at the conclusion of the research effort. Dissertations and technical reports are not usually regarded as formally published documents. They are, nonetheless, very important forms of primary literature, and constitute a sizeable fraction of the total output of scientific literature.

The Research Paper: The research paper published in a refereed primary journal may be said to be the most important basic bibliographic unit, constituting the bulk of the primary literature of science. In a sample study of the literature of computer science, journal articles were found to constitute slightly less than one half of the total volume of primary literature; they were followed by conference papers, books, technical reports, dissertations, and theses, in that order [8]. Like the conference paper, the journal article may be distributed at different stages—as a preprint distributed by the author before the publication

1.3 The Structure of Scientific Literature

of the paper in a journal, as a paper in a printed journal, and as a reprint of the published paper distributed by the author or the publisher.

The total bulk of the primary literature of science and technology consists of a variety of forms of literature besides those already mentioned. The following list represents the wide range of formats available for recording and communicating primary scientific information:

Laboratory notebooks, diaries, notes, medical records

Personal correspondence

Videotapes of experiments, surgical operations, and other scientific or industrial procedures

Graphs, charts, and tables, usually machine-generated during experiments

Transcripts, audio-, or videotapes of lectures and discussions

Motion pictures, slide-tape presentations, or other multimedia presentations depicting experiments or other scientific or industrial procedures

Internal research reports, memoranda, company files

Patent specification

Computer programs (on punched cards, magnetic media, or as computer printout)

Letter to the editor or short communication in primary journal

Preliminary communication in "letters" journal

Preprint or reprint of conference paper

Conference paper and proceedings

Technical report

Thesis or dissertation

Journal article, preprint, or reprint

Newsletter

Standard, specification, or code of practice

House journal

Trade catalog (literature describing products and services, disseminated by manufacturers or distributors)

The characteristics and bibliographic control of some of the principal forms of primary scientific and technical literature will be described in later chapters.

1.3.2 Secondary Sources

The discrete bibliographic units or documents comprising the bulk of the primary literature of science are scattered in a diverse variety of sources published throughout the world in scores of languages. Since the progress of science depends on the utilization of existing scientific knowledge, the tasks of identifying, selecting, and digesting pertinent information from the mass of scientific literature are important phases of the total process of scientific research and development. Surrogation of primary literature, by creating current awareness services, bibliographies, indexes, abstracts, and catalogs, facilitates the identification and selection of pertinent documents appropriate for a given purpose. Repackaging the contents of primary documents into directories, handbooks, yearbooks, and the like provides rapid access to the desired specific piece of information, without the searcher having to wade through an unorganized mass of literature. Digestion and assimilation of scientific knowledge are facilitated by (1) compaction of primary literature into reviews and digests, and (2) integration of new concepts with existing knowledge in treatises, textbooks, and encyclopedias. The processes of surrogation, repackaging, and compaction of the primary literature of science result in the creation of a variety of secondary sources (Figure 1.2).

1.3.3 Tertiary Sources

As discussed elsewhere, scientific literature has been growing at an exponential rate to unmanageable proportions owing to a multitude of factors, including an increase in the number of productive scientists engaged in research and writing, and the current practice of assessing the productivity of a scientist in terms of the number of publications produced. As a result of this proliferation of the primary literature of science, the quantity and diversity of secondary publications have also reached such proportions as to warrant further surrogation of the secondary publications to facilitate identification of appropriate secondary and primary sources for satisfying the information needs of users. The product of this secondary surrogation (Figure 1.2) consists of a variety of tertiary sources such as bibliographies of bibliographies, lists of indexing and abstracting services, directories of directories, and guides to literature. The characteristics and bibliographic control of the principal types of secondary publications will be discussed in later chapters.

Primary, secondary, and tertiary literature may be in any one or more of the following physical media:

Print media (books, journals, and other printed documents)

Near-print media (computer printout; mimeographed, multilithed, or typed documents)

Graphic media (photographs, charts, drawings, etc.)

Microform (microfiche; microfilm strip, reel, or cartridge; microcard; microprint of various sizes and reduction ratios)

1.3 The Structure of Scientific Literature

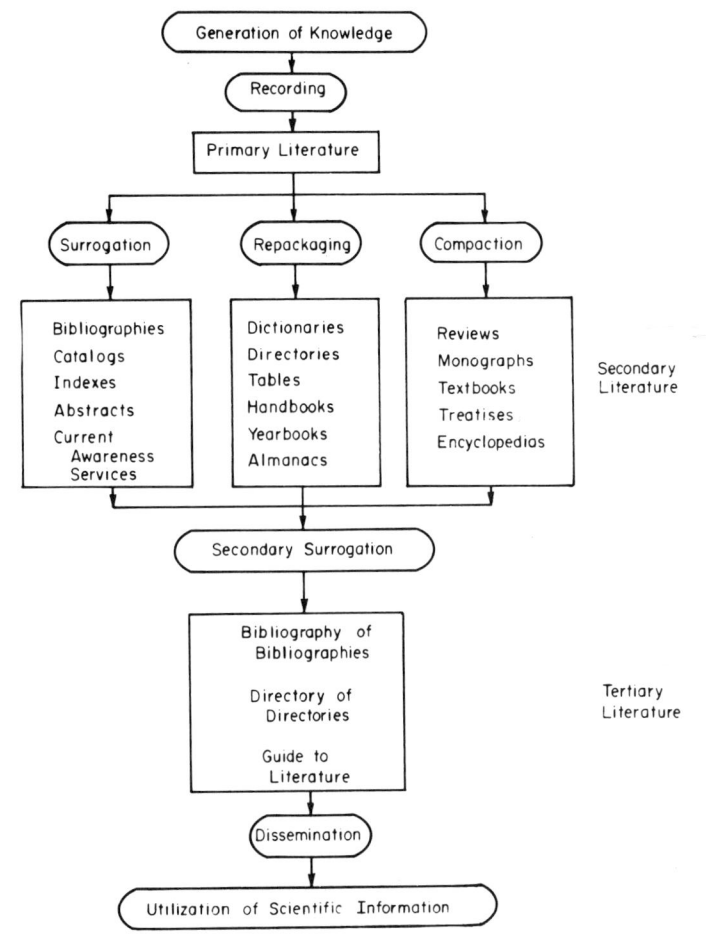

Figure 1.2 The structure of scientific literature. (From *Encyclopedia of Library and Information Science*, vol. 26.)

Machine-readable media (punched card; punched paper tape; magnetic tape, strip, chip, or disk)

Composite media (e.g., aperture card)

The total structure of the literature of science may thus be seen as being made up of three hierarchical levels of publications: (1) primary publications in which new concepts are recorded and disseminated; (2) secondary publications which are derived from the primary publications by the processes of surrogation, repackaging, and condensation; and (3) tertiary publications which are derived by further surrogation of secondary literature.

1.4 Scientific vs. Technical Literature

There seem to be some qualitative differences not only between the types of literature used by scientists and technologists as input into their research and development endeavors, but also between the types of literature generated by them as the product of their efforts. Derek J. de Solla Price has tried to articulate the distinction between scientists and technologists, but the distinction is admittedly not very clear: "They are quite different social groups, comprising, on the one hand, the people who create new knowledge—the scientists, theoretical and applied—and, on the other, those who make new things, new chemicals, new machines: the engineers and technologists. The dividing line is by no means clear; there are many people with scientific and technical skill and training who make nothing new, adding neither to our knowledge nor to our artifacts, but work, with their know-how, well behind the research front" [9].

Investigations on the use of scientific and technical information have indicated consistent differences in the sources of information used by scientists and technologists. In a massive study of information transfer patterns in industrial organizations, Rosenbloom and Wolek obtained data from 1900 engineers and scientists in 13 establishments of four large corporations, and 1200 members of the Institute of Electrical and Electronics Engineers (IEEE). Engineers were found to depend more heavily on internal (intracorporate) sources of information than on outside sources; scientists tend to make substantially more use of sources outside the corporation [10].

Marked differences are also discernible in the types of literature used by scientists and engineers. For "competence-oriented" communication involving such activities as browsing, reviewing new developments in a field, becoming familiar with a new field, and so on, both scientists and engineers depend heavily on documentary sources. But consistent differences may be seen in the nature of these documentary sources. Whereas the documents read by scientists are almost entirely made up of professional literature such as books and journal articles, those used by engineers are about equally divided between professional literature and technical literature, the latter consisting of trade magazines, technical reports, and trade catalogs. The same difference in the nature of documents used is apparent in "problem-oriented" communication in which documents are used for seeking a specific piece of information rather than for current awareness or background appreciation. Scientists make greater use of professional literature, and engineers consult corporate reports and trade catalogs [11].

1.4 Scientific vs. Technical Literature

Another survey by R. A. Davis involving 1800 engineers showed some very interesting results [12]:

1. Over 83% of all engineers surveyed (and 90.6% of the chemical engineers) used handbooks. But only 33.47% of the engineers (and 51.46% of the chemical engineers) used abstracts and indexes. This indicates that engineers are more concerned with problem-oriented communication (in which information is sought for a specific problem or purpose) than with competence-oriented communication for current awareness or background appreciation in a subject.

2. Material published outside the United States was used by 30.71% of the engineers; only 22.14% of the engineers used translations. Once again this indicates that engineers depend heavily on locally produced materials for satisfying their information needs.

3. Research reports and standards and specifications were used heavily: 62.51% of the engineers said they used standards and specifications, and 58.14% of the engineers used research reports. Again, research reports were used more heavily by chemical engineers than by engineers specializing in other branches.

4. Over 85% of the engineers used manufacturers' catalogs, referred to the catalogs at least once a week, and also kept a personal collection of catalogs.

These data support the view that engineers make extensive use of technical literature including technical reports, standards and specifications, and trade catalogs.

To understand the differences between the literature used by scientists and engineers, one has only to observe the differences in the environments in which the two groups function, and in their goals and attitudes. Whereas a scientist is likely to see himself as a member of a larger group of fellow-scientists who share his research interests and attitudes, regardless of organizational or geographic limitations, an engineer tends to identify himself as a member of a smaller team within his immediate working environment. A scientist demonstrates a greater allegiance to the larger social system of science, transcending institutional or even national boundaries [13,14]. But engineers and technologists engaged in similar pursuits in different corporations tend to see each other as competitors rather than as collaborators in a larger community of engineers.

The outlook of an engineer in a corporation is characterized by the norm of allegiance to his group within the corporation and not by the norm of autonomy which is typical of a scientist [15]. He cannot speak freely on sensitive issues, especially if his company is in some way involved in the issue. When the company engineer participates in professional activities such as serving on a standards committee for formulating a technical standard, he seems to do so as an individual engineer rather than as a company representative. Both the Society of Automotive Engineers (SAE) and the American Society of

Mechanical Engineers (ASME) claim that their members serving on standards committees do not represent their companies [16]. It is also true that corporations often encourage their engineers to participate in professional society activities. Many companies pay membership dues, provide paid time for attending society meetings, and give financial support to attend conferences for presenting papers.

There are also consistent differences in the factors that influence the choice of R & D projects by scientists and engineers. The engineer is more or less compelled to choose application-oriented projects that can generate a measurable or tangible outcome. In corporations doing contractual research for other agencies, the projects are suggested by the sponsoring agencies, and the engineer has very limited freedom in choosing projects. Scientists seem to have more freedom in the selection of research projects; organizational constraints exert a less dominant influence than personal interests and preferences in their choice of research projects. In a survey of 57 academic scientists in six science departments of a university, the scientists were asked how important freedom was in research. Thirty-five of the fifty-seven respondents felt that for maximum effectiveness of research, unlimited freedom in the choice of problems was necessary. Fourteen of these scientists considered that such freedom was absolutely essential, while others felt that freedom was "quite necessary," or "merely desirable" [17]. The following three comments typified the views of those academic scientists who considered that unlimited freedom in research was indispensable:

> I think that the moment somebody walks in this door and says, "Here is a problem that the department would sort of like to have you think about. Think about it," I would feel unhappy. My happiness would actually be shot to hell.
>
> (A mathematician)

> I think in my own case that, if I were not permitted to decide what sort of problems I wish to investigate and how I might pursue them, I would not care for research at all, and would take up farming or something.
>
> (A biochemist)

> Independence of what I could do has been a very important factor always. Independence in that I could go any place. I could refuse an offer. I could take up any problem that I wished, or drop it. This demand for personal independence applies to research, but it also applies to everything else, too.
>
> (A physicist)

These differences in the attitudes and organizational settings of scientists and technologists are also reflected in the products of their R & D efforts. Typically, the scientist would disseminate the findings of his research to as

1.5 Nonformal Communication

large an audience as possible and as rapidly as possible. Sometimes, preliminary results of research that is still in progress are published as articles in "letters" journals, or as brief communications in primary journals. Results of research are also presented at conferences and disseminated through nonformal channels before they are published in the form of an article in a primary journal. In other words, the scientist considers the wide dissemination of his findings a necessary part of his commitment to science. Also, the scientist's desire to publish his research results widely (by distributing preprints and reprints, for example) stems from his desire to establish priority and to secure peer approval for his work.

In sharp contrast to this approach of the scientist, the technologist is not quite free to communicate the results of his R & D effort, because of the nature of his activity and also because of organizational constraints. In the first place, the end result of the technologist's developmental activity may be a design of a new or improved product, process, or system, which may not be amenable to presentation in the form of an article in a journal. "The names of Wilbur and Orville Wright are not remembered because they published papers. The technologist's principal legacy to posterity is encoded in physical not verbal structure" [18]. Secondly, even if the results of the technologist's work are amenable to presentation in a paper, the information generated may be of a proprietary nature and may have to be kept confidential to protect the interests of the corporation. In the case of sponsored research, confidentiality may be a contractual requirement stipulated by the sponsoring agency. The engineer is more likely to write a technical report to be submitted to the sponsoring agency, or to apply for a patent to protect his invention from unauthorized exploitation. In fact, many industrial corporations routinely require new entrants to sign a document agreeing to assign to the corporation any patent that may result from their work. The reward for the technologist comes mainly from the employer, and is likely to be more tangible in nature than the reward for the pure scientist who feels gratified by recognition and approbation from his peers. "Both scientists and technologists work in fiercely competitive worlds, but in science the competition is among individuals for prestige, and in technology, it is among corporations for profit" [19].

1.5 Nonformal Communication

It is known that a great deal of information transfer takes place directly from the generators of information to its ultimate users in "invisible colleges" and through "technological gatekeepers." The whole phase of surrogation and dissemination through formal channels is thus bypassed. In fact, studies reported by Voigt, Loosjes, and others have shown that libraries and other formally structured environments are not the most favored channels for the transfer of a great deal of nascent information among scientists and technologists [20,21].

The importance of interpersonal channels of communication has been emphasized in many surveys. Project Hindsight, a Department of Defense study of the origins of information and ideas that were important in the development of 20 successful weapons systems, showed that in 70% of the cases, personal contact was the means through which information was introduced into the using organization [22]. In his study of the information-gathering habits of American medical scientists, Herner found that in 46% of the incidents sampled, medical scientists received the idea for their research through interpersonal channels [23]. Rosenbloom and Wolek reported that engineers were more dependent than scientists on interpersonal contacts for acquisition of useful information [24]. The preferred channels for nonformal communication within the scientific community appear to be the following:

Personal discussion with coworkers in the immediate working environment

Discussion with members of outside organizations in a variety of settings ranging from the utterly informal mixing in "hospitality suites" in conferences to more formal colloquia and summer schools

Personal correspondence

Informal presentation of research results to groups within and outside one's own organization

Direct exchange of preprints and reprints among scientists acquainted with each other

Exchange of reprints, preprints, and memoranda in informal information exchange groups

The following are some of the advantages of nonformal communication:

Promptness: Nonformal communication can take place through face-to-face or telephonic conversation, personal correspondence, and preprint exchange, all of which are faster than dissemination through the formal channels.

Selectivity: Formal channels (e.g., the scientific journal) are designed to reach large audiences, and cannot therefore be sensitive to individual needs. Information transmitted through the nonformal channel is specifically for an individual recipient or a small group.

Interactive Communication: In the nonformal context (e.g., telephonic conversation), continuous interaction between the supplier and receiver of information is possible. This feature is very difficult to achieve in the formal channels.

Screening and Evaluation: The supplier of information provides evaluated and predigested information that can be readily used by the recipient. Books

1.6 Information Exchange Groups

and other products of the formal channel often contain raw data which have to be interpreted, synthesized, and evaluated before they can be used.

Transmission of Sensitive Information: In the nonformal mode, a scientist may not hesitate to communicate opinions and experiences which are too personal or information which is too sensitive to be communicated through the formal channels. For example, a scientist who has had bad experience with the products or services of a company may not venture to publish his experience in the correspondence columns of a journal for fear of inviting unpleasant reactions, but he does not usually mind sharing his thoughts with fellow scientists during nonformal communication.

Personal Appeal: Scientists communicating in the nonformal mode can establish a personal rapport among themselves; this is difficult to achieve in the formal channels.

The methodologies and implications of nonformal communication have been discussed in recent literature on science communication [25-31].

1.6 Information Exchange Groups

In the 1960s an interesting experiment was supported by the National Institutes of Health of the United States to facilitate rapid communication of research information among scientists by creating a formal organization for the exchange of preprints. The experiment, hailed as "one of the most revolutionary innovations in the history of science communication," consisted of setting up a series of Information Exchange Groups (IEGs) for different fields of inquiry. Membership in the group was free and open to any scientist actively engaged in research, and each group had a chairman who would ensure smooth functioning of the group. Any member could submit a written communication for distribution to the other members of the group. The IEG head office in Bethesda, Maryland would make copies of the communications, called IEG Memoranda, and mail them to the group members without charge. There was no restriction on the material submitted for distribution. Copies of papers submitted for publication in primary journals, preliminary reports of unfinished research, comments on other communications, reviews, abstracts, notes on events, and even inquiries were accepted and distributed as IEG Memoranda, without any editorial scrutiny. In all, seven IEGs were established and were in operation between 1961 and 1967.

During the initial years of the project, the IEGs were successful and became quite popular with the participating scientists, mainly because of the speed with which papers could be transmitted to their peers through the IEGs. In fact, the IEGs became so popular that both the membership and the number of communications submitted for distribution grew to unmanageable proportions, and caused the eventual abandonment of the whole project. IEG No. 1, for example, distributed only 27 papers during the first two years of its life, and during the remaining four years, it distributed 800 papers. During the first year, the average membership per group was 56, and the average number of papers distributed was ten. By October 1966, the total membership was 3625 and the number of memoranda circulated rose to an average of 151 per month. During the year 1966, more than 1.5 million copies of some 2000 memoranda were sent out at a cost of $416,000. It was projected that if the IEGs continued to grow at the rate then current, in the next two years the total number of members could reach 14,000 and the total number of copies of memoranda to be distributed would approach the staggering figure of 30 million. The annual cost of maintaining this mammoth organization was variously estimated to be in the range of ten million to one hundred million dollars. There was clearly no possibility of raising this sum of money every year from any source on a continuing basis. Some members expressed serious concern over the overwhelming inundation of memoranda, many of which were of questionable quality and doubtful utility. The uncontrolled growth of the IEGs to unmanageable dimensions caused other problems, including an unacceptable delay in the duplication and transmission of the memoranda, a suspected lowering of the standards of scholarship presumably due to the absence of any editorial scrutiny, and a strong wave of opposition from a group of editors of primary journals. Thus, despite its initial success, the IEG project was called off by the National Institutes of Health in 1967. The main conclusion of the experiment was that the IEG concept was workable, and the IEGs could be a valuable adjunct to complement the primary journal provided that compact groups could be built around well-defined problems or phenomena under active investigation by a small number of scientists [32].

1.7 Use of Scientific Literature

Creating and using scientific literature are important components of the activities of scientists. In an operations research study of the dissemination of information, presented at the International Conference on Scientific Information, Halbert and Ackoff reported that research scientists spend, on an average, about one half of their total time in scientific and business communication, which includes both receiving and disseminating information [33]. According to other estimates, the average professional person spends from 25 to 75% of his time trying to keep abreast of new developments in his field. "A working scientist spends up to one third of his time searching for information, and the cost of this search represents one fifth of all the money allocated to science" [34].

1.7 Use of Scientific Literature

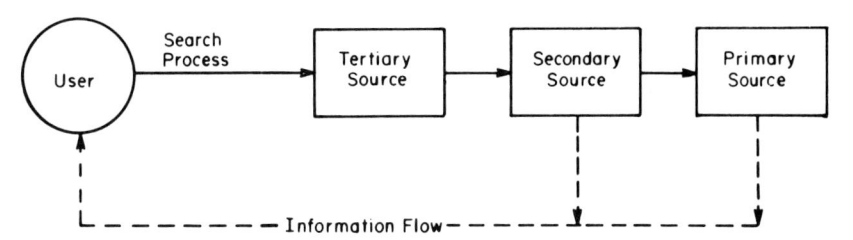

Figure 1.3 The search process and the direction of information flow. (From *Encyclopedia of Library and Information Science*, vol. 26.)

When the user of scientific information desires to acquire the literature pertinent to his needs, the process of literature search normally begins with a tertiary source (e.g., an appropriate guide to the literature of the discipline in question), and ends when a sufficient number of primary documents relevant to the question have been accumulated, as shown in Figure 1.3.

The end user, or the information professional assisting the end user, first consults a tertiary publication to identify appropriate secondary tools relevant to the search process. This is the first phase in the literature seeking process, and the product of this phase is a set of secondary publications. Some of the secondary publications such as directories, encyclopedias, or handbooks directly yield the desired information and there is no need to go further in the search process. Other types of secondary publications such as abstracting and indexing services, catalogs, and bibliographies are simply intermediate search tools that serve as a key to the primary literature. The second phase of the search process then consists of scanning the secondary sources to identify appropriate primary documents which are expected to yield the information that is being sought. The next phase consists of acquiring the primary documents selected, and then using the documents to obtain the desired information. The final product of the search (e.g., specific information obtained from directories or handbooks, or list of references selected from indexing and abstracting services) is usually screened and evaluated by the information professional for relevance to the inquiry before the product is delivered to the end user. These are the basic phases involved in the process of seeking and using scientific literature. In actual practice, this process can be quite complicated, involving such activities as question negotiation, query formulation using a controlled vocabulary, searching one or more computerized databases using an appropriate access language, translating foreign language documents or their abstracts into the user's language, and so on.

The actual process of utilizing scientific literature is one of the less well understood phases of scientific communication. Research on the expressed and implied needs and behavior patterns of information users has been going on to understand the user's motivation and methodology for seeking and using information [35]. Melvin J. Voigt has identified three types of information needs of scientists and engineers [36]. These may be defined as follows:

The Current Approach: This is the need for information about current R & D activities and their technical and socioeconomic implications, in one's own field of specialization as well as in peripheral fields.

The Everyday Approach: This is the need for specific piece of information essential to the day-to-day work of scientists and engineers. This need for a specific piece of information or data, a method, an equation, etc., is felt by scientists and engineers in the course of their daily work.

The Exhaustive Approach: This is the need to find and check through all of the relevant information existing on a given subject in order to determine the current state of the art in a given subject field, problem, or technology. This need arises when a researcher starts work on a new investigation, and when he reports on the results of an investigation in the form of a paper, a patent application, a technical report, or a dissertation.

Voigt has also identified the various channels of communication usually resorted to for the satisfaction of these needs. Nonformal channels are favored for satisfying the current approach and the everyday approach. In-house and external current awareness services and primary journals are used for keeping abreast of current developments. For finding specific data in the everyday approach, next to consultation with colleagues, various kinds of reference books such as dictionaries, directories, handbooks, and yearbooks are used. Indexing and abstracting services and bibliographies are used for exhaustive literature searches. Availability of large-scale machine-readable bibliographic databases for both offline and online searching has greatly enhanced the ease and rapidity with which retrospective searches can be made.

References

1. J. R. Porter, The scientific journal—300th anniversary, *Bacteriological Reviews*, 28(3):211-230 (September 1964).
2. *The Royal Society Scientific Information Conference, 21 June-2 July 1948. Report and Papers Submitted* (London: The Royal Society, 1948), p. 15.
3. Elmer Hutchisson, The role of international scientific organizations in improving scientific documentation, *Physics Today*, 15:24-26 (September 1962).
4. Zhores A. Medvedev, *The Rise and Fall of T. D. Lysenko* (Garden City, N.Y.: Doubleday, 1971). Translated by I. Michael Lerner.
5. Pauline Atherton, *The Role of "Letters" Journals in Primary Contributions of Information: A Survey of Authors of Physical Letters.* Report AIP/DRP-64-1 (New York: American Institute of Physics, 1964).

References

6. Pauline Kean and Jalrath Ronayne, Preliminary communications in chemistry, *Journal of Chemical Documentation*, 12(4):218-220 (November 1972).
7. K. Subramanyam and Constance J. Schaffer, Effectiveness of "letters" journals, *New Library World*, 75:258-259 (December 1974).
8. K. Subramanyam, Core journals in computer science, *IEEE Transactions on Professional Communication*, PC-19(2):22-25 (December 1976).
9. Derek J. de Solla Price, Measuring the size of science, *Proceedings of the Israel Academy of Sciences and Humanities*, 4(6):98-111 (1969).
10. Richard S. Rosenbloom and Francis W. Wolek, *Technology and Information Transfer. A Survey of Practice in Industrial Organizations* (Boston, Mass.: Harvard University, Graduate School of Business Administration, 1970).
11. Rosenbloom and Wolek (Ref. 10), pp. 39-42.
12. R. A. Davis, How engineers use literature, *Chemical Engineering Progress*, 61(3):30-34 (March 1965).
13. Norman W. Storer, Research orientations and attitudes towards team work, *IRE Transactions on Engineering Management*, EM-9(1):29-33 (March 1962).
14. Janice M. Ladendorf, Information flow in science, technology, and commerce, *Special Libraries*, 61:215-222 (May/June 1970).
15. William Evan, Role strain and the norm of reciprocity in research organizations, *American Journal of Sociology*, 68:346-354 (November 1962).
16. David Hemenway, *Industrywide Voluntary Product Standards* (Cambridge, Mass.: Ballinger Publishing Co., 1975), p. 85.
17. S. S. West, The ideology of academic scientists, *IRE Transactions on Engineering Management*, EM-7(2):54-62 (June 1960).
18. Donald G. Marquis and Thomas J. Allen, Communication patterns in applied technology, *American Psychologist*, 21:1052-1060 (1966).
19. Ladendorf (Ref. 14), p. 216.
20. Melvin J. Voigt, *Scientists' Approaches to Information* (Chicago, Ill.: American Library Association, 1961).
21. Th. P. Loosjes, *On Documentation of Scientific Literature* (London: Butterworths, 1973).
22. C. W. Sherwin and R. S. Inemson, *First Interim Report on Project Hindsight* (Summary) (Washington, D.C.: Office of the Director of Defense Research and Engineering, October 13, 1966).
23. S. Herner, The information gathering habits of American medical scientists, in *Proceedings of the International Conference on Scientific Information, Washington, D.C., November 16-21, 1958* (Washington, D.C.: National Academy of Sciences—National Research Council, 1959), vol. 1, pp. 277-285.
24. Richard S. Rosenbloom and Francis W. Wolek, *Technology, Information, and Organization: A Report to the National Science Foundation* (Boston, Mass.: Harvard University, Graduate School of Business Administration, 1967).

25. Diana Crane, *Invisible Colleges* (Chicago, Ill.: University of Chicago Press, 1972).
26. Susan Crawford, Informal communication among scientists in sleep research, *Journal of the American Society for Information Science*, 22:301-310 (September-October 1971).
27. William D. Garvey and Belver C. Griffith, Informal channels of communication in the behavioral sciences: Their relevance in the structuring of formal or bibliographic communication, in *The Foundations of Access to Knowledge: A Symposium*, edited by Edward B. Montgomery (Syracuse, N.Y.: Syracuse University, 1968), pp. 129-151.
28. Carnot E. Nelson and Donald K. Pollock, eds., *Communication Among Scientists and Engineers* (Lexington, Mass.: D. C. Heath & Co., 1970).
29. Robert R. Korfhage, Informal communication of scientific information, *Journal of the American Society for Information Science*, 25:25-32 (1974).
30. Derek J. de Solla Price, Some remarks on elitism in information and the invisible college phenomenon in science, *Journal of the American Society for Information Science*, 22:74-75 (March-April 1971).
31. Gerald Zaltman, A note on an international invisible college for information exchange, *Journal of the American Society for Information Science*, 25:113-117 (March-April 1974).
32. K. Subramanyam, Information Exchange Groups: An experiment in science communication, *Indian Librarian*, 29(4):159-164 (March 1975).
33. Michael H. Halbert and Russell L. Ackoff, An operations research study of the dissemination of information, in *International Conference on Scientific Information* (Ref. 23), pp. 97-130.
34. George Schussel, Advent of information and inquiry services, *Journal of Data Management*, 7(9):24-31 (September 1969).
35. Susan Crawford, Information needs and uses, in *Annual Review of Information Science and Technology* (White Plains, N.Y.: Knowledge Industry Publications, Inc., and American Society for Information Science, 1978), vol. 13, pp. 61-81.
36. Voigt (Ref. 20), pp. 20-33.

2

SCIENTIFIC SOCIETIES

2.1 Introduction

Developments in the production, organization, and use of scientific literature are inseparably connected with the history of scientific societies. Scholarly societies were in existence even in the earliest civilizations of China, Greece, Egypt, India, and Arabia. These societies were associations of scholars who shared their work, results, and aspirations with others of similar interests. In ancient Greece, there were many schools lead by stalwarts such as Thales (ca. 640-546 B.C.) and Pythagoras (ca. 582-497 B.C.); each of these eminent scholars gathered around him a number of students of mathematics and natural science. In the Academy of Plato (427-347 B.C.), zoology, botany, geography, mathematics, astronomy, and philosophy were studied. These schools were organizations for both learning and research, and in this sense may be considered to be the forerunners of modern scientific societies.

The academies of Alexandria founded by the Ptolemies were residential research institutions where renowned men of science were accommodated and supported by the kings, and thus were able to devote all their energies to study and research in the physical and medical sciences. Among the lasting contributions of the Alexandrian academies may be counted *Elements of Geometry* by Euclid and *On Floating Bodies* by Archimedes, which have been the foundations of subsequent developments in these branches of science. The history of these early scholarly societies and of their bibliographical contributions has been documented by Thornton and Tully in their excellent treatise *Scientific Books, Libraries and Collectors* [1].

Scientific societies remained relatively dormant between the suppression of the Platonic schools in A.D. 529 and their revival during the Renaissance. It is believed that the societies continued their deliberations secretly because of opposition from government and religious authorities. During the 16th and 17th centuries, Francis Bacon and other natural philosophers believed that concerted action of scientists was necessary to promote the experimental method of scientific inquiry. In his book entitled *The New Atlantis* propounding the experimental method of research, and also in his other writings, Francis Bacon stressed the necessity for an organization to coordinate the work of individual scientists. Scientists in Europe met, often secretly, to share and discuss their findings. Discussion on experiments and discoveries were often recorded, and copies of these minutes were sent to other centers. Latin was the language commonly used for such scholarly communication. These assemblies led to the establishment of many scientific societies, including the Royal Society in England and the Académie des Sciences in France—two of the greatest academies in the history of science.

2.2 The Royal Society

In the middle of the 17th century, small groups of scholars and philosophers began to meet in various places, including taverns, in London to discuss the experimental method of scientific inquiry propounded by Francis Bacon. These groups, which later became known as the "Invisible Colleges," could not meet openly and regularly because of the civil strife in England. After the civil war ended, these natural philosophers decided to establish a formal constitution, and on November 28, 1660, drew up a memorandum of association. This was the foundation of one of the greatest scientific societies, the Royal Society. On July 15, 1662 the Society received the Royal Charter from King Charles II. Lord Brouncker (the first president), Christopher Wren, Robert Hooke (the first curator), Henry Oldenburg, John Evelyn, Robert Boyle, and Edmond Halley (after whom the Halley's Comet is named), were some of the eminent scientists who founded the Royal Society [2]. In the roster of the presidents of the Royal Society may be seen the names of some of the most distinguished scientists of the world: Sir Humphry Davy, Sir George Gabriel Stokes, Lord Kelvin, Lord Rayleigh, and Sir Joseph John Thomson, to name only a few. Among the fellows of the Society may be counted such illustrious names as Michael Faraday, Charles Darwin, John Dalton, and Clerk Maxwell.

Isaac Newton was elected to the Royal Society in 1671. Newton was indifferent to the publication of his monumental work *Principia*, which has been acclaimed as one of the greatest scientific works ever published. It was published in 1687 by the special efforts of Edmond Halley, himself an outstanding astronomer and geophysicist, who was then assistant secretary of the Royal Society. The manuscript is now preserved by the Society as its most precious scientific treasure. The Society's Newton Collection also contains the reflecting telescope made by his own hand in 1671. Isaac Newton himself became the president of the Society in 1703 and held that office until his death in 1727.

2.2 The Royal Society

Some of the publications of the Royal Society are: *Philosophical Transactions of the Royal Society of London* (1665-), one of the oldest scientific journals still being published; *Proceedings of the Royal Society* (1832-); and a number of monographs such as John Evelyn's *Sylva, or a Discourse of Forest Trees* (1664), the first book printed by the Royal Society; Robert Hooke's *Micrographia, or Some Physiological Descriptions of Minute Bodies Made by Magnifying Glasses* (1665); and Newton's *Philosophiae Naturalis Principia Mathematica* (1687). The Society also published two monumental bibliographies of scientific literature: *The Catalogue of Scientific Papers* and the *International Catalogue of Scientific Literature*. These will be discussed in a later chapter on the bibliographic control of scientific literature.

The Royal Society's Scientific Information Conference held in 1948 may be said to be a landmark event in the history of scientific literature. This international conference was convened as a result of a recommendation from the Royal Society Empire Scientific Conference, 1946. The aim of the conference was to examine the possibility of improvements in the existing methods of production, collection, indexing, abstracting, and distribution of scientific literature. Many important recommendations were made and published in the report of the proceedings of the conference [3].

The Royal Society Scientific Information Conference was followed by the International Conference on Scientific Information, held in Washington, D.C., November 16-21, 1958. This Washington Conference was sponsored by the National Science Foundation, the National Academy of Sciences, the National Research Council, and the American Documentation Institute. The deliberations of the Conference were divided into the following seven areas:

1. Literature and reference needs of scientists: knowledge now available and methods of ascertaining requirements
2. The function and effectiveness of abstracting and indexing services
3. Effectiveness of monographs, compendia, and specialized centers; present trends and new and proposed techniques and types of services
4. Organization of information for storage and search; comparative characteristics of existing systems
5. Organization of information for storage and retrospective search; intellectual problems and equipment considerations in the design of new systems
6. Organization of information for storage and retrospective search; possibility for a general theory
7. Responsibilities of government, professional societies, universities, and industry for improved information services and research

The texts of the papers presented, along with a summary of discussion in each of the areas, were published in a two-volume proceedings [4].

2.3 Scientific Societies in the United States

The Boston Philosophical Society, established in 1683, was probably the first scientific society to be organized in the American colonies. But this society did not function for long. The American Philosophical Society was established in Philadelphia in 1743 by Benjamin Franklin. The members of this society discussed questions of natural philosophy, history, and politics, and carried out investigations in botany, medicine, mineralogy and mining, mathematics, chemistry, industrial arts, and agriculture. The *Transactions of the American Philosophical Society* began in 1771.

John Adams established the American Academy of Arts and Sciences in Boston in 1780. The objectives of this Academy were [5]:

> To promote and encourage the knowledge of the antiquities of America and of the natural history of the country, and to determine the uses to which the various natural productions of the country may be applied; to promote and encourage medical discoveries, mathematical disquisitions, philosophical inquiries and experiments; astronomical, meteorological and geographical observations, and improvements in agriculture, arts, manufactures and commerce, and in fine, to cultivate every art and science which may tend to advance the interest, honor, dignity and happiness of a free, independent, and virtuous people.

The *Memoirs* of the Academy started in 1785.

Many specialized societies at the state level mushroomed in the last quarter of the 18th century: the Medical Society of New Jersey (1776), the Massachusetts Medical Society (1781), the College of Physicians of Philadelphia (1787), the Philadelphia Society for Promoting Agriculture (1785), the Massachusetts Society for Promoting Agriculture (1792), the Chemical Society of Philadelphia (1792), and others. The Academy of Natural Sciences of Philadelphia, started in 1812, published the *Journal of the Academy of Natural Sciences* (1817-) for over a century. The Boston Society of Natural History, established in 1830, has been one of the most active local scientific societies. It is known for its library, its museum, and its journal, the *Boston Journal of Natural History*.

A number of major national societies were established in the United States in the 19th century. Mention should be made of the American Medical Association (1847), the American Dental Association (1859), the American Society of Civil Engineers (1852), the Smithsonian Institution (1846), the American Association for the Advancement of Science (1847), and the National Academy of Sciences (1863). The principal publications of the National Academy of Sciences are: the *Memoirs* (1866-), the *Biographical Memoirs* (1877-), the *Annual Reports* (1863-), the *Proceedings* (1863-1894), and the *Proceedings, New Series* (1915-).

2.3 Scientific Societies in the United States

Bates has noted three main developments during the period from the close of the Civil War to the close of the First World War [6]. These are: a tendency toward specialization, a drift in the direction of national centralization within the specialities, and formation of technological societies in response to the demands of the machine age. The American Chemical Society, which has assumed world leadership in chemical information handling and production of primary and secondary journals in chemistry, was established in 1876. Many national-level engineering societies came into existence during this period: the American Society of Mechanical Engineers (1880), the American Institute of Electrical Engineers (1884), the Society of Automotive Engineers (1904), the American Institute of Chemical Engineers (1908), and the Institute of Radio Engineers (1912). The American Institute of Electrical Engineers and the Institute of Radio Engineers have since merged to become the Institute of Electrical and Electronics Engineers (IEEE).

The trend toward specialized societies at the national level has continued in the recent decades. The Electron Microscope Society of America (1942), and the Society for Experimental Stress Analysis (1943) are typical examples of specialized societies.

Toward the close of the 19th century, it was found that professional engineering societies of national scope grew into centralized monolithic organizations, and could not be sensitive to the interests of remotely stationed members. In the early decades of the 20th century, a large number of regional associations of practising engineers sprang up in different states and regions. Professional engineers could belong to these local societies and actively participate in their deliberations. Some national associations recognized this need and established local chapters. The local chapters started publishing their own periodicals in addition to those published by the national associations. The publications of the regional societies and local chapters are generally intended to provide a forum for communication among members on a variety of professional and related matters. The emphasis is on news about persons and events, job opportunities, relationships between government and industries, employer-employee relations, and professional "shop talk," rather than on research investigations. Some of the earliest known publications of regional professional engineering associations are listed below:

1880 *Michigan Engineer.* Michigan Engineering Society, Detroit, Mich.
1915 *Minnesota Engineer.* Minnesota Federation of Engineering Societies
1917 *Cleveland Engineering.* Cleveland Engineering Society, Cleveland, Ohio
1925 *Illinois Engineer.* Illinois Society of Professional Engineers, Springfield, Ill.
1926 *Baltimore Engineer.* Engineering Society of Baltimore
1930 *Texas Engineer.* American Society of Civil Engineers, Texas Section

1936 *Detroit Engineer.* Engineering Society of Detroit
1939 *New Jersey Professional Engineer.* New Jersey Society of Professional Engineers, Trenton, N.J.
1942 *Texas Professional Engineer.* Texas Society of Professional Engineers, Austin, Tex.
1943 *Georgia Engineer.* Georgia Architectural Engineering Society, Atlanta, Ga.
1948 *Georgia Professional Engineer.* Georgia Society of Professional Engineers, Atlanta, Ga.
1948 *Midwest Engineer.* Western Society of Engineers, Chicago, Ill.

A detailed account of the historical development of American scientific societies may be found in *Scientific Societies in the United States* by Ralph S. Bates [6]. For current information on scientific and technical societies, the following directories are useful:

Encyclopedia of Associations (Detroit, Mich.: Gale Research Co., 1956-)

National Trade and Professional Associations of the United States and Canada, and Labor Unions, 1978. 13th annual edition (Washington, D.C.: Columbia Books, 1978).

World Guide to Scientific Associations and Learned Societies, 2nd edition (New York: R. R. Bowker, 1978).

More descriptive information on these and other similar directories can be found in Chapter 13 (Directories and Yearbooks).

2.4 Publications of Scientific Societies

The publication activities of scientific societies range from membership directories and newsletters to very large integrated information systems generating a variety of publications, information services, and products, as exemplified by the information processing and publication programs of the American Chemical Society and the American Institute of Physics. The following are the various types of publications and information services offered by scientific societies:

Primary journals. American scientific societies together publish about 40% of the world's significant journals.

Abstracting and indexing services (e.g., *Chemical Abstracts).*

Reviews (e.g., *Annual Reports on the Progress of Chemistry* published by the Chemical Society, London).

2.4 Publications of Scientific Societies

Reference books (e.g., *American Institute of Physics Handbook*).

Newsletters.

Translations (for example, *Russian Journal of Physical Chemistry* published by the Chemical Society, London, is a cover-to-cover translation of *Zhurnal Fizicheskoi Khimii* of the Academy of Sciences of the USSR).

Standards, codes of practice, manuals (e.g., *Boiler and Pressure Vessel Code* of the American Society of Mechanical Engineers)

Conference papers and proceedings.

Monographs and monographic series (e.g., Advances in Chemistry series of the American Chemical Society).

Machine-readable databases (e.g., Chemical Condensates and other databases of the Chemical Abstracts Service of the American Chemical Society).

Audiovisual materials (e.g., IEEE Soundings, a series of quarterly colloquia on cassettes for the current awareness and continuing education of engineers; filmstrips, motion pictures, and slide sets prepared by the American Society of Civil Engineers for use as educational aids).

Primary journals published by scientific societies are of two basic types:

1. Journals containing mainly original research papers and papers presented by members at technical meetings. These are usually entitled *Memoirs, Proceedings, Transactions,* or *Journal* of the society (e.g., *Proceedings of the American Society of Civil Engineers, Transactions of the American Society of Mechanical Engineers,* and *Journal of the American Chemical Society*).
2. Periodicals intended to serve as a forum for communication among members on topics of common interest and to report current events and trends, news about members, and proceedings of business meetings. These are usually entitled *Newsletter* or *Bulletin* of the society (e.g., *Bulletin of the American Society for Information Science,* and *Society of Women Engineers Newsletter*).

The American Chemical Society, the American Institute of Physics, the American Society of Civil Engineers, and the Institute of Electrical and Electronics Engineers are among the largest publishers of primary journals. The American Institute of Physics (AIP), established in 1931, is a federation of several societies such as the American Physical Society, the Optical Society of America, the Acoustical Society of America, and the Society of Rheology, with a combined membership of over 70,000. The numerous primary journals and cover-to-cover translations of Russian physics journals published by AIP and its member societies constitute nearly one third of the total world output of significant journal literature in physics. AIP also produces a bibliographic database called Searchable Physics Information Notices (SPIN).

The publication output of the American Society of Civil Engineers (ASCE) includes the following:

Civil Engineering (monthly journal)

Fourteen primary journals (collectively called *Proceedings of the ASCE)* published by the technical divisions of the Society (e.g., *Journal of the Engineering Mechanics Division, Journal of the Sanitary Engineering Division,* etc.)

Engineering Issues (quarterly journal)

Newsletters published by several of the technical divisions (e.g., *Surveying and Mapping Division Newsletter* and *Waterways and Harbors Division Newsletter)*

Hydrotechnical Construction (English translation of the Russian journal *Gidrotekhnicheskoe Stroitelstvo,* monthly)

A biennial directory

An annual report

Official Register (annual)

Cumulative indexes to all ASCE journals

Audiovisual materials, including motion pictures, filmstrips, slide sets, etc.

Manuals and other miscellaneous publications

The Institute of Electrical and Electronics Engineers, which was formed in 1963 by the merger of the Institute of Radio Engineers and the American Institute of Electrical Engineers, publishes about 36 primary journals, annual convention records, standards, a combined index to its journals, and other ad hoc publications. The publication program of the American Chemical Society (ACS) and its Chemical Abstracts Service is very comprehensive and includes primary journals, secondary journals, handbooks, union lists of periodicals, specialized indexes, and machine-readable databases. These will not be discussed here as there are already very many articles and guides describing the various products and services of the Society [7-9]. Developments pertaining to the information processing and publication programs of ACS are regularly reported in *Chemical and Engineering News, Journal of Chemical Information and Computer Science* (formerly, *Journal of Chemical Documentation*), and other journals of the Society. A comprehensive bibliography of the publications of 369 scientific, engineering, and medical societies has recently been compiled by James M. Kyed and James M. Matarazzo [10].

References

1. John L. Thornton and R. I. J. Tully, *Scientific Books, Libraries and Collectors,* 3rd rev. ed. (London: The Library Association, 1971).

References

2. D. C. Martin, The Royal Society's interest in scientific publications and the dissemination of information, *ASLIB Proceedings*, 9(5):127–141 (May 1957).
3. *The Royal Society Scientific Information Conference, 21 June-2 July 1948. Report and Papers Submitted* (London: The Royal Society, 1948).
4. *Proceedings of the International Conference on Scientific Information, Washington, D.C., November 16–21, 1958* (Washington, D.C.: National Academy of Sciences–National Research Council, 1959), 2 volumes.
5. American Academy of Arts and Sciences, *Memoirs*, a new series, XI (1888), p. 78. Quoted in Ralph S. Bates, *Scientific Societies in the United States* (Ref. 6), p. 10.
6. Ralph S. Bates, *Scientific Societies in the United States*, 3rd ed. (Cambridge, Mass.: MIT Press, 1965), p. 85.
7. Dale B. Baker, Chemical Abstracts Service, *Encyclopedia of Library and Information Science* (New York: Marcel Dekker, 1970), vol. 4, pp. 479–499.
8. Joseph H. Kuney, American Chemical Society Information Program, in *Encyclopedia of Library and Information Science* (New York: Marcel Dekker, 1968), vol. 1, pp. 247–264.
9. Toward a modern secondary information system for chemistry and chemical engineering, *Chemical Engineering News*, 53(24):30–38 (16 June 1975).
10. James M. Kyed and James M. Matarazzo, *Scientific, Engineering, and Medical Societies Publications in Print 1976-1977* (New York: R. R. Bowker, 1976).

3

THE PRIMARY JOURNAL

3.1 Historical Overview

Private correspondence was still the predominant means of scientific communication in the middle of the 17th century when the Royal Society came into existence in England. The idea of a journal to disseminate scientific information was first mooted by Sir Robert Moray, president of the Royal Society, in 1661. Henry Oldenburg, who was elected secretary to the Royal Society in 1663, continued to write long letters to scientists all over the world communicating recent scientific achievements. Among the philosophers and scientists with whom Oldenburg corresponded were: Huygens, Leibnitz, Malphigi, Redi, and Spinoza in Europe; Governor Winthrop of Connecticut in the United States; and Boyle, Halley, Hooke, Lister, Newton, and Wren in England. The correspondence work became so voluminous that a committee had to be formed to share the work of letter writing. As a medium of exchange of scientific information, personal correspondence had many defects:

Much time and effort were needed to write letters.

Letters were personal in tone, and were not sent to those who would disagree with and debate their contents.

Unsound theories were not objectively criticized and rejected.

Questions of priority could not be resolved satisfactorily.

Some writers invented ciphers or systems of shorthand to maintain secrecy.

Many people who were interested in science did not receive letters.

In 1663, Francois Mezeray, historian to the French king, obtained a patent for a literary-scientific periodical. This was the first concrete proposal for a scientific journal, but the project did not materialize for various reasons.

3.1 Historical Overview

A proposal to publish a weekly scientific journal was submitted in 1664 by Sir Denis de Sallo, Counselor of the Court of Parliament under Louis XIV. In August 1664, a privilege was signed establishing *Le Journal des Scavans* (Journal of Learned Men), and the privilege was registered in December 1664. The first weekly issue of the *Journal des Scavans* was published on January 4, 1665. One of the objectives of the *Journal* was "to make known experiments in physics, chemistry, and anatomy that may serve to explain natural phenomena, to describe useful or curious inventions of machines, and to record meterological data" [1]. In 1816, the *Journal des Scavans* was reorganized as the *Journal des Savants*, which continues today as one of the leading literary journals of Europe.

At the time the *Journal des Scavans* was being established in France, plans were under way in England to publish a scientific periodical to report the accounts of scientific experiments, excluding legal and theological matters. Moray, Boyle, Hooke, Oldenburg of the Royal Society, and others developed a plan for such a scientific journal. On March 1, 1664, the Council of the Royal Society ordered "that the *Philosophical Transactions,* to be composed by Mr. Oldenburg, be printed the first Monday of every month, if he has sufficient matter for it, and that the Tract be licensed by the Council of the Society, being first reviewed by some of the members of the same . . ." [2].

The first issue of the *Philosophical Transactions* appeared on Monday, March 6, 1665; it consisted of 16 pages and contained a dedication to the Royal Society, an introduction, nine articles, and a listing of important philosophical books. During 1664-1665 there was a great exodus of people from the city of London because of the plague, and this severely affected the sale of the *Transactions.* Back issues of the journal were destroyed during the great London fire in September 1666. The journal was dormant for a short period during 1676-1683. In spite of these early handicaps, the *Philosophical Transactions* has survived for over three centuries and has published some of the most illustrious scientific papers, including those of Herschel, Priestley, Franklin, Rumford, and Henry Cavendish.

The *Journal des Scavans* and the *Philosophical Transactions* served as models for subsequent scientific periodicals of European societies and academies. One of the first journals that followed this model was the Latin journal *Acta Eruditorium* (1682-) published at Leipzig. Many papers by Leibnitz on calculus were published in this journal. Papers reporting original research in physics, chemistry, biology, and medicine began to appear in specialized primary journals in the last quarter of the 18th century. The early history of scientific and technical periodicals has been studied by Kronick [3]. The historical development of abstracting and indexing journals will be discussed in a later chapter on the bibliographic control of scientific literature.

3.2 Functions of the Primary Journal

Since its inception over 300 years ago, the primary journal has been the most important channel for the formal communication of scientific information. Brown and coworkers have characterized the primary journal as a formal, public, and orderly channel of communication among scientists:

Formal: Papers published in journals can be uniquely identified and cited.

Public: Anyone can submit a paper to a journal for publication, and can obtain a journal upon subscription.

Orderly: The input to a journal is accepted or rejected by the scientific community on the basis of its merit [4].

The primary journal serves three important functions. First of all, it is an official, public record of science. The journal serves as an archival record of scientific scholarship, scrutinized and validated by scientists through a "consensus-forming mechanism" that separates trivia and unsubstantiated claims from tested and validated facts, explanations, and predictions. Refereed papers published in primary journals constitute the basic source material for consolidation and compaction into textbooks, reviews, handbooks, encyclopedias, and similar secondary packages.

Secondly, the primary journal is a medium for disseminating information. Besides the results of R & D activity, the journal conveys a variety of information—historical, social, political, commercial, and pedagogical—of interest to scientists. From its beginnings in 1665 until about the middle of the 19th century, the main function of the primary journal was to serve as an archival record of science. As the number of scientists began to increase during the latter part of the 19th century, more and more scientists began to depend on the journal as a medium through which they could keep themselves abreast of current developments in scientific research.

Lastly, the primary journal is a social institution that confers prestige and rewards on authors, editors, referees, subscribers, and publishers. Published papers are regarded as a tangible measure of a scientist's contribution to the advancement of scientific knowledge, and as a basis for an evaluation of his work by his peers and employers. Publication of R & D results in journals facilitates the establishment of priority and ownership of inventions and ideas. The journal also confers recognition and prestige on editors and referees in view of their participation in the monitoring and validating processes that are so essential to maintain the quality of scientific literature. To the subscriber, the refereed primary journal is a symbol of his professional credentials. The reward that accrues to publishers of primary journals is a combination of prestige and financial returns. Scholarly societies and universities publishing primary journals are impelled by their commitment to the advancement of

3.3 Problems of the Primary Journal

scientific disciplines; commercial publishers publish journals in anticipation of financial rewards.

3.3 Problems of the Primary Journal

Despite its important place in science communication, the primary journal has been the subject of severe criticism on account of its many drawbacks. The efficiency of the journal as a channel for communication of science information has suffered because of its uncontrolled growth, which has resulted in overfragmentation and scattering of primary scientific literature. Because of the diversity of its additional functions, some of which are often mutually conflicting, the journal has become too general and expensive a package, unable to perform any of its functions efficiently. The delay in its production, necessitated by its role as a validated archival record of science, is severely restricting its function as a current awareness tool. The following are some of the problems that have invited repeated criticism.

3.3.1 Proliferation

Concern over the uncontrolled proliferation of journals is not new. "In 1716, apprehension was felt about the overproduction of periodicals whose remainder would have to be bundled up and sold like rotten old cheese" [5]. In 1831, the situation was no different [6]:

> This is the golden age of periodicals. Nothing can be done without them. Sects and parties, benevolent societies, and ingenious individuals, all have their periodicals. Science and literature, religion and law, agriculture and the arts, resort alike to this mode of enlightening the public mind. Every man, and every party, that seeks to establish a new theory, or to break down an old one, commences operations, like a board of war, by founding a magazine. We have annuals, monthlys, and weekly—reviews, orthodox and heterodox—journals of education and humanity, of law, divinity and physics—magazines for ladies and for gentlemen—publications commercial, mechanical, metaphysical, sentimental, musical, anti-fogmatical, and nonsensical . . .

Estimates of the number of scientific and technical periodicals in existence vary from 26,000 to 100,000 titles [7-9]. In his 1961 book *Science Since Babylon*, Derek J. de Solla Price postulated an increase by a factor of ten every fifty years: 10 periodicals in 1750; 100 in 1800; 1000 in 1850; and 10,000 in 1900. Based on this pattern, Price predicted that we might well be "on the way to the next milestone of a hundred thousand such journals" [10]. This estimate of 100,000 scientific and technical periodicals perhaps represents those that were founded, and does not take into account the ones that ceased publication. In his subsequent book published in 1963, Price estimated the number of scientific and technical periodicals actually being published at 30,000 [11].

The fourth edition of the *World List of Scientific Periodicals* (London: Butterworths, 1963-1975) lists 59,961 titles. According to K. P. Barr, only about 24,000 of these titles listed in the *World List* are scientific and technical periodicals currently in existence. Barr's reason for this deduction is as follows [12]:

> Only about 40% of the entries (i.e., about 24,000) represent actual current titles, the remaining entries being titles which have ceased publication, cross-references, and what may be termed *"World List* ghosts," that is, titles for which no dates or holdings are given and which have, in some cases, been carried over from earlier editions. In addition, the term "scientific" is interpreted very loosely in the *World List,* encyclopedias, house journals, and business publications, for example, being included. It thus seems probable that the number of current scientific and technical periodicals included in the fourth edition of the *World List* is considerably less than 24,000.

On the basis of the statistics of periodicals acquired by the National Lending Library (now the British Library Lending Division), Barr concluded that "the number of currently available scientific and technical periodicals which contain material of interest to the practicing scientists and technologists is 26,000" [12].

A more conservative estimate was made by King Research, Inc., in an investigation supported by the National Science Foundation [13]. According to this study, in 1960 there were only 1500 scholarly journals published in the United States. This represented a little over 33% of the world journal literature. The number of such journals published in the United States in 1974 was 1945; this represented about 17% of the world figure. On the basis of this estimate, the number of scientific and technical scholarly journals in existence would be slightly less than 11,500.

These are typical of numerous other estimates of the number of scientific and technical journals. The figures vary greatly depending on the definition of the journal and the method of estimating. Predictions of 100,000 journals are unlikely to be fulfilled. The much-apprehended exponential growth of scientific journals is no longer a serious threat, and the growth of scientific journals appears to be approaching a steady state. This belief is based on a close examination of the four editions of the *World List of Scientific Periodicals* [14].

The proliferation of scientific journals may be ascribed to various causes:

> Increase in R & D activity as a basis for national defense, space exploration, and industrial and economic development.

3.3 Problems of the Primary Journal

Increase in the number of scientists and technologists active in R & D and in publishing. Price contends that 80 to 90% of all scientists that have ever lived are alive now [15].

Importance attached to publications as a measure of a scientist's stature by his peers and employers. This phenomenon is often spoken of as the "publish-or-perish syndrome." A survey of biological scientists by the National Academy of Sciences in 1967 showed that, on an average, each scientist contributes four distinct publications every year to the growing body of scientific literature [16].

Increasing specialization and compartmentalization of science and technology.

Developments in high-speed printing technology.

Proliferation of journal literature is due not only to the birth of new journals, but also to the "splitting" of a journal into several sections that eventually become separate journals. For example, the *Transactions of the American Society of Mechanical Engineers,* started as a single journal in 1878, has expanded into the following 13 quarterly primary journals:

1. *Journal of Engineering for Power*
2. *Journal of Engineering for Industry*
3. *Journal of Heat Transfer*
4. *Journal of Basic Engineering*
5. *Journal of Applied Mechanics*
6. *Journal of Lubrication Technology*
7. *Journal of Dynamic Systems, Measurement and Control*
8. *Journal of Engineering Materials and Technology*
9. *Journal of Fluids Engineering*
10. *Journal of Pressure Vessel Technology*
11. *Journal of Biomedical Engineering*
12. *Journal of Mechanical Design*
13. *Journal of Solar Energy Engineering*

Similarly, *IEEE Transactions* is a family of some 36 different journals.

The problem of journal proliferation is further aggravated by a new crop of scientific journals launched by commercial publishers, who have been encouraged both by the publish-or-perish syndrome of scientists and their employers and by the zeal of many librarians who feel compelled to acquire almost everything that is published as a serial, regardless of quality or need. Because of their allegedly lax refereeing standards, the quality of such journals has been questioned by some scientists. In 1973, a concerned group of eleven international scientists published an open letter to the world scientific community urging them not to patronize or publish in these new commercial journals [17].

The increase in the number of journals is accompanied by an increase in the size of the individual journal. For example, the average length of an article in the *Journal of Organic Chemistry* was 4.0 pages in 1964; it increased to 4.4 pages in 1968 and to 4.7 pages in 1972. This increase in the average length of articles was attributed not to verbosity, but to the generation of more data from more sophisticated and thorough experimentation [18].

The number of papers has also registered a similar growth over the years [19]. In a study of the numbers of manuscripts received for publication in *Analytical Chemistry* during the period 1961-1979, the editors of this journal found that the number of manuscripts received seemed to follow the general trends in scientific research activity in the United States (Figure 3.1). The highest point (1140 manuscripts) was reached in 1964, the period when R & D activity was considered to have reached a peak. The fewest manuscripts (706) were received in 1970, coinciding with the period of curtailed emphasis on science in general. Since then the number of manuscripts has again been rising steadily (well over 1000 in 1979), with the exception of 1971 and 1972 when there was a decline—presumably because the journal published the proceedings of two international symposia on chromatography during those years [20].

The proliferation of journals has jeopardized their capacity to transmit information efficiently and rapidly. Scientists have to wade through a large mass of literature to keep themselves informed of current developments in their fields of specialization and in peripheral fields.

3.3.2 Scattering of Journal Literature

Directly related to the problem of proliferation of journals is the phenomenon of scattering of articles on a given subject in a great many journals. Citation studies have established that in any given subject, a substantial proportion of the articles are concentrated in a relatively small number of journals, and the remaining articles are scattered in a very large number of journals peripheral to or outside the subject. This phenomenon of scattering of journal articles was first investigated by S. C. Bradford [21, 22]. It is common knowledge that a large number of articles on librarianship and information science are found in journals in the areas of physics, chemistry, biology, engineering, psychology, sociology, and general science. It is said that the literature on turtles, for example, can be found in some 600 journals [23].

In a study of British journals by Martyn and Gilchrist, 94% of citations were found to have come from a mere 9% of British scientific journals [24]. This means that the remaining 6% of the articles were scattered in a very large number of journals. In the same study, the number of core scientific journals of the world (sufficient to meet about 95% of the demand for literature) was estimated to be between 2300 and 3200.

3.3 Problems of the Primary Journal

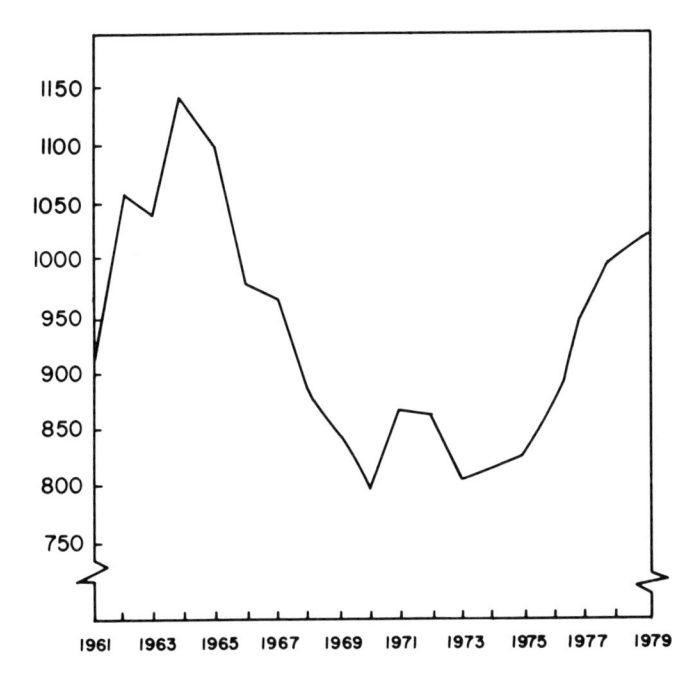

Figure 3.1 Number of manuscripts received by *Analytical Chemistry*. [Reprinted with permission from Research papers in *Analytical Chemistry* (Editor's column), *Analytical Chemistry*, 51(9):1009A-1010A. Copyright 1979 American Chemical Society.]

Some degree of dispersion of articles in various journals may be desirable to promote cross-fertilization of ideas and serendipitous discoveries. "Were the scientist to have restricted access to a few selected but highly homogenous journals, he might not be exposed to other topics, which, though not falling in an expressed area of interest, might be of inestimable benefit for their educational and stimulus value. Apparently unrelated methods or facts have been the source of many scientific insights" [25]. But the disadvantages of the dispersion of papers in numerous journals have far outweighed the advantages of dispersion. The twin problems of proliferation of journals and dispersion of papers are of concern to authors, bibliographers, publishers of secondary services, and users of scientific literature. Editors of abstracting and indexing services and bibliographers have to scan large numbers of journals including those in fields peripheral or unrelated to their areas of interest to be reasonably sure of comprehensiveness. A typical scientist usually scans six to eight journals for current awareness. The effect of individual scientists to remain well informed of current developments in their fields of endeavor is always beset with the frightening possibility of their missing items that may be of crucial importance to their research.

3.3.3 Delay in Publication

Because of the elaborate editorial and refereeing processes, sometimes involving extensive and repeated revision of the manuscript, the time lag between the submission of the first manuscript and the eventual publication of the paper in a journal may range from six months to a couple of years. The growing number of manuscripts that have to be processed also adds to the delay in publication. In 1965, *Physical Review* received 2600 manuscripts of which 2100 (about 80%) were accepted for publication. Of these, about 1000 were accepted by referees with only minor corrections, not requiring return of the manuscript to the authors; about 700 were accepted after more or less straightforward corrections that did not need reexamination by the referees; about 300 had to be reexamined by the referees before final acceptance, and about 70 had to be sent to more than one referee before acceptance [26].

Garvey and coworkers have shown that, on the average, work reported in journal articles was begun 28 months prior to publication, was completed 15 months prior, and was written up and submitted eight months prior [27]. Delay in publishing research results may also result when authors have to hawk their manuscripts from journal to journal until a willing editor can be found. An interesting example of this phenomenon has been described by Shephard [28]. An article describing the successful use of a little-known drug in the treatment of a case of poisoning by *Amanita verna* mushrooms was rejected for various reasons by four major American medical journals, before it was eventually accepted, in its original form, by the author's state medical journal. The timelag between the original poisoning episode and the publication of the paper was 20 months, of which seven months were contributed by the four rejections. In the meantime, the news of the case and its treatment had reached the medical community through the *New York Times* and other newspapers. Because of the delay inherent in the journal publication process, scientists resort to a variety of other channels to disseminate the results of their research; these channels include preprint distribution, conferences, and occasionally the mass media.

Ziman believes that the time required for publishing the results of research is not an unreasonable proportion of the total time required for the research process itself: "... the various stages of hypothesis, design of apparatus, experiment, testing, confirmation, critical analysis, informal discussion, writing up, and so on, take months or years to complete, so that the interval of about four months between the receipt of a typescript and its publication in a reputable journal is not a significant proportion of the time required to make a discovery" [29]. This view is in sharp contrast to the frequent complaints of delay in the publication of journals. Ziman further contends that the existing system of publication can respond speedily enough when necessary [29]:

3.3 Problems of the Primary Journal

The conventional system does not lack the means of quick publication when this is called for. For nearly a century, *Nature* has published "letters" reporting important new discoveries within a few weeks of notification—for example, the first observation of a pulsar was announced on February 24, 1968, in a letter dated February 9. Given that word of the discovery had already been spread by telephone, teleprinter, daily newspaper, jet plane, and first class railway carriage, to everyone who had a professional interest in the subject, this seems quite fast enough for a definitive formal announcement. Let us also recall, more modestly, that only about one paper in ten thousand is so startling as to set the scientific world a-jangling, so that a very small number of quick-publication journals would be quite enough—provided that all the little shephard boys can be taught not to cry "Wolf!" too often.

3.3.4 Diversity of Roles

The primary journal has been too general a package that has tried to be all things to all people. It attempts simultaneously to fulfill many diverse roles:

An archival record of tested and validated knowledge

A current awareness tool for announcement of new discoveries

An instrument through which scientists establish claims for priority to their discoveries and build up their professional stature

A book-reviewing medium to aid librarians in book selection

A channel for the dissemination of a variety of commercial, technical, personnel, and miscellaneous information

Some of these roles are mutually incompatible. For example, the journal's role as an archival record necessitates thorough—and time-consuming—editing, refereeing, and evaluation processes, whereas rapid publication (which may inhibit elaborate refereeing and editing) is indicated if the journal is to be an effective current awareness tool. The journal has not been very effective in any of its diverse and sometimes mutually conflicting functions. According to Rossmassler, "the system does a good job of assembling knowledge into packages which are about 90% mismatched to the needs of users" [30]. This means that users have to scan more and more material to get the same amount of information that is ultimately useful to them.

Also, because of the narrow specialization of scientists and the general nature of the journal, only a small proportion of the contents of a journal is useful to individual specialist readers. In a survey of the users of the *Journal of Organic Chemistry*, Kuney and Weisgerber found that the average subscriber read at least a part of 17% of the articles, and half or more of only 10% of the articles [31]. Elsdon-Dew estimates that: (1) an article in a highly specialized periodical is of interest to only 10% of the workers in the subject area covered by the journal; (2) an article in a general periodical may interest only 2% of its readers; and (3) an article in a local publication may interest only 0.25% of scientists in its field [32].

3.3.5 Increasing Costs

The journal is becoming increasingly expensive. Publication costs are steadily going up, and so are subscription prices. Subscription prices of U.S. journals have been increasing at an average rate of 10-13% per year. During the period 1970-1975, the average price of U.S. periodicals has increased more steeply than other inflation indicators such as the U.S. consumer price index and materials expenditures in academic libraries. Some examples of escalating journal prices are shown in Table 3.1.

The annual surveys of periodical prices reported in *Library Journal* each summer indicate that subscription prices of U.S. journals are increasing at an average of 13% per year, and that chemistry and physics journals are the most expensive ones. The average price of chemistry and physics journals in 1979 was $118.33; this is nearly four times the average price of an American periodical ($30.37) in 1979. The average prices of periodicals in several subject areas during 1978 and 1979 are shown in Table 3.2.

Data in Table 3.2 clearly indicate that periodicals in science, medicine, and engineering are more expensive than those in the social sciences and humanities. The problem of rising costs of journals is equally vexatious to scientists, librarians, and publishers. The high price of scientific journals is rapidly placing the journal beyond the financial means of the individual scientist. The problem is

Table 3.1 Examples of Escalating Journal Prices

Title of journal	Annual subscription[a] 1970	1975	Percent increase
Biochemica et Biophysica Acta (Springer)	495	1551	213
Coordination Chemistry Reviews (Elsevier)	25	136	444
Inorganica Chemica Acta (Elsevier Sequoia)	26	235	804
International Journal of Theoretical Physics (Plenum)	26	135	419
Journal of Theoretical Biology (Academic)	80	234	193

[a] Annual subscription prices paid by the University of Pennsylvania library.
Source: Ref. 33, reprinted with permission.

3.3 Problems of the Primary Journal

Table 3.2 Survey of Periodical Prices

Subject	Average price ($) 1978	Average price ($) 1979	Percent increase
Chemistry and physics	108.22	118.33	9.34
Medicine	57.06	63.31	10.95
Mathematics, botany, geology, and general science	54.16	58.84	8.64
Engineering	39.77	42.95	8.00
Zoology	37.05	40.15	8.37
Psychology	34.21	38.10	11.37
Home economics	21.67	23.21	7.11
Sociology and anthropology	21.58	23.70	9.82
Business and economics	21.09	22.97	8.91
Journalism and communications	19.95	23.86	19.60
Education	19.49	21.61	10.88
Library science	19.34	20.82	7.65
Law	18.74	20.98	11.95
General interest periodicals	17.26	18.28	5.91
Political science	15.62	17.47	11.84
Industrial arts	15.48	17.65	14.01
Fine and applied arts	14.82	17.42	17.54
History	13.71	14.67	7.00
Labor and industrial relations	13.24	15.74	18.88
Literature and language	12.84	13.84	7.79
Agriculture	12.48	14.16	13.46
Philosophy and religion	11.66	13.25	13.64
Physical education and recreation	10.79	12.27	13.72
Children's periodicals	6.34	6.70	5.68
United States periodicals	27.58	30.37	10.12

Source: From Ref. 34. Reprinted from *Library Journal,* September 1, 1979. Published by R. R. Bowker Co. (a Xerox company). Copyright © 1979 By Xerox Corporation.

further compounded by the dispersion of articles and the increasingly diverse nature of the contents of journals. It has been pointed out that a scientist subscribing to a journal is forced to pay for 20 or 30 papers which do not concern him in order to get the one or two papers he wants.

The impact on libraries is no less ravaging. Libraries have been attempting to meet this situation in a variety of ways, among which are: transferring book budgets to serials, cooperative acquisition, participation in networks for resource sharing, and, when inevitable, cancellation of subscriptions to journals. The advent of a new journal entitled the *De-Acquisitions Librarian* (now known as *Collection Management*) reflects the present reaction of librarians to the rising costs of journals.

Scientific and technical journals published by commercial publishers are said to cost 5 to 15 times as much as those published by societies [35]. In a recent study supported by the National Science Foundation, it was discovered that commercial publishers of journals had made an operating profit of about 14% before tax deduction [36]. This is not a very high profit margin in the commercial sector, if it is recognized that this profit margin must account for capital improvements, risk investment, interest on debts, dividends to stockholders, and taxes. Federal, state, and local taxes may together account for nearly 50% of the gross profit.

Editorial and composition costs constitute a substantial part of the total cost of journal production. These are fixed costs and are independent of the number of copies of the journal produced. Hence, a decrease in the number of subscribers inevitably leads to raising of subscription rates by publishers. An increase in subscription prices tends to be followed by a decrease in the number of subscribers. A publishing company caught up in this vicious cycle may easily find itself in a situation in which "an increased subscription rate will lead to a decreased rather than an increased income from subscriptions" [37]. It should be noted here that the economics of magazine publishing is altogether different, because the main source of revenue for the magazine is advertisements, not subscriptions.

Declining incomes from subscriptions, advertisements, page charges, and subsidies have forced journal publishers to increase subscription rates and reduce or eliminate discounts usually allowed to subscription agents. The number of individual subscribers to scholarly journals has shown a downward trend in recent years, presumably because individual subscribers can no longer afford the increased subscription rates. In order to offset this trend, and also to compensate for the possible decline in subscriptions due to the anticipated photocopying in libraries, journal publishers have increased institutional subscription rates more steeply than individual subscription rates. Other measures, such as limiting the size and frequency of journals, generation of spinoff products from existing databases, and changes in printing and distribution methods, have not had much impact on the economics of the primary journal.

To cope with the increased cost of publication, some publishers have been asking for "voluntary" page charges, a concept pioneered by the American Institute of Physics [38, 39], and subsequently endorsed by the United States Federal Council for Science and Technology. The SATCOM Report has further recommended payment of page charges to support the primary publication of research results [40]. This recommendation was made by SATCOM in recognition of the idea that information transfer is an integral part of the R & D effort and that research is not complete until the results of research are disseminated through appropriate channels of communication. Gannett has estimated that revenues from page charges pay for anywhere from 25% to over 50% of the total publication costs of roughly one half of all U.S. scientific journals [41]. The former United States Atomic Energy Commission alone was contributing over $400,000 in page charges annually to the American Institute of Physics journals.

In a study of 20 physics journals, Matarazzo showed that journals with page charges have lower subscription prices, and that increases in their subscription rates in relation to increases in the number of pages have been modest [42]. Although page charges are said to be voluntary, it is suspected that some societies give a higher priority in their publication schedule to papers covered by page charges than to those not so covered, thus bringing indirect pressure on contributors (or their employers or sponsors) to pay page charges.

3.4 Current Trends in Journal Publishing

Several innovations have been suggested and tried with varying degrees of success to offset the problems discussed in the preceding sections. Some of these are simple modifications that can be incorporated into the existing structure of the journal to improve the speed of dissemination of scientific information. Other suggestions have been more drastic. It has been suggested, for example, that the primary journal, in its present format, is totally ineffective and should be replaced with a system for distribution of "separates" or single articles, or with an "electronic journal." Several measures aimed at improving the speed of dissemination of scientific information and the economics of journal publishing are discussed in this section.

3.4.1 Computer-Aided Production

Modern methods of composition and printing, increasingly aided by the computer, have considerably speeded up journal publication. Letterpress printing is rendered obsolete by the faster and more economic offset process. Monotype composition is being increasingly replaced by newer methods of composition: typewriter composition (using Varityper, Justowriter, or other similar word processing equipment), computerized photocomposition, and COM (computer output micrographics). The advent of COM has greatly speeded up the production of both microfilm and offset master directly from the computer, thus eliminating many time-consuming intermediate steps in the preparation of the offset master.

3.4.2 Auxiliary Publication

Another solution to the problems of the growing volume of published material and the delay in its publication has been to print only the main text or an abbreviated version of the paper in the journal and to store all auxiliary material (such as supporting data, computer programs and printouts, mathematical derivations, bibliographies) on microfilm or microfiche in a depository for dissemination to users in response to specific requests [43]. The National Auxiliary Publication Service of the American Society for Information Science is an example of this trend. The American Chemical Society, the American Psychological Association, the British Library Lending Division, and the Canadian Research Council are among other organizations that have similar provisions for storage of auxiliary material.

3.4.3 Microform Publication

Several journal publishers, such as the American Chemical Society, the Institute of Electrical and Electronics Engineers, and the American Medical Association, have resorted to microform publication to supplement or supplant the hard-copy editions of their journals. All the primary journals of the American Institute of Physics are also published in a collective microfilm edition entitled *Current Physics Microform*. The journal *Wildlife Diseases* is published only on microcard. Each issue of the *Honeywell Computer Journal* arrives with a microfiche copy of the same issue in a back-cover pocket. Journals on microfiche can be mailed first class, and reach the user faster than paper editions. Although microform editions are compact, portable, and cheaper than paper editions, user acceptance of microforms has been slow. Gannett has reported a survey in which only 1% of the members of the Society of Automotive Engineers and the IEEE preferred microfiche to the printed edition [44]. User acceptance may be improved by the use of microfilm cartridges and motorized reader/printers that facilitate rapid searching and hard-copy printing [45].

3.4.4 Advance Announcement of Contents

In an attempt to perform the current awareness function more efficiently, the editors of many primary journals indicate in advance the contents of future issues of their journals. Papers accepted for publication in the *Journal of the Science of Food and Agriculture* are listed in *Chemistry and Industry*. For some time the American Institute of Physics published advance abstracts of forthcoming papers in *Current Physics Advance Abstracts*. This service has now been discontinued. *Talanta* lists "Papers Received" with the actual date of receipt of the manuscript indicated in parentheses.

Another variation of this practice is to provide information on concurrent publications in other journals, usually of the same group. *IEEE Spectrum* has

3.4 Current Trends in Journal Publishing

two regular features: One, entitled IEEE Tables of Contents for Current and Future Publications, provides details of articles in concurrent or future issues of other IEEE journals. The second feature is entitled Future Special Issues. The monthly journal *Physics Teacher* runs a regular feature, Read it in AJP, listing articles in the recent or concurrent issues of the *American Journal of Physics* with brief annotations.

Since 1969 the American Chemical Society has been providing a semimonthly Single Article Announcement Service. This is an alerting service that enables users to scan the tables of contents of 18 primary journals published by the ACS and to order copies of desired articles.

Yet another approach to improve the current awareness function of the journal is to publish a "preview journal" containing summaries of future articles. *Biochemica et Biophysica Acta* publishes author summaries of forthcoming papers several months before the papers are published. A compilation of these advance summaries is sold as a separate journal entitled *Biochemica et Biophysica Acta Previews*. Preview journals and advance announcement of forthcoming papers in journals facilitate rapid dissemination of current information and alert scientists to journal articles that they might otherwise overlook.

3.4.5 "Letters" Journals

An entirely new breed of journals, known as short communication journals or "letters" journals, has come into existence exclusively for the rapid publication of preliminary results of research. The articles published in such journals are usually short, and receive minimal or no editing. Some letters journals (e.g., *Tetrahedron Letters*) use the author's copy for printing, thus considerably speeding up the production process. The institution of the short communication journal followed a steep increase in the proportion of preliminary communication in regular primary journals publishing full-length articles. In the six years preceding 1972, the ratio of preliminary communications to articles increased by 40% in the *Journal of the American Chemical Society,* and by 200% in the *Journal of the Chemical Society* (London) [46]. The idea was to use these short communication journals solely for the publication of preliminary findings of outstanding importance to the scientific community, with the implicit understanding that these preliminary communications would later be followed up by full-length papers incorporating the complete details of the research project. But this objective does not seem to have been fully achieved. An investigation by Kean and Ronayne showed that many papers published in letters journals were short, definitive accounts of "dead end" research with little or no followup potential, rather than preliminary communications of outstanding importance or general interest. Only 29% of British papers in *Chemical Communications* and 20% of those in *Tetrahedron Letters* were found to have been subsequently followed up and published as full-length papers [47]. Only about 50% of the communications published in *Physical Review Letters* are later published as full reports.

The letters journal, however, seems to have succeeded in appreciably speeding up dissemination of results of research by reducing or eliminating editorial and refereeing processes and by employing faster production methods. The time lag of publication in *Chemical Physics Letters* is said to be 14 days from the date of acceptance of the manuscript [48]. *Tetrahedron Letters* claims to publish communications within four weeks. In a comparative study of *Analytical Chemistry* and *Analytical Letters,* the mean time lags were found to be 191 days and 33 days respectively [49]. This time lag represented the interval between the receipt of the manuscript in the editorial office and the actual distribution of the printed journal. The longest time lag in *Analytical Chemistry* was one year and in *Analytical Letters* was two months. Thus, even in the worst case, the publication delay in the letters journals was far shorter than that in the standard primary journal publishing refereed, full-length papers.

3.4.6 The Synopsis Journal

The idea of publishing scientific papers at two different levels of completeness was proposed by N. W. Pirie during the Royal Society Scientific Information Conference in 1948 [50]. This suggestion was based on the premise that there are basically two types of users who read a scientific paper: those who are interested in the details of the investigation and its methodology, and those who are interested only in the conclusions of the investigation. It was further contended that readers of the second type far outnumbered those of the first type, and that a short version of the paper, containing a summary of the investigation, should be prepared to meet the needs of the majority of readers, although the preparation of two versions of the paper would place an additional burden on the author. During the International Conference on Scientific Information in 1958, J. D. Bernal made a similar suggestion: ". . . instead of the present intermediate length paper of ten to twenty pages, it would be better to have a short, pointed paper of some two pages in the form of what has been called an informative abstract. This would be supplemented by a longer, more detailed paper, not printed and published, but available in duplicated, microfilm, or other modern method of reproduction, to all those thought to be interested in it or who requested it" [51].

An interesting variation involving three levels of publication was suggested by Phipps in 1959 [52]. He proposed that each author prepare (1) a full length report of his research, (2) a two-page summary, and (3) an abstract for deposit in a central office. All of these papers would be assigned the same code number. Scientific journals would print summaries of most papers and full texts of selected papers. Abstracts would be sent to journal subscribers, perhaps on edge-notched cards. Subscribers could obtain copies of desired papers from the central office.

In 1968, the American Chemical Society examined a proposal to publish its *Journal of Organic Chemistry* in two editions: a complete edition with full experimental data and details for libraries, and a condensed version with only the main findings and limited data for general circulation. A survey was conducted to assess subscribers' reaction to this proposal. The survey led the Society to the conclusion that "such a system, while perhaps needed, could not unilaterally be adopted by *Journal of Organic Chemistry* without loss in favor" [53]. The ACS has recently resumed its dual basic journal experiment with financial support from the National Science Foundation's Division of Science Information. The purpose of this experiment is to determine whether dual journals can provide a less expensive method of communicating research findings that will also be more useful to individual scientists. In this experiment, the *Journal of the American Chemical Society* is being published in two versions. One, called the summary version, contains synopses of research papers, prepared by the authors themselves. This version, which is primarily intended for individual subscribers, also contains book reviews and communications to the editor. In the second version, called the archival edition, which is primarily meant for libraries, the author's typewritten manuscript is reproduced at a reduced size. The ACS is conducting studies to assess users' reaction to this dual journal concept and the economics of publication [54].

The synopsis journal now under experimentation in Europe represents another attempt at multilevel journal publishing. The West German *Chemie-Ingenieur-Technik* prints only synopses of some of its technical articles; the original manuscript is available on microfiche upon request. The normal publication time of 9-12 months in the conventional journal format is claimed to have been reduced to less than three months in the synopsis structure [55]. A similar project is being planned by the Chemical Society in the United Kingdom. The synopsis journal serves both the archival function and the current awareness function of the traditional primary journal.

3.5 Alternatives to the Scientific Journal

In the previous section we have discussed several innovations that are being tried to improve the speed and economics of journal publishing without changing the basic structure of the traditional primary journal. From time to time, suggestions have been made to alter the traditional format of the primary journal and to replace it with alternative mechanisms for the dissemination of scientific information. The following are some of the proposals suggesting drastic modifications or alternatives to the primary journal that have been put forward during the last three decades:

> Organization of information-exchange groups for the public distribution of preprints [56, 57]

On-demand distribution of author-prepared summaries and/or full papers following computerized matching of user interest profiles and subject headings assigned by authors to their papers [58]

Repackaging of primary journals into "user journals" or "superjournals" for particular user groups [59]

Establishment of separate radio stations and/or television stations for broadcasting science reports

Distribution of research reports solely on tape recordings

Replacement of the primary journal by the individual paper (or "separate") as the primary unit for distribution

Extensively documented reports of these and similar proposals have been made by Phelps [60] and Hills [61]. By far the most momentous alternative to the scientific journal yet suggested is the substitution of the individual paper or "separate" as the primary unit of distribution. Perhaps the earliest proposal to replace the primary journal by collections of separates was made by Pownall in 1926 [62]. During the following decades, a number of quite similar schemes were proposed for the establishment of regional, national, or international centers for the distribution of individual scientific papers as separates. Watson Davis suggested, for example, the establishment of a national center that would take over the functions of scientific publishing and bibliography. Authors would submit manuscripts with brief summaries to this center, rather than to journals, for reviewing by referees. Accepted papers would be typed on special paper suitable for photographic duplication. The author-prepared summaries would be published in a series of weekly or monthly journals devoted to different fields. Scientists could scan the summary journals and order full texts of desired papers. The national center would make copies of papers only as needed to fill the orders received from readers of the summary journals.

Davis' plan was published in 1939 as an appendix to J. D. Bernal's *Social Functions of Science* (London: George Routledge, 1939), in which Bernal had suggested a similar scheme for distribution of separates. Bernal later presented his "Provisional Scheme for Central Distribution of Scientific Publications" at the Royal Society Scientific Information Conference in 1948 [63]. Bernal's scheme called for the establishment of a number of National Distributing Authorities (NDA) which, working in conjunction with various scientific societies, would be responsible for the publication and distribution of scientific papers. The author would send the manuscript, along with an abstract, to the NDA. The paper would then be referred to a review panel for acceptance. Titles of accepted papers would be published in a weekly list, and the paper itself would be published by the NDA as a preprint for distribution to individuals and libraries as well as to foreign NDAs for local distribution in other countries.

3.5 Alternatives to the Scientific Journal

Bernal claimed that such a centralized system of distribution of separates would be more advantageous than conventional journal publishing in terms of improvements in speed, convenience, comprehensiveness, and economy. Bernal's plan was the subject of much debate and criticism both during and after the Royal Society Scientific Information Conference. Ten years later, in the 1958 Washington Conference, Bernal's own thinking on his earlier proposal was different [64]:

> ... I had been so much impressed, through the experience of my own work, with the importance of reprints that I had proposed a scheme for substituting a rational distribution of these for the traditional scientific periodicals. This scheme roused much feeling and was even castigated in a *Times* leader as "Professor Bernal's insidious and cavalier proposals." However, the result of the pilot survey showed me that scientists as a whole did not work the way I did, but rather made use primarily of libraries where the disadvantages of the bound periodicals largely disappeared. Consequently, I immediately abandoned my original proposals and publicly withdrew them at the Conference.

Similar schemes for the distribution of separates as a possible alternative or supplement to the scientific journal were proposed by Atherton Seidell, M. B. Visscher, T. S. Harding, Zeliaette Troy, Bernard Berelson, P. W. Wokes, Friedrich Kaysser, J. W. Kuipers, J. H. Wilson, D. A. Brunning and others. These and several other proposed alternatives to the scientific journal have been documented and reviewed by Phelps [60]. In the same paper, Phelps has also documented a number of objections to the various preprint distribution schemes. The main objections to any proposal that seeks to replace the scientific journal with preprint distribution are the following:

1. Preprint distribution is impractical; no scientist could read all the papers relevant to his work, even if they were made available to him.
2. Distribution of separates militates against standardization of scientific terminology, especially in the biological sciences.
3. Scientists would not support any plan that would diminish the importance of scientific societies as the guardians of high standards in scientific communication.
4. Distribution of separates in response to requests from users would place an additional work load on users, who would have to first read title lists or abstracts of papers and then place orders for full copies of desired papers. This arrangement would also cause inconvenience to scientists and delay their access to scientific papers.
5. Maintaining and using files of separates in libraries would be more difficult than handling bound volumes of journals.
6. Selective distribution of separates to individual scientists would call for the development and adoption of a classification scheme or a scheme of

categories. It is obvious that no one scheme of classification or categorization would be acceptable to all or even the majority of the various national centers responsible for the distribution of separates in different countries.
7. Distribution of separates cannot facilitate casual reading and browsing, which are important to scientists. It would hinder serendipitous discoveries and cross-fertilization of ideas, and would accelerate excessive fragmentation of science.
8. Scientific societies advocate freedom of scientists to pursue their inquiries in any field of their choice and publish their findings in any of a variety of ways. Centralized distribution of separates would tend to minimize the potential of scientific societies in protecting the heritage of scientific freedom.
9. Centralized agencies for distribution of separates would not be economically viable since they could not derive any revenue from advertisements. Such agencies would have to be supported by government. Also, mailing separates could be costlier than mailing journals. Consequently, libraries and individual scientists might find acquisition of separates to be more expensive than subscription to journals.
10. The scientific journal is identified with a field of scientific inquiry or profession, and by virtue of this identity, it confers status and prestige on authors, editors, and referees. The preprint or separate does not have the individuality, continuity, or prestige of the refereed scientific journal.

Several scientific societies such as ASCE, ASME, SAE, and IEEE have experimented with distribution of separates to replace or supplement their journals. In an attempt to overcome delay in the review and publication of papers, the ASCE discontinued its *Proceedings* in February 1950 and began to distribute the proceedings papers as numbered separates. This arrangement was found to be unsatisfactory, and in January 1956, the Society introduced its division journals. Individual papers, however, are still obtainable as reprints.

The IEEE tried an experimental reprint supply service called the IEEE Annals as an adjunct to its journals. Articles accepted for publication in the IEEE journal were mailed to the subscribers after matching their interest profiles with document profiles provided by indexers. In addition, contents of about 36 IEEE journals were also regularly announced in the *IEEE Spectrum*, and users could order reprints of desired articles before or after their publication in the IEEE journals. The IEEE Annals was not expected to replace the regular IEEE journals; it was set up as a parallel service in a "quasi-competing mode," and was expected to be economically viable. The IEEE experiment was modeled after the Mathematical Offprint Service of the American Mathematical Society, and was recently discontinued [65].

3.6 The Future of the Scientific Journal

After a detailed review of the various alternatives to the scientific journal, Phelps concluded that a system of separates distribution was not a satisfactory solution to the problems of the primary journal [60]:

> The experience of the three societies which tried and abandoned the distribution of separates as an alternative to journal publication, the experience of the Engineering Societies Library in handling separates, and the published literature critical of proposals to replace the periodical by separates convince us that a system of separates distribution is not a practical solution to problems of scientific communication.

3.6 The Future of the Scientific Journal

Garvey and Gottfredson feel that the scientific journal has "taken on new functions which are more important for the promotion of scientists than science" [66]. Scientists depend heavily on nonformal channels of communication, and the journal is no longer the only medium for disseminating current scientific information. Scientists use the journal largely for transferring scientific information from the nonformal to the formal domain and for gaining visibility for themselves. Consequently, journals attract a very large number of articles which stand as isolated packets of unintegrated information. The impact of such isolated packets on the totality of scientific information is at best uncertain. Garvey and Gottfredson propose that the existing journal system be changed to a delayed integrative journal system to serve the archival function [66]:

> These future journals will no longer accept piecemeal articles. Rather, they will require that such articles' publication be delayed until a coherent series of research works can be synthesized into a single major article. Journals will serve mainly the functions of integrating and storing information in the archives of science.

Although many changes have taken place in the journal publication process in the recent past, as discussed in an earlier section, the basic structure of the primary journal has remained substantially unaltered. In this section, we shall consider some developments and proposals that have the potential of drastically altering the format of the primary journal.

3.6.1 Integrated Journal Publishing

Most of the proposed alternatives to the primary journal considered so far have not passed beyond the proposal stage. There is, however, one system that has reached a fully operational stage. The International Research Communication System (IRCS), described by Eakins [67], is an integrated information system in which a single input can generate a variety of different outputs. Papers in the whole field of biomedicine are submitted to a single editorial office, and after scrutiny and acceptance by referees and editors, are first published as separates. Each month, a subject-classified list of titles of papers accepted for publication during the month is compiled. At the end of each month, a number of specialist journals are published by selecting appropriate papers pertaining to the subject of the specialist journals. A paper can appear in more than one specialist journal; typically, a paper appears in three specialist journals. In addition, selected papers of wide interest to the biomedical community are published in a general biomedical journal entitled the *IRCS Journal of International Research Communications*. Finally, all the papers accepted by the system are microfilmed and published in the *IRCS Library Compendium*. Speed, flexibility, and predictability are said to be the main advantages of this system. Subscribers can obtain specialist journals of their choice and thus keep nonpertinent material to a minimum. It is also possible for subscribers to obtain copies of only desired papers and not subscribe to any journal.

3.6.2 The Editorial Processing Center

The most important single force that is sure to influence the basic structure of the primary journal is the computer. The American Chemical Society, the American Institute of Physics, and other journal publishers have been developing computer-based techniques for keyboarding manuscripts, typesetting, microform publication through COM, and index production [68-73]. But the computer has not so far altered the basic structure of the primary journal; the machine has been used mainly to expedite certain operations (e.g., typesetting) which were formerly done manually.

Use of the computer in journal publishing is probably so limited because the scale of operation of the typical scientific journal publisher is not large enough to justify investment in new technology. Although an enormous quantity of scientific and technical literature is published each year, it is produced by a large number of publishers. Computerization becomes worthwhile if a number of small journal publishing operations can be combined to achieve a scale large enough for cost-effective investment in computer-based publishing technology. The Editorial Processing Center (EPC) has been suggested as a possible step in this direction [74-76]. The EPC is conceived as a mechanism for combining a number of small publishing operations into an integrated publication-dissemination system on a scale large enough to make computerization worthwhile, at the same time allowing each editor full autonomy and control over his publications. The EPC is a centralized, computer-based system that can take over much of the routine, programmable tasks now performed by authors, editors, and referees, and allow them to concentrate on their nonprogrammable intellectual functions. The EPC is expected to facilitate a wide range of functions including data capture, text editing, typesetting, page composition, manuscript tracing, subscription fulfillment, index preparation, reviewer file maintenance, and information retrieval.

3.6 The Future of the Scientific Journal

In a typical EPC setting, the author of a paper prepares his manuscript using special quality paper and typeface for optical character recognition (OCR) input. The author then marks the manuscript for the attention of the editor of his preferred journal and mails it to the EPC. The entire manuscript is recorded into the EPC computer by means of an OCR scanner. From this stage onwards, the editor, the referees, and the author can interact through the EPC computer. The final product of this interaction is the refereed and accepted manuscript on magnetic tape, ready for photocomposition. Copies of the tape may also be sent to abstracting and indexing services and to information analysis centers for creating secondary services and products. The main advantages of the EPC are the following:

> Operating economies resulting from (1) elimination of manuscript rekeyboarding and galley proof corrections; (2) computerization of housekeeping operations such as routine correspondence, filing, and follow-up; and (3) computer support of referee selection, manuscript revision, and copy editing
>
> Faster publication of papers
>
> Potential for online linkages to editors, referees, authors, and even to readers
>
> Faster production of secondary services and products (e.g., abstracting and indexing services, reviews, bibliographies, etc.)

With financial support from the National Science Foundation, the Aspen Systems Corporation has set up an experimental EPC. The American Society for Information Science, the American Society for Microbiology, the Entomological Society of America, and the American Society of Biological Chemists will each produce one journal through the experimental EPC, to study the practical and economic feasibility of such an integrated approach to journal production.

3.6.3 The Electronic Journal

Herschman has speculated that in the foreseeable future the primary journal, as it is known to us now, may be replaced by a computerized central depository with a network of consoles serving the various groups of users: authors, editors, referees, information analysts, evaluators, and readers [77]. In 1970, a committee of the National Academy of Sciences predicted that in about three decades, the primary journal would be rendered obsolete by a network of consoles that would facilitate intercontinental interactive communication [78]. These proposals might appear futuristic at first sight. But along these very lines, a concrete proposal has been made for what has been called "the electronic journal" in an investigation carried out at the University of Toronto with financial support from the National Science Foundation [78]. The electronic journal is conceptually similar to the EPC, but would consist of a distributed electronic network that could provide online access to scientific papers by everyone concerned: authors, editors, referees, and users (readers). In addition to performing, in an online mode, the basic functions of manuscript input, reviewing and editing, and housekeeping operations, the electronic journal would provide an extended spectrum of services such as browsing facility, retrospective searching, commentaries, user scratchpads and filing systems, and data manipulation capability. It would be possible for the user, for example, to manipulate the data presented in a paper, draw graphs and histograms, and have the results displayed on the terminal. The user would be able to communicate with other users over the network through computer conferencing. Authors in different geographical areas could collaborate in preparing articles. The electronic journal could also carry news items and even advertisements.

Besides being faster than the traditional printed journal, the electronic journal can provide international access to scientific papers in an online mode, and can also capture "fugitive materials" such as unpublished technical reports. It is also expected to be more cost-effective than an equivalently large paper journal. The approximate cost of an electronic journal publishing about 40,000 pages per year has been estimated thus: The total startup costs would be about $20 million, and the operating costs about $10 million a year. It is expected that an electronic journal of this size could probably pay for itself out of income from "subscriptions" [79].

Schemes such as the electronic journal might soon become technologically feasible and economically viable. The chief impediments to the successful implementation and acceptance of the electronic journal appear to be predominantly sociological and technological in nature. The traditional scientific journal has become so firmly rooted within the scientific and technical communication system that any alternative that does not have the individuality and prestige-giving power of the primary journal may not be easily accepted by the scientific community. The scientific societies are also likely to resist implementation of such a system on the ground that it would diminish their independence and importance. The high degree of standardization and conformity in publication practices that such a centralized publishing system might impose may also not be acceptable to journal publishers. For widespread acceptance of the centralized computerized networks typified by the electronic journal, the advantages gained by the scientific community must far outweigh the disadvantages perceived by the individual scientists and the scientific societies. The scientific community seems to prefer the refereed journal as a means of establishing priority and winning prestige and recognition. According to Herschman, technology and tradition are the two main drawbacks of his "journal of the future" [77]:

> Its drawbacks are, of course, technology (the access must be really universal) and tradition (aesthetics). The visible evidence of one's progeny is not

there. This does not mean that the copy which any user sees, either on his scope, or as printed matter, won't be typographically excellent—it will, but it won't be there to cover your walls and heft in your hands and loving finger. All that can be said to this is that when the technology is ready, perhaps we will be also.

3.7 Bibliographic Control of Journals

Numerous directories, catalogs, and union lists exist for the bibliographic control of journals and other types of serials. The bibliographic control of the papers published in journals is achieved largely through abstracting and indexing services. These will be discussed in detail in a later chapter on the bibliographic control of scientific and technical literature.

3.7.1 Directories of Serials

The *Ulrich's International Periodicals Directory* (1932-), and *Irregular Serials and Annuals: An International Directory* (1972-), both published biennially by R. R. Bowker Company, New York, are noted for their comprehensive listing of serials of all kinds. The 18th edition of *Ulrich's International Periodicals Directory* (1979-80) provides data on some 62,000 current periodicals published throughout the world. Each entry contains the following data elements as applicable:

Title and subtitle of the periodical

Dewey Decimal Classification number

Country code

International Standard Serial Number (ISSN)

Language

Year first published

Frequency of publication

Corporate author

Publisher's name and address

Editor's name

Index information

Micropublisher

Abstracting and indexing services in which covered

Former title, if any

The entries are arranged under 250 subject headings ranging from Adventure and Romance through Biology, Chemistry, Engineering, Physics, Transportation, and Women's Interests. Most subject headings have several subheadings. Abstracting and indexing services, publications of international organizations, and cessations are listed in separate sections. An alphabetical title index facilitates location of desired titles in the directory. Only periodicals which are currently in print and are published more frequently than once a year are described in this directory.

Irregular Serials and Annuals (5th edition, 1978), another Bowker serials directory, lists 32,500 currently published serials and continuations such as conference proceedings, transactions of societies, annual reviews, yearbooks, and monographic series—all of which constitute the twilight area between books and periodicals. Entry format and arrangement are similar to those in the *Ulrich's International Periodicals Directory*.

Ulrich's Quarterly (New York: R. R. Bowker, March 1977-) is an inter-edition supplement to both *Ulrich's International Periodicals Directory* and *Irregular Serials and Annuals*. Each issue of the quarterly supplement lists some 2500 new titles of periodicals, irregular serials and annuals, title changes, and cessations.

Yet another Bowker directory of journals is *Magazines for Libraries* (3rd edition, 1978) edited by Bill Katz and Berry Richards. This is an annotated list of some 6500 periodicals (including scholarly scientific journals) considered to be "most useful for the average public, academic, elementary, secondary, or special library" (preface). The entries are arranged under subject categories ranging from Aeronautics and Space Science to Marine Science, Men, . . . Pets, Philosophy, Photography, Physics, and so on. Abstracting and indexing services are listed in a separate section at the beginning.

The *Standard Periodicals Directory* (New York: Oxbridge Communications, Inc., 1964-) contains an extensive listing of U.S. and Canadian periodicals. The sixth edition (1978) provides data on 68,720 periodicals. Other types of directories of journals include those published by libraries (e.g., *Journals Available in the ERDA Library)* and by publishers of secondary services (e.g., *Serial Sources for the BIOSIS Data Base)*. The following is an illustrative list of directories of journals and other serials.

1. *Ayer Directory of Publications.* Philadelphia, Penn.: Ayer Press, 1869-, annual.

2. BioSciences Information Service of Biological Abstracts, Chemical Abstracts Service, and Engineering Index, Inc. *Bibliographic Guide for Editors and Authors.* Washington, D. C.: American Chemical Society, 1974. This is a list of about 27,700 serials covered by BIOSIS, CAS, and EI.

3.7 Bibliographic Control of Journals

3. *Catalogue of Scientific Serials of all Countries Including the Transactions of Learned Societies in the Natural, Physical and Mathematical Sciences, 1633-1876.* Samuel H. Scudder. Cambridge, Mass.: Library of the Harvard University, 1879 (Reprinted by Kraus Reprint Corporation, New York, 1965. 358p).
4. *Directory of Canadian Scientific and Technical Periodicals.* Ottawa: National Research Council of Canada, 1969.
5. *Directory of Japanese Scientific Periodicals, 1967.* Tokyo: National Diet Library, 1967. 660p.
6. *Guide to Scientific Periodicals: An Annotated Bibliography.* Maureen J. Fowler. London: Library Association, 1966. 318p.
7. *Half a Century of Soviet Serials, 1917-1968.* Rudolph Smits, comp. Washington, DC: Library of Congress, 1968. 2 v. 1661p.
8. *Journals Available in the ERDA Library.* Washington, DC: U.S. Energy Research and Development Administration (Department of Energy), 1976.

    Lists over 1600 journals. New titles are listed in the *Weekly Accessions List.*
9. *National Directory of Newsletters and Reporting Services.* 2nd ed. Robert C. Thomas, ed. Detroit, Mich.: Gale Research Co., 1978-.

    To be published in four parts, with approximately 750 entries in each part.
10. *Publications Indexed for Engineering (PIE).* New York: Engineering Index, Inc., 1974.

    Provides bibliographic information for 2138 serials including conference proceedings that are abstracted and indexed by Engineering Index, Inc.
11. *Serial Sources for the BIOSIS Data Base.* Philadelphia, PA: BioSciences Information Service, 1978-, annual.

    The 1978 edition has 14,207 entries, of which 8580 are currently active serials. Supersedes: *BIOSIS List of Serials.*
12. *World List of Aquatic Sciences and Fisheries Serials Titles.* Rome: Food and Agriculture Organization of the United Nations, 1975.

    Contains bibliographic information on 1237 serials.

3.7.2 Union Lists of Serials

Union lists of serials are indispensable bibliographic tools for ascertaining the availability of serials held in various libraries. There are probably several thousands of local union lists each listing the holdings of a small number of libraries.

The *Union List of Serials in Libraries of the United States and Canada* (3rd edition, 1965), published by the H. W. Wilson Company, is one of the most important international union lists. The preface to this edition contains a brief historical account of union lists of serials. This five-volume work was prepared under the sponsorship of the Joint Committee on the Union List of Serials with the cooperation of the Library of Congress. The third edition describes the holdings of 156,449 serial titles (which began publication before 1950), held in 956 libraries in the United States and Canada. Serials which began after December 31, 1949 are covered in a quarterly supplement entitled *New Serial Titles: A Union List of Serials Commencing Publication After December 31, 1949*, compiled and edited by the Serial Record Division of the Library of Congress. A 21-year cumulative edition entitled *New Serial Titles: A Union List of Serials Commencing Publication After December 31, 1949. 1950-1970 Cumulative* (Washington, DC: Library of Congress; New York: R. R. Bowker, 1973), as well as a subject guide to this cumulative edition have also been separately published. In the *New Serial Titles 1950-1970 Subject Guide* (New York: R. R. Bowker, 1975), entries are arranged alphabetically under country within 225 subject headings. The subject headings are arranged in the sequence of the Dewey Decimal Classification schedule. An alphabetical subject index of 1600 terms to the classification scheme is also provided to facilitate specific subject access to the serial titles.

The following are two additional examples of noteworthy international union lists of serials:

1. *British Union Catalogs of Periodicals.* Hamden, Conn.: Shoe String Press, 1968. 4v.
2. *World List of Scientific Periodicals Published in the Years 1900-1960.* 4th ed. London: Butterworths, 1963-1975, 3v. 1824p.

Both these union lists are supplemented by *New Periodical Titles* (London: Butterworths, 1964-), published quarterly with annual cumulation.

Chemical Abstracts Service Source Index, 1907-1974 Cumulative (CASSI) is in fact a union list of serials covered in *Chemical Abstracts.* CASSI provides complete bibliographic identification of all source documents ever monitored by Chemical Abstracts Service since 1907. Additionally, it facilitates access to these publications by giving holdings information in some 400 libraries in the United States and overseas. The 1907-1974 cumulative edition in two volumes contains 53,593 entries including numerous cross references. It is also available in computer-readable form from Chemical Abstracts Service, The Ohio State University, Columbus, OH 43210. The list includes publications covered by BIOSIS, Engineering Index, Inc., and the Institute for Scientific Information. Thus the scope of CASSI extends far beyond the field of chemical sciences, into the physical, engineering, and biological sciences. A list of 1000 serials

3.7 Bibliographic Control of Journals

most frequently cited in *Chemical Abstracts* and a directory of participating libraries are also included in the 1907-1974 cumulative edition. CASSI is kept up to date by a quarterly supplement. About 5000 new entries are added each year through the quarterly supplement.

3.7.3 Cumulative Indexes to Journals

Current issues of primary journals are valuable as sources of current information during the first few months after publication. The volume of use of journals then falls off more or less rapidly, depending on the "half-life" of the periodicals in a given discipline. In general, journals in the fast-moving fields such as electronics, space travel, and most branches of engineering become obsolete more rapidly than those in slow-moving fields such as botany, geology, and mathematics. Once the journal issues have served their current awareness function, they become "back files" or bound volumes. But much of the materials published in refereed, scholarly journals is of lasting value, and will continue to be sought frequently during retrospective literature searches. To facilitate identification and retrieval of specific information in primary journals, most journals publish indexes at the end of each volume. There is considerable variety in the kinds of indexes provided, and in their depth and quality. Subject and author indexes are the most common types of indexes provided. The volume indexes are usually printed in the last issue of each volume, or in one of the early issues of the succeeding volume. Sometimes the indexes are issued separately.

Many publishers of primary journals collate the volume indexes to their journals and publish cumulated indexes covering 10, 15, or 20 years. While 5 or 10 year cumulations are common, there are some that cover 50 or even 100 years. The importance of such cumulated indexes for retrospective literature searching cannot be overemphasized. This practice of publishing cumulated indexes to one or a group of journals appears to be particularly characteristic of scholarly societies. The American Society of Civil Engineers publishes annual subject and author indexes to all of its regular publications in the annual volumes of its *Transactions*. These annual indexes cover all the articles published in the monthly journal *Civil Engineering* and in the technical journals of the Society's 14 divisions. Besides these annual indexes, the ASCE has periodically published various cumulative indexes to its *Transactions*. These are:

1. *ASCE Index to Transactions*, vol. 1-83 (1867-1920)
2. *ASCE Index to Transactions*, vol. 84-99 (1921-1934)
3. *Cumulative Index to ASCE Publications: Proceedings*, 1950-1959 (vol. 76-85); *Transactions*, 1935-1959 (vol. 100-124); *Civil Engineering*, 1930-1959 (vol. 1-29)

4. *Cumulative Index to ASCE Publications: Journals (Proceedings)*, 1960-1969 (vol. 86-95); *Transactions,* 1960-1969 (vol. 125-134); *Civil Engineering,* 1960-1969 (vol. 30-39).
5. *Cumulative Index to ASCE Publications: Journals (Proceedings)*, 1970-1974 (vol. 96-100); *Transactions,* 1970-1974 (vol. 135-139); *Civil Engineering,* 1970-1974 (vol. 40-44)

The Society of Analytical Chemistry, London, published a cumulative index to the first 20 volumes (1877-1896) of its journal, *The Analyst;* since then, decennial cumulated indexes to the journal have been published.

The ASME publishes cumulative indexes to all of its technical literature in response to user needs ascertained through a survey of index users. A 60-year index covering the period 1880-1940 and a ten-year index for 1941-1950 were both superseded by the *Seventy-Seven Year Index 1880-1956* published in 1957. This publication is in two parts, covering the years 1880-1939 and 1940-1956. A further 14-year cumulative index to papers published in the *Transactions of the ASME* (1957-1970) was recently published by the Society.

The Engineer Index, 1956-1959 (London: Morgan Brothers, 1964) is an example of a cumulative index published by a commercial publisher. This was intended to be a cumulative index to the 200 volumes covering the first 100 years of the journal's life (1856-1956). Since the preparatory work took five years, it was possible to include the volumes published up to the end of 1959. However, this is not a simple collation of the individual volume indexes: "Had this been done, it would have resulted in an index of no less than 3000 pages and, as indexing methods had changed during the period, it would have been difficult to use" (foreword). A completely new index of names and subjects was prepared by omitting certain materials which were not of permanent interest or which could be traced through other sources. The types of material excluded from this cumulative index are: patent specifications, standards and codes of practice, review articles, leading articles reflecting contemporary opinion, correspondence, book reviews, very short articles, etc.

The following are additional examples of cumulated indexes to primary journals:

1. *A.I.Ch.E. Publications: Combined Cumulative Index, Subject and Author, 1955-1972.* Tokyo: Nichigai Associates, in association with the American Institute of Chemical Engineers, 1973. 620p.
2. *Analytical Index to the Publications of the Institution of Civil Engineers, January 1970-December 1974.* London: Institution of Civil Engineers, 1975. 149p.
3. *Journal of Applied Physics and Applied Physics Letters. Combined Cumulative Index-Subject and Author, 1962-1973.* New York: Nichigai Associates, 1974.

4. *Physical Review and Physical Review Letters. Combined Cumulative Subject Index, 1951-1973.* New York: Nichigai Associates, 1974.
5. *Scientific American. Cumulative Index, 1948-1971: An Index to the 284 Issues from May 1948 Through December 1971, Inclusive.* New York: Scientific American, 1972.

3.7.4 Special Issues and Supplements

Many journals issue special issues and supplements periodically to focus attention on events or topics of current relevance or to provide information that cannot be included in the regular issues. Several technical journals such as *Chemical Week, Chemical Engineering, Electronics,* and *Nuclear News* issue annual buyers' guides. These special issues are actually directories describing products, equipment, and their manufacturers and distributors. Every year in November the American Association for the Advancement of Science (AAAS) publishes a directory of scientific instruments as a special issue of *Science.* In April of each year, the ACS publishes a special issue of *Analytical Chemistry* dedicated to review articles. Directory issues of technical magazines are discussed in detail in a later chapter on trade catalogs. Information on special issues and supplements as well as indexes and their cumulations can be found in *Guide to Special Issues and Indexes of Periodicals,* 2nd ed. by Charlotte M. Devers, Doris B. Katz, and Mary M. Regan (New York: Special Libraries Association, 1976), 289p.

References

1. J. R. Porter, The Scientific journal—300th anniversary, *Bacteriological Reviews,* 28(3):215-216 (September 1964).
2. Porter (Ref. 1), p. 221.
3. David A. Kronick, *A History of Scientific and Technical Periodicals. The Origins and Development of the Scientific and Technical Press, 1665-1790,* 2nd ed. (Metuchen, N. J.: Scarecrow Press, 1976).
4. W. S. Brown, J. R. Pierce, and J. F. Traub, The future of scientific journals, *Science,* 158(3805):1153-1159 (1 December 1967).
5. Charles M. Gottschalk and Winifred F. Desmond, World-wide census of scientific and technical serials, *American Documentation,* 14(3):188-194 (July 1963).
6. Periodicals, *Illinois Monthly Magazine,* 1:302-303 (1831), quoted in Gottschalk and Desmond (Ref. 5), p. 193.
7. Charles H. Brown, *Scientific Serials: Characteristics and Lists of Most Cited Publications in Mathematics, Physics, Chemistry, Geology, Physiology, Botany, Zoology and Entomology.* ACRL Monograph No. 16 (Chicago: Association of College and Research Libraries, 1956).

8. Purpose in publication, *Nature,* 191(4788):527-530 (5 August 1961).
9. P. B. Mangla, Scientific literature and documentation, *Herald of Library Science,* 3(4):286-292 (October 1964).
10. Derek J. de Solla Price, *Science Since Babylon* (New Haven, Conn.: Yale University Press, 1961), p. 95.
11. Derek J. de Solla Price, *Little Science Big Science* (New York: Columbia University Press, 1963).
12. K. P. Barr, Estimates of the number of currently available scientific and technical periodicals, *Journal of Documentation,* 23(2):110-116 (June 1967).
13. D. W. King and others, *Statistical Indicators of Scientific and Technical Communication, 1960-1980.* PB 254 060 (Washington, D.C.: U.S. Government Printing Office, 1976).
14. K. Subramanyam, The scientific journal: A review of current trends and future prospects, *Unesco Bulletin for Libraries,* 29(4):192-201 (July-August 1975).
15. Price, *Little Science Big Science* (Ref. 11), p. 1.
16. Demise of scientific journals, *Nature,* 228:1025-1026 (12 December 1970).
17. Too many chemistry journals, *Chemical & Engineering News,* 51:44-43 (10 December 1973).
18. James A. Moore, An inquiry on new forms of primary publications, *Journal of Chemical Documentation,* 12:75-78 (May 1972).
19. Simon Pasternack, Is journal publication obsolescent? *Physics Today,* 19: 38-43 (May 1966).
20. Research papers in *Analytical Chemistry* (Editor's column), *Analytical Chemistry,* 51(9): 1009A-1010A (August 1979).
21. S. C. Bradford, On the scattering of scientific subjects in scientific periodicals, *Engineering,* 137:85-86 (1934).
22. S. C. Bradford, *Documentation,* 2nd ed. (London: Crosby Lockwood, 1953, p. 18.
23. Denis J. Grogan, *Science and Technology: An Introduction to the Literature,* 2nd ed. (Hamden, Conn.: Shoe String Press, 1973), p. 147.
24. J. Martyn and A. Gilchrist, *Evaluation of British Scientific Journals* (London: ASLIB, 1968).
25. Harold P. Van Cott and Albert Zavala, Extracting the basic structure of scientific literature, *American Documentation,* 19(3):247-262 (July 1968).
26. Pasternack (Ref. 19), pp. 40-41.
27. William D. Garvey and others, Some comparisons of communication activities in the physical and social sciences, in *Communication Among Scientists and Engineers,* edited by Carnot E. Nelson and Donald K. Pollock (Lexington, Mass.: D. C. Heath & Co., 1970), p. 63.
28. D. A. E. Shephard, Some effects of delay in publication of information in medical journals, and implications for the future, *IEEE Transactions on Professional Communication,* PC-16:143-147; 181-182 (1973).

References

29. J. M. Ziman, Information, communication, knowledge, *Nature*, 224: 76-84 (25 October 1969).
30. Stephen A. Rossmassler, Scientific literature in policy decision making, *Journal of Chemical Documentation*, 10:163-167 (August 1970).
31. Joseph A. Kuney and William H. Weisgerber, System requirements for primary journal systems: Utilization of the *Journal of Organic Chemistry*, *Journal of Chemical Documentation*, 10:150-157 (August 1970).
32. R. Elsdon-Dew, The library from the point of view of the research worker, *South African Libraries*, 23:51-54 (October 1955).
33. Richard DeGennaro, Escalating journal prices: Time to fight back, *American Libraries*, 8(2):69-74 (February 1977).
34. Norman B. Brown and Jane Phillips, Price indexes for 1979: U.S. periodicals and serial services, *Library Journal*, 104(15):1628-1633 (1 September 1979).
35. Moore (Ref. 18), p. 75.
36. Bernard M. Fry and Herbert S. White, *Economics and Interaction of the Publisher-Library Relationship in the Production and Use of Scholarly and Research Journals*. November 1975. (Springfield, Va.: National Technical Information Service, Order No. 249108).
37. Louis P. Hammett, Choice and chance in scientific communication, *Chemical & Engineering News*, 39:94-97 (10 April 1961).
38. H. A. Barton, The publication charge plan in physics journals, *Physics Today*, 16:45-57 (June 1963).
39. H. William Koch, Publication charges and financial solvency, *Physics Today*, 21:126-127 (December 1968).
40. *Scientific and Technical Communication. A Pressing National Problem and Recommendations for its Solution* (Washington, D.C.: National Academy of Sciences, 1969), p. 121; Recommendation C8, p. 66.
41. Elwood K. Gannett, Primary publication systems and services, in *Annual Review of Information Science and Technology* (Washington, D.C.: American Society for Information Science, 1973), vol. 8, p. 262.
42. James M. Matarazzo, Scientific journals: Page or price explosion? *Special Libraries*, 63:53-58 (February 1972).
43. W. Davis, Developments in auxiliary publication, *American Documentation*, 2(1):7-11 (January 1951).
44. Gannett (Ref. 41), p. 261.
45. Lee N. Starker, User experiences with primary journals on 16mm microfilm, *Journal of Chemical Documentation*, 10:5-6 (February 1970).
46. Moore (Ref. 18), p. 77.
47. Pauline Kean and Jalrath Ronayne, Preliminary communications in chemistry, *Journal of Chemical Documentation*, 12(4):218-220 (November 1972).
48. Denis J. Grogan, *Science and Technology: An Introduction to the Literature*, 3rd ed. (London: Clive Bingley, 1976), p. 157.
49. K. Subramanyam and Constance J. Schaffer, Effectiveness of "Letters" journals, *New Library World*, 75:258-259 (December 1974).

50. N. W. Pirie, Note on the simultaneous publication of papers at two different levels of completeness, in *The Royal Society Scientific Information Conference, 21 June-2 July 1948. Report and Papers Submitted* (London: The Royal Society, 1948), pp. 419-422.
51. J. D. Bernal, The transmission of scientific information: A user's analysis, in *Proceedings of the International Conference on Scientific Information, Washington, D.C., November 16-21, 1958* (Washington, D.C.: National Academy of Sciences-National Research Council, 1959), vol. 1, pp. 77-95.
52. T. E. Phipps, Jr., Scientific communication, *Science*, 129(3342):118 (16 January 1959).
53. Moore (Ref. 18), p. 76.
54. Rebecca L. Rawls, ACS conducting dual basic journal experiment, *Chemical & Engineering News*, 54(24):28-29 (7 June 1976).
55. D. H. Barlow, A & I services as database producers: Economic, technological and cooperative opportunities, *ASLIB Proceedings*, 28(10):325-337 (October 1976).
56. A debate on preprint exchange, *Physics Today*, 19:60-73 (June 1966).
57. K. Subramanyam, Information Exchange Groups: An experiment in science communication, *Indian Librarian*, 29(4):159-164 (March 1975).
58. Brown, Pierce, and Traub (Ref. 4), pp. 1157-1159.
59. Kenneth D. Carroll, Development of a national information system for physics, *Special Libraries*, 61:171-179 (April 1970).
60. Ralph H. Phelps, Alternatives to the scientific periodical: A report and bibliography, *Unesco Bulletin for Libraries*, 14(2):61-75 (March-April 1960).
61. Jacqueline Hills, *Review of the Literature on Primary Communication in Science and Technology* (London: ASLIB, 1972), pp. 10-11.
62. J. F. Pownall, *Organized Publication* (London: Elliott Stock, 1926) cited in Phelps (Ref. 60), pp. 63-64.
63. J. D. Bernal, Provisional scheme for central distribution of scientific publications, *Royal Society Scientific Information Conference* (Ref. 50), pp. 253-258.
64. Bernal, Transmission of scientific information, (Ref. 51), p. 92.
65. Jordan J. Baruch and Nazir Bhagat, The IEEE Annals: An experiment in selective dissemination, *IEEE Transactions on Professional Communication*, PC-18(3):296-308 (September 1975).
66. William D. Garvey and S. D. Gottfredson, Scientific communication as an integrative social process, *International Forum on Information and Documentation*, 2:9-16 (January 1977).
67. J. P. Eakins, Integrated publication system: A new concept in primary publication, *ASLIB Proceedings*, 26(11):430-434 (November 1974).
68. Joseph H. Kuney, New developments in primary journal publication, *Journal of Chemical Documentation*, 10:42-46 (February 1970).
69. Arthur Herschman, Keeping up with what's going on in physics, *Physics Today*, 24:23-29 (November 1971).

References

70. A. W. Kenneth Metzner, Multiple use and other benefits of computerized publishing, *IEEE Transactions on Professional Communication*, PC-18 (3):274-278 (September 1975).
71. Paul A. Parisi, Composition innovations at the ASCE, *IEEE Transactions on Professional Communication*, PC-18(3):244-273 (September 1975).
72. Robert W. Bemer and A. Richard Shriver, Integrating computer text processing with photocomposition, *IEEE Transactions on Professional Communication*, PC-16(3):92-96 (September 1973).
73. Dorothy K. Korbuly, A new approach to coding displayed mathematics for photocomposition, *IEEE Transactions on Professional Communication*, PC-18(3):283-287 (September 1975).
74. Harold E. Bamford, Jr., The editorial processing center, *IEEE Transactions on Professional Communication*, PC-16(3):82-83 (September 1973).
75. Sarah N. Rhodes and Harold E. Bamford, Editorial processing centers: A progress report, *American Sociologist*, 11:153-159 (August 1976).
76. *Editorial Processing Centers: Feasibility and Promise* (Rockville, Md.: Aspen Systems Corporation; Westat, Inc., 1976).
77. Arthur Herschman, The Primary journal: Past, present and future, *Journal of Chemical Documentation*, 10(1):37-42 (February 1970).
78. J. W. Senders, C. M. B. Anderson, and C. D. Hecht, *Scientific Publication Systems: An Analysis of Past, Present and Future Methods of Scientific Communication*. PB 242 259 (Springfield, Va.: National Technical Information Service, June 1975).
79. Herschman (Ref. 77), p. 41.

4
CONFERENCE LITERATURE

4.1 Scientific Conferences

Presentation of papers at local, national, and international conferences has been one of the most important methods of disseminating scientific information. Nascent R & D information can be communicated more rapidly and directly through conference papers than through papers published in journals. On the average, papers are presented at national conferences about one year prior to their publication in journals [1]. Another advantage of the conference paper over the journal article is the possibility of obtaining immediate feedback from the conference participants in the form of questions and comments following the presentation of the paper. Conferences also provide additional opportunities for nonformal communication and for developing and strengthening personal contacts with other scientists engaged in similar pursuits.

Conferences range from small gatherings of local chapters or special interest groups of national societies to large international congresses attended by thousands of delegates from all over the world. Depending upon their purpose and setting, these meetings are variously called conferences, congresses, symposia, seminars, workshops, colloquia or teach-ins, and generate varying quantities of published and unpublished literature. Hundreds of papers are read in some of the larger conferences such as the meetings of the ACS or the ASME. For example, the *Proceedings of the Second International Conference on the Peaceful Uses of Atomic Energy* (Geneva: The United Nations, 1958) contains 2100 scientific papers submitted by participants from 46 countries and six international agencies.

The conference as a forum for information exchange is not a new phenomenon. According to Alexander King, three international conferences were held in 1853, over 100 in 1909, and at least 2000 in 1953 [2]. In 1958, the International Organizations Section of the General Reference and Bibliography Division of the Library of Congress had identified and recorded in its files 1008 multilateral meetings for 1953, and 3249 meetings for 1957. Based on

4.1 Scientific Conferences

these records, which were admittedly incomplete, Kathrine O. Murra estimated the annual number of multilateral conferences at approximately 5000 [3]. According to a more recent estimate, about 10,000 conferences are held each year all over the world [4]. A National Science Foundation survey showed that 94% of professional scientific societies in the United States organize annual meetings at which original research is reported [5].

The following are some of the factors responsible for the enormous increase in the number of conferences in recent years:

Increased tempo of R & D activity throughout the world

Increase in the rate at which new information is generated and exploited

Growing interest in international cooperation in sharing the results of R & D activity

Need for a means of communication that is faster and more direct than the printed medium

Oversaturation and lack of speed of the printed medium as a means of information transfer

It appears that the scientific conference has become an established institution for both formal and nonformal communication in response to a basic need for a faster, more direct, and more vital means of communication than the overloaded and slower traditional means of information transfer.

Conferences are organized for a variety of reasons. In smaller meetings, new knowledge is announced and its implications are discussed in an expert milieu. Some gatherings are mainly instructional in nature. Others are held in order to review the developments, assess the current state of knowledge, and identify goals for future R & D endeavors in a particular branch of learning. Large congresses are sometimes mainly an expression of solidarity of those attending. At such congresses, little completely new material may be presented, although much of the material presented could be novel to most of the participants.

Alexander King has summarized the main functions of conferences thus [6]:

Announcement of new knowledge

Exchange of information and experience

Education

Formulation of problems and situations, especially in interdisciplinary areas

Fact-finding and reporting

Negotiations and policy formulation

Status and ceremonial congregation

The first four functions seem to be the dominant functions of the majority of the scientific conferences. Baum has identified four major types of conferences on the basis of their principal objectives and format [7]:

1. Conferences at which experts in a given field gather to discuss problems of mutual interest. Such meetings are typified by the Gordon Research Conferences, which are week-long discussion sessions. Attendance is by invitation. No written papers are presented and no proceedings are published [8, 9].

2. Current awareness conferences, typified by the meetings of the Federation of American Societies of Experimental Biology. Summaries of progress reports to be presented at the conference are printed in the society journal well before the conference. Speakers are not required to prepare written copies of their papers. They may, however, later publish a formal paper based on the presentation in a journal.

3. Learned society meetings, exemplified by the meetings of the ASME and the American Institute of Electrical Engineers portion of the IEEE. Papers may be submitted as either "conference papers" or as "transaction papers" for screening by experts and subsequent presentation and discussion at the conference. Transactions papers are preprinted and made available to participants before the conference. Conference papers may be submitted for formal publication by the society after the conference.

4. Professional group conferences, organized by a specialized group of a larger professional society. This approach is taken by the Institute of Radio Engineers portion of the IEEE, which consists of a number of decentralized professional groups devoted to the study of specific branches of electronics engineering. The role of the main professional society in such conferences in minimal.

Regardless of their function or format, conferences usually result in a great many publications of various types, distributed before, during, and after the conference. A notable exception to this are the Gordon Research Conferences which do not generate any printed record except the preconference announcements.

4.2 Preconference Literature

Long before the commencement of the conference, announcements, calls for papers, and programs are published in journals and are also mailed to the members of the sponsoring society and to other potential participants. Journals such as *Science, Nature,* and *Physics Today,* and newsletters (e.g., *Information Hotline)* are invaluable sources of advance information on forthcoming conferences. A number of specialized bibliographic tools that provide advance information on forthcoming conferences are listed and described by Jiřina Čermáková [10].

Abstracts of conference papers are often distributed to intending participants or published in journals prior to the conference. The ACS, the A.I.Ch.E., and some other societies publish premeeting abstracts to tell prospective participants what they may expect to hear at the conference, and also to enable those not

4.5 Bibliographic Control of Conference Literature

attending the conference to gain at least a brief view of the proceedings [11]. The SAE, the American Petroleum Institute (API), and some divisions of the ACS issue preprints of meeting papers either as separates or in bound volumes. Authors of conference papers also send copies of their papers to their friends and colleagues before presenting the papers at the conference. If preprints are distributed in advance, it is usual for the speaker to assume that the paper has been read by the delegates. Distribution of preprints of conference papers gives the conference participants an opportunity to study the papers in advance, so that more time can be spent on discussion of the papers at the conference.

4.3 Literature Generated During the Conference

Very often, abstracts and preprints of papers are made available to participants during the conference. Some societies, such as the American Society for Testing and Materials (ASTM) and the SAE, supply conference papers through a mail order service. Additional literature such as copies of opening and closing speeches, keynote addresses, texts of resolutions, and lists of participants is distributed during the conference. These may also be printed in the official organ of the sponsoring society. The American Society for Information Science (ASIS), for example, provides abstracts of papers on hard copy and full texts of papers on microfiche during its annual conferences.

4.4 Postconference Literature

Postconference literature may appear in one of several alternative forms. The complete proceedings of the conference, consisting of edited versions of the papers read, discussions, speeches, minutes, resolutions, and list of attendees, may be published as a monograph or as a special issue of a journal. Alternatively, some or all of the individual contributions presented at the conference may be published, often suitably revised, as papers in primary journals. It is not uncommon for a conference paper to be subsequently issued as a technical report by the agency supporting the research or the organization in which the investigation was carried out.

4.5 Bibliographic Control of Conference Literature

Papers presented at conferences are very important for current awareness because of the currency of the information reported in them. Published volumes of conferences are valuable as authoritative surveys of developing subjects and as reference works. The bibliographic control of conference papers and published proceedings of conferences is beset with many problems. An editorial in *Nature*, contemplating the future place of symposium volumes in scientific literature, listed several defects of published symposium proceedings [12]:

> The publication of symposium proceedings is delayed because of the time involved in obtaining manuscripts from speakers, editorial processing, and the actual production of the printed proceedings volume.
>
> Symposium volumes contain repetitive material already published elsewhere, because of the need to provide background for the benefit of the conference audience.
>
> Many conference papers do not contain adequate bibliographic references.
>
> Many symposium volumes contain contributions that would not have been approved by referees for publication in a journal.
>
> It is difficult to identify and retrieve information buried in symposium volumes.

The last difficulty is the result of several separate problems such as the geographic distribution of conference avenues, linguistic diversity of the proceedings, inadequate bibliographic control of conference literature, delay in the publication of proceedings, lack of adequate indexing in published proceedings, and so on.

Multiplicity of languages is a problem that conference papers and proceedings share with other forms of scientific literature. This problem was highlighted in a survey carried out by the Union of International Associations covering 285 conferences held during 1960-1961 [13]. Table 4.1 shows the frequency with which various languages were used for written communication in the 285 conferences surveyed.

It is apparent from Table 4.1 that many of the conferences were multilingual and accepted papers written in several languages.

Yet another problem is the lack of a uniform policy regarding the subsequent publication of papers presented at conferences. The Gordon Research Conferences "deliberately discourage publication so that participants can report and discuss more freely work that is not sufficiently advanced for wide dissemination" [14]. The erstwhile Institute of Radio Engineers inserted the following note in its journal, concerning papers read at the Convention of the Institute in 1949 [15]:

> No papers are available in preprint or reprint form, nor is there any assurance that any of them will be published in the *Proceedings of the IRE*, although it is hoped that many of them will appear in these pages in subsequent issues.

A list of papers presented at the 1950 winter general meeting of the former American Institute of Electrical Engineers was prefaced with a similar note [16]:

4.5 Bibliographic Control of Conference Literature

Table 4.1 Languages Used in Conferences

Language	Number of conferences
English	250
French	242
German	121
Spanish	47
Italian	24
Russian	12
Dutch	8
Norwegian	7
Esperanto	6
Hebrew	2
Hindustani	2
Portuguese	2
Danish	1
Finnish	1

These papers are not scheduled for publication in *AIEE Transactions* or *AIEE Proceedings,* nor are they available from the Institute.

Although many channels exist for disseminating conference papers before, during, and after the conference, a large proportion of papers read at conferences are never published, and are thus lost to the scientific community. According to one estimate, as much as 25% of conference papers may not appear in print at all [17]. In a paper read at the International Conference on Scientific Information, 1958, Felix Liebesny reported that 48.5% of some 383 contributions presented in four conferences of American societies remained unpublished [18]. Of the papers that were published, about one third appeared in periodicals other than that in which the abstracts of the papers had appeared.

As mentioned earlier, delay in the publication of conference papers is another problem that renders their bibliographic control more difficult. In the above investigation by Liebesny, it was found that 114 papers (nearly 30%) were published within 12 months, 63 were published between 13 and 24 months, and 14 between 25 and 36 months after the conference. Four and one half years had elapsed before two of the papers were seen in print.

In an ASLIB survey of 194 conference proceedings listed in the *British National Bibliography,* 42 proceedings were found to have been published in the same year as the conference concerned, 93 were published in the following year, 46 were issued two years later, and 13 appeared three years or more after the conference [19].

A frequent drawback of published conference proceedings volumes is the lack of appropriate indexes. In the ASLIB study mentioned above, 205 conference reports noted in the *British National Bibliography* in classes 500 (science) and 600 (technology) during 1956-59 were examined for indexes. Of these, 103 publications (50%) had no index at all; 66% had no author index, and 59% had no subject index. Only a quarter of the publications examined had both author and subject indexes. The investigators felt that the lack of an author index in a conference proceedings volume was not a serious disadvantage unless a large number of authors (say, over 25) were involved. But the lack of a subject index is a definite disadvantage, especially in voluminous publications.

One other important problem afflicting the bibliographic control of conference literature is its uneven coverage in abstracting and indexing services. A study of the International Mineral Dressing Congress, Stockholm, September 1957, showed that of 34 papers, 31 were dealt with in *Chemical Abstracts,* six were abstracted in *Bulletin Signaletique,* and seven in the *Journal of the Iron and Steel Institute* (U.K.). *IMM Abstracts* (of the Institution of Mining and Metallurgy, U.K.) abstracted all the 34 papers (32 were abstracted as "preprints" before the conference and the rest were abstracted after the conference). None of the 34 papers was abstracted in Germany or in the USSR [20].

In another study, Hanson and Janes of ASLIB surveyed the coverage given in English-language abstracting journals to the publications of ten conferences consisting of a total of 386 papers [21]. No abstracts were discovered of any of the papers presented in two of the ten conferences. In all, 117 abstracts (representing 30% of the 386 papers searched) were discovered, but only 79 (or 20%) were abstracts of the papers as printed in the conference proceedings; the rest were abstracts of the same papers published additionally elsewhere, or of the reviews of the papers. Nearly half of the 386 papers were neither abstracted nor indexed.

In spite of the unevenness of their coverage, abstracting and indexing services are important bibliographic tools for locating conference papers and proceedings. Some representative examples are: *Applied Mechanics Reviews, Biological Abstracts, Chemical Abstracts, Computing Reviews, Government Reports Announcements and Index, International Aerospace Abstracts,* and *Physics Abstracts.* The last named publication has a separate index to conferences. Reviews of meetings, usually with abstracts of papers, are published in society journals such as those of the ACS (*Chemical & Engineering News, Rubber Age,* etc.) and the ASME (e.g., *Mechanical Engineering). Chemical Abstracts* includes abstracts of papers presented at the meetings of American societies, even if the papers are not subsequently published in a journal [22].

Several specialized bibliographic tools such as *Conference Papers Index* and *Index to Scientific and Technical Proceedings* are available for tracing

4.5 Bibliographic Control of Conference Literature

conference papers and proceedings. *Conference Papers Index* (formerly, *Current Programs)*, is a monthly current awareness service, with quarterly and annual cumulations, published by Data Courier, Inc., Louisville, Ky. It contains listings of papers presented at conferences in the fields of life sciences, medicine, engineering and technology, chemistry, and the physical sciences. The coverage is world-wide. The database, consisting of over half a million bibliographic entries of papers presented since 1973, is not available for online searching.

Index to Scientific and Technical Proceedings is a monthly indexing service started in 1978 by the Institute for Scientific Information. This index to conference proceedings contains listings, by title and author, of all papers presented in the conferences covered. Some 80,000 papers from about 3000 conference proceedings, including those published as journal issues or in book form, are indexed each year. Alternative approaches to the main entries are provided through several indexes: subject category index, permuterm subject index, author/editor index, sponsor index, corporate source index, and meeting location index.

Information on forthcoming conferences may be obtained from the following sources:

1. *Scientific Meetings.* New York: Special Libraries Association, 1957-, quarterly.

    Describes future meetings of technical, scientific, medical, and management organizations and universities.

2. *World Meetings: United States and Canada.* New York: MacMillan Information, 1963-, quarterly.

3. *World Meetings: Outside United States and Canada.* New York: MacMillan Information, 1968-, quarterly.

4. *Yearbook of International Conference Proceedings.* Brussels: Union of International Associations, biennial.

The following sources are useful in locating conference papers and proceedings:

1. The Research Libraries of the New York Public Library and the Library of Congress. *Bibliographic Guide to Conference Publications, 1975.* Boston: G. K. Hall, 1976. 684p.

    Includes publications catalogued during the year by the New York Public Library and additional entries from LC MARC tapes. Conferences in all languages and all subjects are covered.

2. *Index of Conference Publications Received by the BLLD.*
    Boston Spa, England: British Library Lending Division, 1974-, monthly.

    The BLLD attempts to acquire all conference proceedings that are announced in the *World Meetings* series. Monthly issues of the *Index* are

indexed by subject (keywords). Entries give date and place of meeting, title of proceedings, and BLLD classification.
3. *Directory of Published Proceedings. Series SEMT.* Interdock Corporation, 1964-, monthly. Annual cumulation.
4. *Proceedings in Print.* Matapan, N.J.: Proceedings in Print, Inc., 1964-, bimonthly.

Additional bibliographic tools for identifying conference papers and proceedings are listed and described by Short [4] and Čermáková [10]. Trade bibliographies, catalogs and promotional literature of publishers, and announcements and reviews in journals are other sources of information concerning conferences and conference literature.

4.6 UNESCO-FID Recommendations

The quality and bibliographic control of conference literature have been subjects of much concern to both scientists and librarians. The ban on written communications in the Gordon Research Conferences is a strong deterrent to the proliferation of premature papers reporting undigested information on incomplete investigations. The International Federation for Documentation (FID) has very actively promoted measures to improve the accessibility of conference literature. In 1959, the Federation, under contract with UNESCO, conducted a study on "The Content, Influence and Value of Scientific Conference Papers and Proceedings." For this study, the FID consulted the Union of International Associations, the Abstracting Board of the International Council of Scientific Unions, and several other organizations and individuals, and also obtained information from its own member organizations. The findings of this study on scientific conferences have been published in two parts in the *UNESCO Bulletin for Libraries* [5, 20], and also as a separate report [23]. In this report, known as the *Poindron Report,* FID suggested that organizers of scientific conferences take the following steps in order to make the proceedings of the conferences useful to research workers, particularly those who are unable to attend the conference:

1. To publish, if possible, all papers (accompanied in every case by an author's summary) before the conference; to make available, upon request, the text of unpublished papers.
2. To publish the proceedings as soon as possible after the conference, within a year at the latest, and, in this respect, to consider the advantages, from the point of view of diffusion, of publishing them in a periodical. Members of the Abstracting Board of the International Council of Scientific Unions have proposed that no grant be made by an international union or a government to any conference unless its organizers pledge themselves beforehand to publish the proceedings within a year.

3. To send all conference publications, immediately after they appear, to the secretaries of the international bibliographies concerned, so that not only the volumes considered as a whole, but each individual paper, may be indexed or abstracted (using the author's summary).
4. To give all particulars of all conferences in a single calendar.
5. To provide, through some appropriate means, information about conference publications even before they are issued.
6. To list library holdings of conference publications in union catalogs.
7. To promote standardization in respect of terminology of conferences and their publications, the style of conference publications, and rules for cataloguing them.

Laudable as these recommendations are, they are difficult to implement in view of the numerous and diverse groups of people concerned with and participating in the organization of scientific conferences.

References

1. Johns Hopkins University. Center for Research in Scientific Communication, *Reports of studies of the publication fate of materials presented at national meetings (two years after the meeting)*. PB 185 469 (Springfield, Va.: National Technical Information Service, June 1969).
2. Alexander King, Concerning conferences, *Journal of Documentation*, 17(2):69-76 (June 1961).
3. Kathrine O. Murra, Futures in international meetings, *College and Research Libraries*, 19(6):445-450 (November 1958).
4. P. J. Short, Bibliographic tools for tracing conference proceedings, *IATUL Proceedings*, 6(2):50-53 (May 1972).
5. International Federation for Documentation, The content, influence and value of scientific conference papers and proceedings, *Unesco Bulletin for Libraries*, 16(3):113-126 (May-June 1962).
6. King (Ref. 2), pp. 71-72.
7. Harry Baum, Documentation of technical and scientific meetings, *Proceedings of the American Documentation Institute. Parameters of Information Science*, Philadelphia, Penn. October 5-8, 1964 (Washington, D.C.: American Documentation Institute, 1964), pp. 243-246.
8. W. George Parks, Gordon Research Conferences: Program for 1964, *Science*, 143(3611):1203-1205 (13 March 1964).
9. Dael Wolfle, Gordon Research Conferences, *Science*, 148(3670):583 (April 30, 1965).
10. Jiřina Čermáková, International scientific congresses and conferences: Calendars, bibliographies of congress proceedings and conference technique handbooks, *Annals of Library Science and Documentation*, 19(3):104-113 (September 1972).

11. Eileen F. Dirksen, Meetings and their publications, *Journal of Chemical Documentation*, 5(3):124-125 (August 1965).
12. Symposium volumes, *Nature*, 214(5083):46 (April 1, 1967).
13. *International Associations*, No. 8:484 (1960), cited in Ref. 5.
14. Denis J. Grogan, *Science and Technology: An Introduction to the Literature*, 2nd edition (Hamden, Conn.: Shoe String Press, 1973), p. 172.
15. *Proceedings of the IRE*, 37(2):160 (1949).
16. *Electrical Engineering* (New York), 69(3):250 (March 1950).
17. Short (Ref. 4), p. 50.
18. Felix Liebesny, Lost information: Unpublished conference papers, *Proceedings of the International Conference on Scientific Information, Washington, D.C., Nov. 16-21, 1958* (Washington, D.C.: National Academy of Sciences-National Research Council, 1959), pp. 475-479.
19. C. W. Hanson and M. Janes, Lack of indexes in reports of conferences: Report of an investigation, *Journal of Documentation*, 16:65-70 (June 1960).
20. International Federation for Documentation, Availability of scientific conference papers and proceedings, *UNESCO Bulletin for Libraries*, 16(4):165-176 (July-August 1962).
21. C. W. Hanson and M. Janes, Coverage by abstracting journals of conference papers, *Journal of Documentation*, 17:143-149 (September 1961).
22. C. Alan Moore, Preprints: An old information device with new outlooks, *Journal of Chemical Documentation*, 5(3):126-128 (August 1965).
23. UNESCO. *Scientific Conference Papers and Proceedings* (Paris, 1963).

5

DISSERTATIONS, THESES, AND RESEARCH IN PROGRESS

5.1 Dissertations

The doctorate degree is awarded upon successful completion of supervised research and the formal presentation of the results of research in a dissertation. Each dissertation should deal with some aspect of a subject not previously treated. Doctoral dissertations are therefore an important source of original information. Before embarking on research for the doctoral dissertation, the researcher invariably conducts a comprehensive literature survey; this is done not only to ascertain the present state of knowledge in the field of inquiry, but also to make sure that no one else has made a prior investigation on the proposed research topic. The results of this comprehensive literature survey are presented as a separate chapter in the dissertation in the form of a state-of-the-art review, accompanied by an exhaustive bibliography. Thus, dissertations can also be used as a secondary source for locating reviews and bibliographies.

In 1861, Yale University awarded America's first three earned doctorates in psychology, physics, and classics. Since then, through 1970, American universities have awarded some 340,000 doctoral degrees. One half of these degrees were awarded in the last nine years of this period. According to one projection, some 350,000 to 400,000 doctoral degrees are expected to be awarded during the decade 1971-1980. Slightly more than one half of these will be in science and engineering, including mathematics and the social sciences [1]. During the academic year 1976-1977, American and Canadian universities reported 32,705 doctoral dissertations. Of these, 15,552 (or 47.55%) were in the physical, earth, and biological sciences; the remaining 17,153 were in the social sciences and humanities. A ten-year summary of doctoral degrees granted by American universities is shown in Table 5.1.

Although a large number of doctoral dissertations are written each year, not much use is made of them by scholars and students. Considerable effort, time, and resources are expended in doing research for the doctoral degree and in

Table 5.1 Doctoral Degrees Granted by American Universities

Disciplines	1968	1969	1970	1971	1972	1973	1974	1975	1976	1977
Philosophy	291	303	357	395	375	503	494	441	452	387
Religion	337	330	316	331	419	487	468	602	621	588
Humanities	2187	2847	3076	3651	3735	4528	3950	3885	3839	3481
Social sciences	6876	8316	9938	10,664	11,574	13,104	13,044	12,986	13,322	12,697
Physical sciences	7013	8207	9008	9204	8488	8901	8185	8336	7924	7144
Earth sciences	478	610	635	826	884	1149	1072	1094	951	943
Biological sciences	4593	5605	6242	6540	6383	7676	7608	7468	7691	7465
Totals	21,775	26,218	29,572	31,611	31,858	36,348	34,821	34,812	34,800	32,705

From Ref. 2, by permission.

5.2 Master's Theses

presenting the results of research in a dissertation. According to one estimate, the average cost of a science doctorate is $62,000. This cost represents only the resources expended by the university, and does not include the time invested by the doctoral student [3]. Yet, gauged by the number of citations generated in published literature, doctoral dissertations produce a miniscule impact on the progress of science. In a 3698-item bibliography on the communication of scientific and technical information published by Rutgers University Press, only nine items were indexed as dissertations or theses [4]. In a study of 6838 bibliographic references cited in computer science periodicals, dissertations and theses accounted for only 2.79% of the cited references [5]. Calvin Boyer has suggested the following as the probable reasons for the low volume of use made of doctoral dissertations [3]:

Many dissertation authors do not publish the contents (or a portion) of the completed dissertation in the open literature. Contrary to popular belief, most dissertations do not find their way into print in the form of monographs or as parts of serials; 23% of chemistry dissertations and 51% of psychology dissertations did not yield subsequent publications.

Immediate physical access to most dissertations is relatively unavailable to potential users because libraries do not acquire dissertations systematically. In spite of improved bibliographic access through the *Dissertation Abstracts International* and physical access to copies of dissertations through University Microfilms International, "there is no available evidence to suggest that libraries acquire dissertations in substantially greater numbers now than before the advent of the improved services."

Comments frequently made by students, faculty, and scholars would be somewhat like this [3]:

The last source of information that I would consult for material is *Dissertation Abstracts*. Probably, any material identified would not be held by our library and, usually, the cost, nuisance, and time required to obtain the dissertation are too great. Therefore, *Dissertation Abstracts*, and consequently dissertations, will remain of little value to me as sources of information.

Comments of this nature indicate that, in general, researchers typically use only those materials that are easily accessible, tend to overlook those that are difficult to identify, and ignore those that are difficult to acquire.

5.2 Master's Theses

Research at the master's level is not always a requirement, and the policies of universities and those of individual departments within universities are quite divergent. The present trend appears to be to do away with research and theses as requirements for the master's degree. As in the case of doctoral dissertations, lists of theses published periodically by individual universities are important bibliographic resources for identifying master's theses. A good tertiary source that lists bibliographies of master's theses is the *Guide to Lists of Master's Theses* by Dorothy M. Black (Chicago: American Library Association, 1965). This 144-page guide consists of two sections: (1) lists of master's theses in specific subject fields, and (2) lists of master's theses of specific institutions.

There are two major recurring bibliographies of master's theses at the national level. *Master's Theses in the Pure and Applied Sciences Accepted by Colleges and Universities of the United States and Canada* (New York: Plenum Press, 1957-) is an annual list of master's theses completed at accredited colleges and universities. "Mathematical and most life sciences have been excluded from this publication. . . .Biochemistry, biophysics, and bioengineering are included in the coverage when titles in these areas are reported together with chemistry, physics, and engineering and not as a separate discipline" (from the contents page). The first 12 volumes (1957-1968) were produced and published by the Thermophysical Properties Research Center, School of Mechanical Engineering, Purdue University, Lafayette, Indiana. Volumes 13 through 17 (1969-1973) were produced by the Center, but printed and distributed by Xerox University Microfilms. Canadian theses have been included beginning with volume 18. Plenum Press has been the publisher since volume 18 (1974).

The second major bibliography of master's theses is *Master's Abstracts: A Catalog of Selected Master's Theses on Microfilm* (1962-) published quarterly by University Microfilms International. This abstracting journal was instituted at the instance of the National Science Foundation and the Association of Research Libraries in the wake of increased interest in research at the master's degree level during the mid-1950s. The abstracts are brief (about 150 words), and the coverage is not comprehensive, as many universities do not send copies of master's theses to University Microfilms International. Each issue of the abstract journal has subject and author indexes, which are cumulated in the last issue of each volume. The indexes are also cumulated and published separately every five years.

5.3 Foreign Dissertations and Theses

With the exception of a few countries like the United States and England, national bibliographic control of dissertations is not well organized. Even in the Soviet Union, where vigorous attention is given to other types of bibliographies, the task of compiling bibliographies of dissertations appears to have met with less favor. Generally, national bibliographies cover only dissertations published in monographic form. Unpublished dissertations have to be traced through

5.3 Foreign Dissertations and Theses

other bibliographies and lists of dissertations issued by universities. National bibliographies of dissertations of many countries are listed in general guides to literature such as A. J. Walford's *Guide to Reference Material,* 3rd edition (London: The Library Association, 1973), and Eugene Paul Sheehy's *Guide to Reference Books,* 9th edition (Chicago: American Library Association, 1975).

In England, the ASLIB has been producing an annual index since 1953: *Index to Theses Accepted for Higher Degrees in the Universities of Great Britain and Ireland.* This index lists some 7500 dissertations and master's theses annually, arranged by degree-awarding school and then by discipline. Some doctoral dissertations accepted at British universities during and after 1970 and held at the BLLD are announced in the *BLLD Announcement Bulletin.* This is a guide to British reports, translations, and theses published monthly by the BLLD. Documents listed in the *Bulletin* may be acquired on loan from the BLLD.

Typical of university lists is London University's *Theses and Dissertations Accepted for Higher Degrees,* 1953-. Cambridge University and Oxford University also publish similar serial bibliographies of dissertations.

The official bibliography of French dissertations is an annual publication entitled *Catalogue des Thèses de Doctorat Soutenues Devant les Universités Francaises,* 1884/85-. Until 1959, the title of this publication was *Catalogue des Thèses et Ecrits Académiques.* French dissertations are also listed periodically in a supplement to *Bibliographie de la France-Biblio,* which is a continuation of *Bibliographie de la France* (1811-1971).

Since 1945, all Russian dissertations have been deposited in the State V. I. Lenin Library, Moscow. These are listed in a quarterly publication entitled *Katalog Kandidatskikh i Doktorskikh Dissertatsii Postupivshikh v Biblioteku imeni V. I. Lenina i Gosudarstvennuyu Tsentral'nuyu Nauchnuyu Meditsinskuyu Biblioteku* (Moscow, 1958-). This is the principal current bibliography of master's theses and doctoral dissertations deposited with the V. I. Lenin Library and the State Central Library of Medicine. Eleanor Buist's article in the *Library Quarterly* (April 1963) contains a comprehensive description of the national bibliographic organization of dissertations in the Soviet Union [6]. As examples, other national bibliographies of dissertations may be mentioned: *Canadian Theses* (National Library of Canada, 1962-annual), and *A Bibliography of Doctoral Dissertations Accepted by Indian Universities 1857-1970* (Inter-University Board of India and Ceylon, New Delhi, 1972-).

Copies of foreign dissertations may be obtained through the Center for Research Libraries, Chicago. The Center acquires doctoral dissertations required by its member libraries. However, dissertations from Britain, Canada, the United States, and the Soviet Union, and master's theses, are not handled by the C

5.4 Bibliographic Control of Dissertations and Theses

Most abstracting and indexing journals cover dissertations, but the coverage is far from comprehensive. Individual universities periodically issue lists of dissertations approved. For example:

> Cambridge University. *Abstracts of Dissertations Approved for the Ph.D., M.Sc., and M.Litt. Degrees in the University of Cambridge* ... *1925/26-1956/57.*

This has been continued by:

> Cambridge University. *Titles of Dissertations Approved for the Ph.D., M.Sc., and M.Litt. Degrees in the University of Cambridge, 1957/58-*, Annual.

A list of "Serial Publications Listing or Abstracting Dissertations" is printed in the preliminary pages of *American Doctoral Dissertations.*

The earliest systematic attempt to list doctoral dissertations on a national level was made by the Library of Congress. *List of American Doctoral Dissertations Printed* was initiated in 1912 as an annual publication. Only dissertations printed as monographs or as parts of scholarly journals were included in this list, and unpublished dissertations were not listed; the assumption was that all dissertations would eventually be printed. The series was discontinued with the 1938 issue, printed in 1940.

Doctoral Dissertations Accepted by American Universities (1933/34-1955/56) was an annual bibliography, without abstracts, compiled for the Association of Research Libraries and published by the H. W. Wilson Company. The title of this series changed to *Index to American Doctoral Dissertations* (1955/56-1963/64). This was followed by *American Doctoral Dissertations*, published annually by Xerox University Microfilms for the Association of Research Libraries [7]. *American Doctoral Dissertations* is an annual listing of all doctoral dissertations (including a number of dissertations not abstracted in *Dissertation Abstracts International)* accepted by American and Canadian universities. The list is compiled from the commencement programs issued by universities, and the entries are arranged by subject categories and degree-granting institutions. *American Doctoral Dissertations* is published on a school-year basis.

By far the most comprehensive abstracting service for dissertations is the *Dissertation Abstracts International* published by University Microfilms International (formerly known as Xerox University Microfilms), Ann Arbor, Michigan [8]. This is a monthly compilation of abstracts of doctoral dissertations submitted to University Microfilm International by more than 415 cooperating institutions in the United States, Canada, and a few European countries. However, the coverage of American and Canadian dissertations is not as complete as *American Doctoral Dissertations* [9]. *Dissertation Abstracts International*

5.4 Bibliographic Control of Dissertations and Theses

was begun in 1938 as *Microfilm Abstracts,* and was known as *Dissertation Abstracts* from 1952 until 1969. Its current title reflects the expansion of its coverage to include dissertations from European universities. Since 1966, the abstract journal has been appearing in two sections: Section A, The Humanities and Social Sciences, and Section B, the Sciences and Engineering. Each entry includes the title of the dissertation, author's name, university, date, name of the supervisor, an informative abstract, number of pages, and order number. The abstracts are arranged under subject headings. Each issue has a subject index and an author index. The indexes cumulate annually. Full texts of dissertations may be purchased from University Microfilms International either on microfilm, microfiche (from 1976 onwards), or as enlarged prints.

A cumulated index to the first 29 volumes of *Dissertation Abstracts* was published by Xerox University Microfilms in 1970. This consists of subject and author indexes (in 11 volumes) to all dissertations abstracted in *Microfilm Abstracts* (1938-1951) and *Dissertation Abstracts* (1952-1969). In 1973, Xerox University Microfilms published a monumental index to over 400,000 dissertations from nearly 400 institutions, covering a period of 111 years. The *Comprehensive Dissertation Index* (CDI) is a computer-generated index by keywords and authors, and attempts to list all dissertations accepted at universities of the United States from 1861 onwards. Numerous Canadian and foreign dissertations are also indexed, but no claim of completeness of listing is made for universities outside the United States. The *Index* is in 37 volumes, grouped under disciplines, as follows:

Volumes	Disciplines covered
1-4	Chemistry
5	Mathematics and statistics
6-7	Astronomy and physics
8-10	Engineering
11-13	Biological sciences
14	Health and environmental sciences
15	Agriculture
16	Geography and geology
17	Social sciences
18-19	Psychology
20-24	Education
25-26	Business and economics
27	Law and political science
28	History
29-30	Language and literature
31	Communication and the arts
32	Philosophy and religion
33-37	Author index

Entries in the CDI include title, author, degree, date of degree, institution granting the degree, number of pages in the dissertation, reference to *Dissertation Abstracts International, American Doctoral Dissertations,* or other source of information, and order number when available from University Microfilms International. The *Index* is kept up to date by annual supplements. The 1975 supplement in five volumes indexes some 36,760 dissertations. CDI completely replaces earlier cumulative indexes to dissertations such as the *Retrospective Index* to the first 29 volumes of *Dissertation Abstracts,* the Wilson indexes entitled *Doctoral Dissertations Accepted by American Universities,* 1933/34-1954/55, and the Library of Congress *List of American Doctoral Dissertations Printed,* 1912-1938. The CDI database is also available for online searching.

The Retrospective Index to Theses of Great Britain and Ireland, 1716-1950 (Oxford: European Bibliographical Center-Clio Press, 1976), edited by Roger R. Bilboul, is another massive retrospective index to some 40,000 theses. The index consists of a subject sequence and an author sequence; each sequence is arranged alphabetically. There are five volumes:

Volume 1: Social sciences and humanities. Approximately 13,000 theses. 393p.

Volume 2: Applied sciences and technology including agriculture. Approximately 7000 theses. 159p.

Volume 3: Life sciences, including biological and medical sciences. Approximately 8500 theses.

Volume 4: Physical sciences. Approximately 5000 theses. 99p.

Volume 5: Chemical sciences. Approximately 7500 theses. 251p.

In addition to these general indexes to dissertations and lists of dissertations issued periodically by the degree-granting universities, a third category of bibliographies consists of specialized listings of dissertations in specific disciplines. The following examples are representative of many bibliographies of this kind:

Bibliography of Theses in Geology, 1965-66. Dederick C. Ward and T. C. O'Callaghan. Washington, D.C.: American Geological Institute, 1969. 255p.

Bibliography of Theses in Geology: United States and Canada, 1967-1970. Dederick C. Ward. Boulder, Geological Society of America, 1973. 274p.

Dissertations in Physics: An Indexed Bibliography of all Doctoral Theses Accepted by American Universities, 1861-1959. M. Lois Marckworth. Stanford University Press, 1961. 803p.

Dissertations in Physics lists 8216 dissertations, including the very first doctoral dissertation accepted by an American university: *"Having Given the Velocity and Direction of Motion of a Meteor on Entering the Atmosphere of the Earth, to Determine its Orbit about the Sun, Taking into Account the Attractions of Both these Bodies,"* by Arthur Williams Wright (Yale University, 1861).

An important tertiary guide that describes bibliographies of dissertations and theses is the following:

A Guide to Theses and Dissertations: An Annotated Bibliography of Bibliographies. Michael M. Reynolds. Detroit, Mich.: Gale Research Co., 1975. 599p.

This is a classified bibliography of bibliographies of dissertations and master's theses, both completed and in progress. The guide describes over 2000 bibliographies of dissertations.

Guide to Bibliographies of Theses, United States and Canada, 2nd edition, compiled by Thomas R. Palfrey and Henry E. Coleman, Jr. (Ann Arbor, Mich.: University Microfilms, 1969, 54p.) is a much smaller tertiary guide.

5.5 Research in Progress: SSIE

Since a doctorate is awarded for original work, it is important for a doctoral aspirant to ensure that no one else has completed or is currently conducting an investigation on the proposed research topic. Research intelligence at the wavefront of knowledge is essential not only to prevent unintended duplication of research, but also to exploit the results of research already completed or in progress. Doctoral research currently in progress is listed in the *Directory of Graduate Research* (1953-), a biennial publication of the ACS Committee on Professional Training. The subtitle of the 1975 edition reads thus:

Faculties, Publications, and Doctoral Theses in Departments or Divisions of Chemistry, Chemical Engineering, Biochemistry, and Pharmaceutical and/or Medicinal Chemistry at Universities in the United States and Canada.

Scientific Research in British Universities and Colleges (1951/52-), an annual directory published by the Department of Education and Science and the British Council, provides details of research projects currently in progress in some 160 academic institutions and 104 government and nonacademic institutions in the United Kingdom. Research projects of doctoral candidates are also included.

A few primary journals (e.g., *Chemistry in Canada)* and secondary services (e.g., *Bibliography and Index of Geology)* also list dissertations recently completed and dissertation proposals approved.

A unique source of current information on ongoing research projects is the Smithsonian Science Information Exchange (SSIE). It was established in 1949 as the Medical Sciences Information Exchange under the aegis of the National Academy of Sciences-National Research Council to serve as a clearing house for exchange of research information among several federal agencies concerned with research in the medical sciences. In 1953, the unit was transferred to the Smithsonian Institution as a nonprofit corporation and was redesignated as the Biosciences Information Exchange. During the subsequent years, the scope of its activities expanded to include physical and engineering sciences, and consequently the name was changed in September 1960 to Science Information Exchange of the Smithsonian Institution [10]. SSIE's aim is to "facilitate the planning and management of scientific research by furnishing information about research in progress to scientists, research program managers, and research administrators" [11].

SSIE receives input on research in progress from more than 1300 federal agencies, state and local organizations, foundations, associations, and universities, and processes information on over 100,000 research projects annually. Its computerized database, from which a number of products and services are generated, contains records on more than 200,000 ongoing or recently completed research projects in all branches of pure and applied sciences and engineering, including agriculture, medical sciences, and social sciences. The basic record, called the "Notice of Research Project," contains the following items of information on research projects: Supporting organization; grant or contract number; name and address of performing organization; name, department, and speciality of the principal investigator; names of coinvestigators; project title; period covered; funding; and, in most cases, a 200-word technical description of the project. Project summaries are indexed to an average of 14 subject categories, each of which may include as many as five hierarchical subcategories. An online search system was implemented in 1973. The following are among the products and services of SSIE:

> Custom searches of the database for Notices of Research Project on specific subjects, organizations, geographic areas, or any combination of these.
>
> Selective dissemination of information services for users with individualized interest profiles
>
> *SSIE Newsletter*, published ten times a year.
>
> Catalogs containing summaries of current research projects. The following is an example:
>
> *Science and Technology Research in Progress, 1972-73.* (Orange, N.J.: Academic Media, in cooperation with the SSIE, 1973). This 12-volume catalog contains brief descriptions of over 125,000 current research projects in physics, chemistry, mathematics, engineering, earth and space sciences, as well as medical, biological, agricultural and behavioral sciences.

In addition, the SSIE database on magnetic tape may be acquired, with or without the subject index, on a one-time or periodic-update basis. The database is also available for online searching through remote terminals.

Information on ongoing agricultural and related research projects may be obtained from the Current Research Information System (CRIS) of the U.S. National Agricultural Library. The CRIS database is available for online searching, and contains information on ongoing and recently completed agricultural and related research sponsored or conducted by the U.S. Department of Agriculture, the State Agricultural Experiment Stations, State Forestry Schools, and other cooperating institutions. The database is updated monthly for addition of new project information, and annually for deletion of information on projects terminated [12].

References

1. Dael Wolfle and Charles V. Kidd, The future market for Ph.D.'s, *Science*, 173(3999):784-793 (August 27, 1971).
2. University Microfilms International, *American Doctoral Dissertations* (Ann Arbor, Mich., 1977-1978), p. v.
3. Calvin J. Boyer, *The Doctoral Dissertation as an Information Source: A Study of Scientific Information Flow* (Metuchen, N.J.: Scarecrow Press, 1973.)
4. Rutgers University Press, *Bibliography of Research Relating to the Communication of Scientific and Technical Information* (New Brunswick, N.J., 1967), pp. 630, 723.
5. K. Subramanyam, *A bibliometric investigation of computer science journal literature.* Ph.D. Dissertation, University of Pittsburgh, 1975.
6. Eleanor Buist, Soviet dissertation lists since 1934, *Library Quarterly*, 33(2):192-207 (April 1963).
7. Julie L. Moore, Bibliographic control of American doctoral dissertations, *Special Libraries*, 63(5/6):227-230 (May/June 1972).
8. Patricia M. Colling, Dissertation Abstracts International, *Encyclopedia of Library and Information Science* (New York: Marcel Dekker, 1972), v. 7, pp. 238-240.
9. Julie L. Moore, Bibliographic control of American doctoral dissertations, *Special Libraries*, 63(7):285-291 (July 1972).
10. William H. Fitzpatrick and Monroe E. Freeman, Science Information Exchange: The evolution of a unique information storage and retrieval system, *Libri*, 15(2):127-137 (1965).
11. Monroe E. Freeman, Science Information Exchange as a source of information, *Special Libraries*, 59(2):86-90 (February 1968).
12. National Agricultural Library: CRIS-Online, *NFAIS Newsletter*, 19(2):6 (April 1977).

6
PATENTS

6.1 Introduction

A patent is a protection granted by the government to an inventor to prevent unauthorized exploitation of his invention. The subject matter of a patent is an invention. The precise nature of the protection granted should be clarified: What is granted to the inventor (or his heirs or assigns) is not the right to make, use, or sell, but the right to exclude others from making, using, or selling the invention. In the United States, patents are granted by the United States Patent and Trademark Office, an agency of the Department of Commerce of the Federal Government; the duration of the protection is 17 years. The right conferred by the patent grant extends throughout the United States and its territories and possessions. In return for the protection granted by the state, the inventor is required to file a description of the invention in the form of a patent specification.

To be eligible for patent protection, an invention must be a new and useful process, machine, manufactured product, or chemical composition or substance, or a new and useful improvement to any of these. By "useful" is meant the condition that the invention serve a useful purpose and be operable. Patents have been granted to very novel and curious inventions. On March 10, 1896, J. C. Boyle was granted U.S. patent 556,248 for an automatic hat tipper. This mechanism enables a gentleman to keep his hands in his pocket while still performing the chivalrous duty of tipping his hat. A mere idea or a suggestion, or a machine that does not actually operate or perform the intended function cannot be patented. Alleged inventions of perpetual motion machines are not considered patentable. Also, under the Atomic Energy Act of 1954, inventions useful solely in the utilization of special nuclear material or atomic energy for atomic weapons cannot be patented.

A patent should be distinguished from a trademark and a copyright. All these are protections granted by the state, but are of different kinds. A trademark is the name or symbol used with a product to indicate its source or origin,

6.2 The United States Patent and Trademark Office

and to distinguish it from the goods of others. Trademarks may be registered in the Patent Office. Trademark registration excludes others from using the registered name or symbol, but does not prevent others from making and selling the same goods under a different name or symbol. Copyright protects the writings of an author against copying. Copyright protects the form of expression rather than the subject matter of the writing. A description of a machine could be copyrighted as a writing. This would only prevent others from copying the description; it does not prevent others from writing another description of the machine or from making and using the machine. Literary, dramatic, musical, and artistic works may be copyrighted. Copyrights are registered with the Copyright Office in the Library of Congress, Washington, D.C. 20540.

The history of patents has been traced back to the ancient Greek civilization of the third century B.C. [1]. But the oldest patent system with a continuous history to the present time is the British patent system, which is said to have been started from the Statute of Monopolies of 1623. The patents granted until that time were really monopolistic grants to royal court favorites of the sole right to deal in certain commodities or trades. In fact, the word "patent" means "open," and is used as an abbreviation of "royal letters patent"—an open letter from the state to all subjects announcing privileges. Patents became a form of scientific literature when the Patent Law Amendment Act of 1852 in England directed that all patents subsequently granted should be printed. At that time all the earlier patents starting from Number 1 of 1617 (for "A Certain Oyle to Keep Armour and Armes from Rust and Kanker") were also printed [2].

6.2 The United States Patent and Trademark Office (PTO)

Before the Constitution of the United States was adopted, the various American colonies granted patents to settlers for methods of making salt and agricultural products. The first patent in Colonial America was granted by the Massachusetts General Court in 1641 to Samuel Winslow for making salt. A number of patents were granted for salt making (presumably by different processes) in other States: Plymouth (1641), Massachusetts (1652), Virginia (1660), New York (1661), Connecticut (1691), and South Carolina (1725).

The Constitution of the United States contains a specific provision (Article 1, Section 8, Clause 8) empowering the Congress to enact laws relating to patents: "Congress shall have the power ... to promote the progress of science and useful arts, by securing for limited times to authors and inventors the exclusive right to their respective writings and discoveries." Under the first patent act approved by George Washington on April 10, 1790, the power to grant patents was vested in a board of three members consisting of the Secretary of State, the Secretary for the Department of War, and the Attorney General. The first Board of Examiners consisted of Thomas Jefferson (Secretary of State), Henry Knox

(Secretary of War), and Edmund Ralph (Attorney General). Each patent had to be signed by the President and the Secretary of State, and the administration of the patent system was entrusted to the Department of State. Thus Thomas Jefferson became the first administrator of the U.S. patent system. The first patent under the Act was granted on July 30, 1790 to Samuel Hopkins for "Making Pot and Pearl Ashes."

The Patent Act of 1836 provided for the establishment of the Patent and Trademark Office (PTO) under the direction of a Commissioner of Patents. The PTO was tranferred from the Department of State to the Department of the Interior in 1849, and became a part of the Department of Commerce in 1925. The present patent laws were enacted in 1870 and revised in 1952.

The role of the PTO is to encourage the development of business and industry in the United States. Toward this end, the PTO performs the following functions:

Provide patent protection for inventions

Register trademarks for corporate and product identifications

Advise and assist other government agencies in matters pertaining to inventions, patents, and technology utilization

Publish and distribute patents

Maintain a public search room where patents and related records are accessible to the public

Collect, classify, and disseminate the technological information disclosed in patents

A list of publications of the PTO is appended at the end of this chapter.

Examination of patent applications is the largest and most important function of the PTO. It has approximately 3000 employees, and about one half of these are examiners and other professional staff with technical and legal education. Each year the PTO receives over 100,000 applications, and grants about 70,000 patents. So far, the PTO has granted about 4.1 million patents.

The current numbering system for U.S. patents was initiated in 1836. Till then some 4000 patents had been issued. The number of patents granted each year has been increasing. The one millionth patent was issued on August 8, 1911–75 years after the beginning of the new numbering system. The second and third millionth patents were issued at equal intervals of about 25 years: Patent No. 2,000,000 was issued on April 30, 1935, and patent No. 3,000,000 on September 12, 1961. It took only 15 years to issue the next million patents. In a special introduction to the December 28, 1976 issue of the *Official Gazette*, which included patent No, 4,000,000 ("A Process for Recycling Asphalt-Aggregate Compositions"), the Commissioner of Patents and Trademarks wrote:

6.3 Foreign Patents

> The increasing rate of patenting reflects the accelerating pace of our technological progress. . . . Patents encourage the development of technology by providing incentives to make inventions, to invest in research and development, to put new or improved products and processes on the market, and to disclose inventions that otherwise would be kept as trade secrets. I am confident that the patent system will continue to help bring forth technology to satisfy our nation's needs—needs such as new energy sources, a cleaner environment, and better food and medical care. . . . It is fitting at the close of our nation's 200th birthday year to pay tribute to the system that has contributed so greatly to our technological strength. . .

Patent specifications are published on the day they are granted. Printed copies of patents are sent to about 30 patent depositories in the United States and also to foreign patent offices under exchange agreements. Patent depositories in the United States are listed at the end of this chapter. Additional copies of patents are sent on a standing order basis to subscribers who wish to receive all new patents issued in a specified class or subclass of the patent classification system. Individual copies or subclass collections of copies are supplied by the PTO in response to specific orders for these from the public; the price is 50 cents per copy. Every year, the PTO distributes about ten million copies of patents to the public.

The size of the patent specifications varies from a couple of pages to a few hundreds of pages or even more. The British patent 1,108,800 for an IBM computer has 1139 pages and 419 sheets of drawings [3]. Microfiche copies of U.S. patents and the weekly journal *Official Gazette* can be obtained from the firm Research Publications, Inc., 12 Lunar Drive, New Haven, CT 06525.

A science library, a search room, and a record room, all located in the PTO at 2021 Jefferson Davis Highway, Arlington, VA 22202, are open to the public. The science library has 200,000 books, 2000 journals on subscription, and 28,000,000 foreign patent documents from 52 countries. All U.S. patents granted since 1836 are available for examination in the search room. One set of patents is arranged in numerical order by patent number, and a second set is filed according to the U.S. Patent Classification System. The record room contains files of records and papers pertaining to all U.S. patents. These are available for public inspection.

6.3 Foreign Patents

Patent laws and procedures differ from one country to another. National patent laws specify the nature of inventions that can be patented, the conditions under which patents are granted, and the procedure for securing a patent grant. An introduction to the patent procedures may be found in *Foreign Patents: A Guide to Official Patent Literature* by Francis J. Kase (Dobbs Ferry, N.Y.:

Oceana Publications, 1972). The British patent procedure allows the inventor to file a provisional patent application, even when the invention is still not complete in its projected final form; this provisional specification is helpful to the inventor in establishing his claim for priority. A complete specification has to be filed within one year from the date of filing the provisional specification. Each year some 50,000 patent applications are examined in the British Patent Office, and often two or more years elapse between the filing of the application and its acceptance. When the inventor files an application to protect his invention, the patent office gives a serial number to the application and announces the title of the invention in the weekly *Official Journal (Patents)*. The application is then examined by the patent office examiners, and, if found acceptable, it is given a seven-digit number which uniquely identifies the patent, and the acceptance is announced in the *Official Journal*. The patent specification is then printed and put on sale. If there is no objection from anyone challenging the invention during the next three months, the patent document is stamped with the official seal and then the patent is said to have been granted. This document becomes the "royal letters patent"—a legal document on parchment, separate from the printed specification. The protection in England is for 16 years from the date of filing the complete specification, and an annual renewal fee has to be paid after the first four years. Approximately 45,000 patents are published annually in Britain.

Unlike the copyright protection, patent protection is not international, and inventors have to obtain patent protection in many countries by filing patent applications in the respective national patent offices. Equivalent or corresponding patents for the same invention in chemistry and related fields can be traced through the Patent Concordance of *Chemical Abstracts,* which covers over 50,000 patents annually from many countries of the world.

6.4 Patents as a Source of Technological Information

About 400,000 patents are issued each year worldwide; of these, about 70,000 are U.S. patents. Since inventors are required to provide a complete description of the invention in order to obtain a patent grant, it is conceivable that descriptions of almost all new technology can be found in patents. Despite the wealth of technological information contained in patents, several studies have indicated that patents are not used extensively by scientists, engineers, and technologists as a source of technical information [4,5]. Terapane has suggested the following as possible reasons for the meager use of patents [6]:

Researchers may be unaware of the kind of information that patents contain, or they may not know how to obtain patents.

The information in patents may seem out of date because of the delay—often several years—between development of an invention and its acceptance and publication as a patent.

Inefficiencies in abridgements (abstracts) and indexes hinder certain types of searches.

The peculiar language (technical-legal jargon) in which patents are written is difficult for researchers to comprehend. Foreign language patents present a similar barrier.

A more serious deterrant to the use of patents is the widely held belief that technical information contained in patents will eventually be republished in technical magazines and journals. Evidence reported in literature does not support this contention. Felix Liebesny and coworkers found that out of a random sample of 1058 British patents, only 61 (5.77%) were substantially republished in journal articles, books, and other forms of nonpatent literature [7]. According to Vcerasnij, only 5-10% of new technological solutions offered by patents are published subsequently in nonpatent literature [8].

In another study of 435 randomly chosen U.S. patents, Terapane found that only 16% of patented technology was substantially disclosed in published nonpatent literature, and 13.3% was partially disclosed. An article was considered to "substantially" disclose the technology described in a patent if 75% or more of the patented technology was reported; disclosure of less than 75% of patented technology was considered "partial" disclosure. The remaining 70.7% of patented technology was not at all disclosed in subsequently published nonpatent literature [9]. Thus, if patents relevant to a given research investigation are not identified and used, a great deal of technical information contained therein is either likely to be irretrievably lost or may have to be regenerated at considerable expense of research resources and time.

6.5 Bibliographic Control of Patents

New patents granted by the national patent offices are announced in their official journals (e.g., the *Official Journal* of the British Patent Office). In the United States, the *Official Gazette* is the official journal of the PTO relating to patents and trademarks. It has been published weekly (every Tuesday) since 1872. Each weekly issue of the *Official Gazette* contains abstracts of about 1500 newly granted patents. The abstracts include a selected drawing from the patent where applicable. The weekly issues of the *Official Gazette* have an index of inventors and a subject index in which patent numbers are listed under the class and subclass numbers of the U.S. Patent Classification System. These indexes are cumulated and published annually as the *Index of Patents*. Part I of of the *Index of Patents,* List of Patentees, is an alphabetical listing of all patentees and assignees who received patents during the year. Part II, Index to Subjects of Invention (classification of patents), lists all patents granted during the year under the class and subclass numbers of the U.S. Patent Classification System. The *IFI Assignee Index to United States Patents* is a similar index published quarterly (with annual cumulation) by IFI/Plenum Data Company, Arlington, VA 22202.

Each abstract in the *Official Gazette* is assigned a class number according to the U.S. Patent Classification System. The *Manual of Classification* of the PTO contains a list of about 350 main classes and 100,000 subclasses of the U.S. Patent Classification System, along with an alphabetical index of about 50,000 entries [10]. The *Manual* is a loose-leaf book with periodic supplements, and can be obtained on an annual subscription basis from the Superintendent of Documents, Washington, DC 20402.

The *Index to U.S. Patent Classification* is an alphabetical index to the *Manual of Classification.* The *Index* consists of alphabetical listing of subject matter headings, and serves as an initial entry point into the classification system by providing specific class and subclass numbers for subject matter headings. The class and subclass titles found in the *Manual* are not always fully indicative of the scope of each subclass. More descriptive definitions for each class and subclass may be found in a separate publication entitled *Classification Definitions.* The *Manual of Classification,* the *Index to U.S. Patent Classification,* and *Classification Definitions,* together constitute a set of three important tools of the U.S. Patent Classification System. The following concordance is useful in finding International Patent Classification numbers corresponding to U.S. Patent Classification numbers.

> United States. Patent Office. *Concordance: U.S. Patent Classification to International Patent Classification.* Washington, D.C.: U.S. Government Printing Office, 1969. 194 p.

Several abstracting and indexing services, such as those published by INSPEC (International Information Services for the Physics and Engineering Communities) cover patents, but the coverage of patents in secondary services is, in general, very uneven. In a smaple study of U.S. and foreign patents covered in *Chemical Abstracts,* Oppenheim obtained following results [11] in Table 6.1.

Oppenheim's results were similar to those obtained by D. Kaye in 1965. Kaye examined the coverage in *Chemical Abstracts* of five random samples of 50 patents from the United States, West Germany, United Kingdom, and Belgium. In the case of the United States and West German patents, the *Chemical Abstracts* coverage was 100%; 70% of the British patents and only 2% of the Belgian patents were found to have been covered in *Chemical Abstracts.* Thus, the coverage of patents in *Chemical Abstracts* is uneven, although the abstracting

6.5 Bibliographic Control of Patents

Table 6.1 Coverage of Chemical Patents in *Chemical Abstracts*

Sample of patents examined	Patents covered in *Chemical Abstracts* Number	Percentage
567 U.K. patents on 20-keto steroids	472	83.2
809 U.K. patents on 11-keto steroids	786	97.2
94 U.S. patents on aminopyridines	92	97.9
62 U.S. patents on halogenated hydrocarbons	58	93.5
113 West German pharmaceutical patents	109	96.5
133 Japanese pharmaceutical patents	69	51.9
463 Belgian pharmaceutical patents	7	1.5

Source: Data from Ref. 11 with permission.

service endeavors to cover patents of chemical and chemical engineering interest exhaustively from 20 countries (including the United Kingdom, Japan, Belgium, and West Germany), and selectively from 11 other countries [12].

Each weekly issue of *Chemical Abstracts* has a numerical patent index and a patent concordance. In the numerical patent index, patents are indexed first alphabetically by country, and then under each country numerically by patent number; the patent numbers are linked to the appropriate abstract number.

The patent concordance links corresponding patents. Each time Chemical Abstracts Service receives a patent that corresponds to another patent abstracted earlier in *Chemical Abstracts*, the corresponding patent received later is not abstracted again; instead it is linked to the earlier corresponding patent and its abstract number, as shown in Figure 6.1.

The Austrian patent 340018 corresponds to Belgian patent 852569; an abstract of the Belgian patent has appeared in *Chemical Abstracts*, volume 88, abstract No. 75397k. When the corresponding Austrian patent 340018 is received at the chemical Abstracts Service, it is not abstracted again; instead, it

Patent number	Corresponding patent	CA Ref. number
AUSTRIAN 340018	Belg 852569	88, 75397k

Figure 6.1 Sample entry from *Chemical Abstracts* Patent Concordance.

is linked with the corresponding Belgian patent abstracted earlier and its abstract number, as shown in Fig. 6.1.

Both the numerical patent index and the patent concordance are cumulated semiannually in the *Chemical Abstracts* volume indexes, and again in the quinquennial collective indexes.

Henry M. Woodburn's *Using Chemical Literature: A Practical Guide* (New York: Marcel Dekker, 1974) is one of several guides that are helpful in searching the patent indexes and other indexes of *Chemical Abstracts*.

There are a few secondary services, other than the *Official Gazette* of the PTO, that exclusively cover patents. The *NASA Patent Abstracts Bibliography* (1972-) is a semiannual recurring bibliography of patents owned by the National Aeronautics and Space Administration (NASA). The following publications are other examples of this type:

API Patent Alert. New York: American Petroleum Institute, 1972-, weekly. (Formerly, *American Petroleum Institute Abstracts of Refining Patents*.)

Polymer Science and Technology—Patents (POST-P). Columbus, Ohio: Chemical Abstracts Service, 1967-, semimonthly. Since 1971, this is available only on computer-readable magnetic tape.

Uniterm Index to Chemical Patents and Defensive Publications. New York: IFI/Plenum Publishing Corporation, 1972-.

The IFI/Plenum Data Company in Arlington, Va., has developed a computerized database service entitled CLAIMS (*C*lass Code, *A*ssignee, *I*ndex, *M*ethod Search). The database is updated monthly and contains information relating to over 350,000 U.S. chemical and chemically related patents issued since January 1950. The database is searchable online through Lockheed's DIALOG system.

Derwent Publications Limited, of London, specializes in publishing patent abstract services. Each year, abstracts of some 60,000 patents from 24 countries are published in Derwent publications. Some of these publications are: *British Patent Abstracts* (all subjects); *British Patent Report* (chemical patents only); *Soviet Inventions Illustrated* (chemical, electrical, and mechanical and general patents); *Japanese Patents Gazette* (chemical patents only), and so on. Derwent's computerized databases (entitled *World Patents Index*, *World Patents Abstracts*, and *Central Patents Index*) cover patents issued in over 20 industrial nations and are available for online searching.

Some primary journals and technical magazines (e.g., *Journal of Applied Chemistry, Modern Plastics, New Scientist, Production Engineer*) also announce new patents, sometimes with abstracts, as a regular feature. Latest developments in the field of patent information are reported in a new quarterly journal entitled *World Patent Information*, sponsored jointly by the Commission of the European Communities and the World Intellectual Property Organization.

Publications of the Patent and Trademark Office

Patents: Printed copies of any patent, identified by its number, may be purchased from the Patent and Trademark Office, Crystal Plaza, 2021 Jefferson Davis Highway, Arlington, Va., at a cost of 50¢ each, postage free. Plant patents cost $1.00 each, and design patents 20¢ each.

Official Gazette of the United States Patent and Trademark Office: Published weekly since January 1872. Orders for subscription or for single copies should be sent to the Superintendent of Documents, U.S. Government Printing Office, Washington, D.C. 20402.

Index of Patents: This is an annual index to the *Official Gazette,* and can be obtained from the Superintendent of Documents.

Index of Trademarks: An annual index of registrants of trademarks; sold by the Superintendent of Documents.

Manual of Classification: A loose-leaf book containing a list of all the classes and subclasses of inventions in the Patent and Trademark Office classification system, a subject matter index, and other information relating to patent classification. Substitute pages are issued from time to time. Sold by the Superintendent of Documents.

Classification Definitions: Contains the changes in classification of patents as well as definitions of new and revised classes and subclasses. Sold by the Patent and Trademark Office.

Weekly Class Sheets: Lists showing classification of each patent in the weekly issue of the *Official Gazette.* Sold on annual subscription by the Patent and Trademark Office.

Patent Laws: A compilation of patent laws in force. Sold by the Superintendent of Documents.

Title 37 Code of Federal Regulations: Includes rules of practice for patents, trademarks, and copyrights. Available from the Superintendent of Documents.

Trademark Rules of Practice of the Patent and Trademark Office with Forms and Statutes: Rules governing the procedures in the Patent and Trademark Office in trademark matters and a compilation of trademark laws in force. Sold by the Superintendent of Documents.

General Information Concerning Trademarks: A brief introduction to trademark matters including an account of the workings of the Patent and Trademark Office, instructions for registrants, and definitions of patents, copyrights, and trademarks. Obtainable from the Superintendent of Documents.

Patents and Inventions—An Information Aid for Inventors: Provides information which may help inventors decide whether to apply for patents and aid them in obtaining patent protection and promoting their inventions. Sold by the Superintendent of Documents.

Directory of Registered Patent Attorneys and Agents Arranged by States and Counties: A geographical listing of patent attorneys and agents registered to practice before the U.S. Patent and Trademark Office. Sold by the Superintendent of Documents.

Manual of Patent Examining Procedure: A loose-leaf manual which serves primarily as a detailed reference work on patent examining procedure for the PTO's examining corps. Subscription service, sold by the Superintendent of Documents, includes basic manual, quarterly revisions, and change notices.

Guide for Patent Draftsmen: Describes PTO requirements for patent drawings. Illustrated. Sold by the Superintendent of Documents.

The Story of the United States Patent Office: A chronological account of the development of the U.S. Patent and Trademark Office and patent system and of inventions which had unusual impact on the American economy and society. Sold by the Superintendent of Documents.

Patent Cooperation Treaty: Copy of the Treaty Articles and Regulations available in limited quantities from the PTO.

General Information Concerning Patents: A brief introduction to patent matters, including an account of the workings of the PTO, instructions to patent applicants, definitions of patents, copyrights, and trademarks, and a set of specimen forms useful to patent applicants. Obtainable from the Superintendent of Documents.

List of Patent Depositories in the United States

1. ALBANY, N.Y., State University of New York Library.
2. ATLANTA, Ga., Georgia Tech. Library.
3. BOSTON, Mass., Boston Public Library.
4. BUFFALO, N.Y., Buffalo and Erie County Public Library.
5. CHICAGO, Ill., Chicago Public Library.
6. CINCINNATI, Ohio, Cincinnati Public Library.
7. CLEVELAND, Ohio, Cleveland Public Library.
8. COLUMBUS, Ohio, Ohio State University Library.
9. DETROIT, Mich., Detroit Public Library.
10. KANSAS CITY, Mo., Linda Hall Library.
11. LOS ANGELES, Calif., Los Angeles Public Library.

12. MADISON, Wis., State Historical Society of Wisconsin.
13. MILWAUKEE, Wis., Milwaukee Public Library.
14. NEWARK, N.J., Newark Public Library.
15. NEW YORK, N.Y., New York Public Library.
16. PHILADELPHIA, Penn., Franklin Research Institute.
17. PITTSBURGH, Penn., Carnegie Library of Pittsburgh.
18. PROVIDENCE, R.I., Providence Public Library.
19. ST. LOUIS, Mo., St. Louis Public Library.
20. STILLWATER, Okla., Oklahoma Agricultural and Mechanical College.
21. SUNNYVALE, Calif., Sunnyvale Public Library.
22. TOLEDO, Ohio, Toledo Public Library.

References

1. Herman Skolnik, Historical aspects of patent systems, *Journal of Chemical Information and Computer Sciences*, 17(3):119-121 (August 1977).
2. Denis J. Grogan, *Science and Technology: An Introduction to the Literature*, 3rd edition (London: Clive Bingley, 1976), p. 252.
3. Grogan (Ref. 2), p. 256.
4. J. S. Gilmore and others, The channels of technology acquisition in commercial firms, and the NASA dissemination program NASA CR-790 (Denver Research Institute, June 1967).
5. N. B. Hannay and others, *Cost-Effectiveness of Information Subsystems*. (Subcommittee on the Economics of Chemical Information, Committee on Corporation Associates, American Chemical Society, May 1969.)
6. John F. Terapane, A unique source of information, *Chemtech*, 8(5):272-276 (May 1978).
7. Felix Liebesny, J. W. Hewitt, P. S. Hunter, and M. Hannah, The scientific and technical information contained in patent specifications. The extent of time factors of its publication in other forms of literature, *The Information Scientist*, 8(4): 165-177 (December 1974).
8. Rostislav P. Vcerasnij, Patent information and its problems, *UNESCO Bulletin for Libraries*, 23(5):234-239 (September-October 1969).
9. Terapane (Ref. 6), pp. 273-274.
10. Kendall J. Dood, The U.S. Patent Classification System, *IEEE Transactions on Professional Communication*, PC-22(2):95-100 (June 1979).
11. C. Oppenheim, The patent coverage of *Chemical Abstracts*, *The Information Scientist*, 8(3):133-137 (September 1974).
12. Introduction to Patent Concordance in *Chemical Abstracts* volume indexes, v. 89, December 31, 1978.

7

TECHNICAL REPORTS

7.1 Introduction

The technical report may be thought of as an accepted bibliographic format for the dissemination of technical information generated through R & D effort in the same way that the scholarly journal article is the accepted channel for communicating the results of scientific research. The English word "report" is derived from the Latin word *reportare* which means "to bring back." This implies that a report contains information produced in response to a specific request or need and submitted to the individual or agency making the request or commissioning an investigation. C. P. Auger offers the following definition of a report [1]:

> A report is a document which gives the results of or the progress with research and/or development investigation. Where appropriate it draws conclusions and makes recommendations, and is initially submitted to the person or body for whom the work was carried out. Commonly a report bears a number which identifies both the report and the issuing organization.

A very large number of technical reports are issued each year. Estimates of the annual output of reports have ranged from 50,000 to 500,000 [2,3]. About 80 to 85% of the world output of report literature is produced in the United States.

7.2 History of Report Literature

Neil Brearley suggests that technical reports predate scientific journals and that "scientists were exchanging reports with one another long before scientific communication was institutionalized" [4]. Copernicus, for example, distributed a preliminary draft of his new cosmology to a few selected scientists two decades prior to publishing his monumental work in 1543 [5]. Industrial research

100

7.2 History of Report Literature

laboratories have always used technical reports and memoranda for internal communication. But the history of the technical report as a distinct format for disseminating technical information dates back only to the beginning of the 20th century. The *Professional Papers of the United States Geological Survey* (published from 1902), and the *Technologic Papers of the National Bureau of Standards* (starting from 1910) may be said to mark the beginning of report literature. Since 1928, the *Technologic Papers* have been incorporated into the *NBS Journal of Research*. The earliest reports to be issued in Great Britain were the Reports and Memoranda series of the Advisory Committee for Aeronautics (now called the Aeronautical Research Council) starting from 1909 [6].

The Second World War spurred a great deal of research and development (R & D) activity in the United States, especially in subjects directly or indirectly affecting the war effort. Government expenditure on R & D greatly increased as numerous defense-related projects were sponsored by many government agencies such as the Army, the Navy, and the War Production Department. A separate agency called the Office of Scientific Research and Development (OSRD) was established in June 1941 to mobilize scientific and technical information resources for national defense. Research laboratories in universities and industrial corporations that performed research for the various government agencies were required to submit progress reports to the sponsoring agencies.

With the cessation of hostilities the OSRD was abandoned, but the pace of government-sponsored R & D effort and the generation of technical reports from these projects did not slacken in the years following the War. By 1950, the annual output of technical reports from government-sponsored R & D activity was estimated to be in the region of 75,000 to 100,000. At the close of the Second World War, great quantities of report literature were acquired by the teams of Allied specialists who went to Germany and Japan with the following mission [7]:

1. To capture enemy documents of a scientific and technical nature which might be of use to the Allies. Some 1500 tons of documents and reports thus acquired were known as Captured German Documents (CGD).
2. To interview German and Japanese scientists, engineers, and manufacturers to obtain additional technical data. Results of these interviews and visits to factories were recorded in technical reports.
3. To capture useful equipment and facilities.
4. To capture key scientists.

Of these four objectives, the first two are of bibliographical interest. The findings reported in these captured reports and in other reports generated through sponsored research had to be kept secret for some time. After the War, most of these reports were declassified and released for general access. In June 1945, a Cabinet Committee called the Publications Board (PB) was established by an Executive Order to release scientific and technical information generated during the War. Later in the same year, the Publications Board was also authorized to process reports acquired from enemy sources and from cooperating foreign governments. By early 1946, documents from various sources were pouring into the office of the Publications Board at the rate of 5000 pieces a month. Each of these reports was given a PB accession number. A weekly announcement service entitled *Bibliography of Scientific and Industrial Reports* (BSIR) was established to promote wide dissemination of the reports to industries and businesses for exploitation. Through a series of changes in practically every aspect (including title, sponsor, frequency, indexes, and even volume numbering), this announcement service has evolved into the present *Government Reports Announcements and Index* (GRAI). These changes are depicted in Table 7.1. The reports themselves were stored in the three national libraries. The Army Medical Library (now the National Library of Medicine) received reports on medical and pharmaceutical topics; reports on agriculture, forest products, chemistry, and textiles were sent to the Library of the Department of Agriculture (now the National Agricultural Library); and all the remaining reports were sent to the Library of Congress.

The Publications Board merged with a newly created agency called the Office of Technical Services (OTS), which was established in 1946 under the Department of Commerce to handle the distribution of technical reports. The three national libraries supplied microfilm or photostat copies of reports in response to requests received from the public through the OTS. In a six-month period during 1946, requests for a quarter of a million copies were processed by the libraries. In 1950, the OTS started receiving technical reports on a regular basis from various government agencies such as the National Advisory Committee for Aeronautics (now the National Aeronautics and Space Administration), the United States Atomic Energy Commission (no longer in existence), the Naval Research Laboratory, and the Tennessee Valley Authority, as well as from universities and other agencies performing research under government contract.

In 1964, the Clearinghouse for Federal Scientific and Technical Information (CFSTI) was established under the National Bureau of Standards, and the functions of the OTS were transferred to the Clearinghouse. In 1970, CFSTI was merged with the newly established National Technical Information Service (NTIS) under the Department of Commerce. The present activities and services of NTIS are described in a later section.

Concurrently with the above series of developments leading to the establishment of NTIS, a number of parallel series of events took place, and these culminated in the establishment of various agencies such as the United States Atomic Energy Commission (USAEC), the Defense Documentation Center (DDC), and the National Aeronautics and Space Administration (NASA), all of which have been responsible for the production and distribution of large quantities of

7.2 History of Report Literature

Table 7.1 History of Government Reports Announcements and Index

Dates of publication	Publisher	Title	Frequency	Volumes	Indexes
1/1946-6/1949	OTS	*Bibliography of Scientific and Industrial Reports*	Weekly	1-11	Subject; Patent (German and Japanese)
7/1949-12/1954	OTS	*Bibliography of Technical Reports*	Monthly	12-22	Subject; PB number
1/1955-6/1961	OTS	*U.S. Government Research Reports*	Monthly	23-35	Subject; PB number
7/1961-12/1964	OTS	*U.S. Government Research Reports*	Semimonthly	36-39	Subject; PB number
1/1965-12/1966	CFSTI	*U.S. Government Research and Development Reports*	Semimonthly	40-41	*Government-Wide Index to Federal Research and Development Reports*: subject, personal author, corporate author, contract number, report number
1/1967-1969	CFSTI	*U.S. Government Research and Development Reports*	Semimonthly	67-69	Same as above
1970-2/1971	NTIS	*U.S. Government Research and Development Reports*	Semimonthly	70-71	Same as above
3/1971-3/1975	NTIS	*Government Reports Announcements*	Semimonthly	71-75	*Government Reports Index*: subject, personal author, corporate author, contract number, report number
4/1975-present	NTIS	*Government Reports Announcements and Index*	Semimonthly	75-	Index section combined with the abstracts section

Source: "Technical Literature," *Encyclopedia of Library and Information Science*, vol. 30.

technical reports. Detailed accounts of these historical developments in the United States, Great Britain, and other countries have been presented by Tallman [8,9], Boylan [10,11], and Auger [1]. The activities and services of these and other agencies concerned with the bibliographic control of technical reports are mentioned in a later section.

7.3 Characteristics of Technical Reports

The boundaries of technical report literature are not easy to delineate not only because of the very large number of reports produced by government and private agencies all over the world, but also because of great variations in the nature and quality of their contents. On the heterogeneous nature of report literature, the following observation has been made in the SATCOM Report [12]:

> Other attributes of technical reports as a whole are so heterogeneous that one can find ready examples to support almost any generalization that happens to strike his fancy: that they are too long or too short; badly refereed or well refereed—or not refereed at all; reliable or unreliable; inadequately distributed or too widely distributed; too detailed and technical or not technical enough; too expensively printed or shoddily assembled; a valuable complement to journals or a serious handicap to conventional publication.

The status of technical reports among other more conventional forms of scientific and technical literature has been questioned repeatedly; the following quotation is from an editorial in *Nature* [13]:

> The writing of technical reports has been one of the most rapidly growing components of scientific enterprise.... Some of these are humdrum documents, reviews of the literature in some narrow field, reports on particular experiments or calculations more suitable for back of envelopes than for the solemn stationery in which they are distributed. Some, however, turn out to be important and distinguished contributions to understanding, and the question arises as to how these are eventually to form part of the scientific literature.

Much of the debate recurring in published literature concerning technical reports centers around three themes: (1) the uneven quality of technical reports, (2) the diverse nature of their contents, and (3) their status as formal publications, especially in relation to the scholarly journal articles.

7.3.1 Quality of Technical Reports

Most technical reports are the product of contractual R & D projects, usually sponsored by a government agency. A normal requirement of such contracts is

7.3 Characteristics of Technical Reports

that the agency performing the investigation submit to the sponsoring agency one or more reports on the project within a specified time limit. Preparation and timely submission of the reports are usually the responsibilities of the principal investigator of the project. While many companies performing contractual research engage technical information officers or staff editors for editing technical reports, this is by no means a standard practice. Thus, the uneven quality of technical reports may be attributed to the following factors:

Most technical reports are written by engineers or technologists.

The reports are addressed to the technical experts of the sponsoring agency and not to the entire scientific and technical community.

The time available for the preparation of reports is usually very limited.

Because technical reports are intended to be working documents and not a part of the archival literature of science, and in many cases because of the confidential nature of their contents, reports are not refereed by outside experts.

Technical editing expertise and facilities available for report editing are usually very limited.

The Weinberg Panel recommended that agencies handling large numbers of reports (e.g., the Department of Defense and NASA) establish resident referees to critically review the reports early in the information transfer chain, preferably before the reports are submitted to the agency for distribution [14]. The Panel also pointed out a precedence for such refereeing of technical reports. In 1949, all the captured German documents were screened by hundreds of American experts under contract to the OTS. Only 20% of the reports passed the scrutiny and entered the OTS system. The SATCOM Committee felt that scientific and technical societies should take steps to improve the quality of technical reports [15].

7.3.2 Diversity of Contents

Reports vary greatly in the nature of their contents. Besides the results of R & D, diverse types of material such as literature reviews, bibliographies, compilations of statistical data, catalogs, directories, and conference papers and proceedings appear as technical reports. The following examples illustrate the diverse nature of material contained in technical reports:

NASA TM X-3334 is a compilation of the collected works of Robert T. Jones, issued to commemorate the 65th birthday of this distinguished aeronautical engineer and inventor.

AFGL-TR-76-0306, a report originating from the U.S. Air Force Geophysical Laboratory, Hanscomb Air Force Base, New Hampshire, is the proceedings of the Ninth AFGL Scientific Balloon Symposium held at Wentworth-by-the-Sea,

Portsmouth, New Hampshire, October 20-22, 1976. Texts of 33 papers presented at the symposium are reproduced in this report.

AGARD R-657, "Maximizing Efficiency and Effectiveness of Information Data Banks," by Y. M. Braunstein, contains material from a paper entitled "Public policy and research on the economics of information transfer" presented by the author at the 39th annual meeting of the American Society for Information Science, and subsequently published in the *Proceedings of the 39th ASIS Annual Meeting*, 1976.

AGARD R-658 is a "Catalog of Current Impact Devices."

The subjects covered in technical reports encompass all branches of science, engineering, technology, the social and behavioral sciences, interdisciplinary areas including various aspects of energy and environment, and even some branches of the humanities. J. P. Chillag has cited a recondite example: "If at any time one is interested in papers given by Episcopal cathedral deans on various aspects of the Creation, STAR (*Scientific and Technical Aerospace Reports* of NASA) will be just the place in which to find it (N68-12081)" [16]. The following examples taken from recent issues of *Government Reports Announcements and Index* (GRAI) serve to illustrate the diversity of the contents of technical reports:

AD-036 928/0G1	"Social Structure and Motivational Pattern in an Expressive Medium: American and Mexican Popular Songs" (1978)
COM-75-10484/4G1	"Control of Sex in Fishes" (1975)
AD-A013 001/3G1	"Teaching a Large Russian Language Vocabulary by the Mnemonic Keyword Method" (1975)
PB-243 817/4G1	"Moral Views on Gambling Promulgated by Major American Religious Bodies" (1975)
HRP-0011 408/2G1	"Abortion: Do Attitudes of Nursing Personnel Affect the Patient's Perception of Care?" (1976)
AD-A021 123/5G1	"Women's Rights: Selected References" (1976)
ED-144 541	"Monkey See, Monkey Do" (1978)
PB-253 756/1G1	"Selected Bibliography of Egyptian Educational Materials" (1976)
PB-285 760/5G1	"U.S. Foreign Relations and Multinational Corporations: What's the Connection?" (1978)

7.3.3 The Status of Technical Reports

The uncertain status of technical reports as a form of scientific literature has been described by the Weinberg Panel thus [17]:

7.3 Characteristics of Technical Reports

The documentation community has taken an equivocal attitude toward informal reports; in some cases the existence of these reports is acknowledged and their contents abstracted in the abstracting journals. In other cases informal reports are given no status; they are alleged to be not worth retaining as part of the permanent record unless their contents finally appear in a standard hardcopy journal. Whether this position is tenable even in the basic sciences is open to question; it certainly is no longer tenable in technological development. Here the informal report, rather than the formal paper, has always been the main vehicle of publication.

The Panel further pointed out that the technical report, as originally conceived, was not intended to become a part of the archival record of science; but the growth of government information systems has tended to normalize the report and accord it an archival status although the quality of the report has not improved accordingly.

Editors of many scholarly scientific journals have criticized the uncertain quality and uncontrolled proliferation of technical reports. A certain ambivalence may be observed in the attitude of journal editors toward report literature. On the one hand, many editors have refused to consider manuscripts which had previously been distributed in the form of reports on the grounds that such distribution constitutes prior publication; on the other hand, many editors have discouraged their authors from citing reports in their manuscripts on the grounds that reports are not "published" literature and are not easily accessible to readers [18]. This ambivalent attitude of journal editors was also reflected in an opinion survey conducted by the Task Group on the Role of the Technical Report in Scientific and Technological Communication set up by the Committee on Scientific and Technical Information (COSATI) of the Federal Council for Science and Technology. While most of the editors generally agreed that limited dissemination of information in the form of technical reports did not constitute prior publication, many editors qualified this opinion and said that they would deem a report as a regular publication and would not consider it for publication in their journals if the report had been announced in abstracting services or if copies of the report were available from a clearinghouse or from the originating laboratory.

It is generally believed that much of the material presented in technical reports is subsequently republished in other formats such as journal articles. A survey of nearly 2500 unclassified technical reports carried out by the Library of Congress led to the following conclusions [19]:

About 60 to 65% of unclassified technical reports contain publishable information.

About one-half of such publishable information is published in the open literature within two to three years.

About a fifth of the reports that contain publishable information are not published in the open literature for at least several years; much of them are probably never published.

One technical report rarely leads to exactly one journal article. Very often a single journal article contains information drawn from a number of reports. Also, the information contained in one report may be published in several journal articles.

The following are some of the reasons why information contained in a large number of technical reports is not published in the open literature:

The information contained in technical reports is likely to be highly specialized, and addressed to the technical experts of the sponsoring agency, and therefore not of much interest to the larger scientific and technical community.

Unlike basic research, applied R & D projects often result in the design, development, or improvement of a process or product, and such results are not always amenable to publication in a journal article.

Information contained in classified reports is proprietary or confidential, and the disclosure of such information in open literature is prohibited, at least until the report remains classified.

Many technical reports would need condensation, extensive rewriting, and editing before they could be submitted for publication in a journal. Authors of technical reports, who are likely to be engineers or technologists, may not have the time, training, or inclination to "clean up" the report and make it ready for submission to a journal.

Editors of many scholarly journals are not favorably inclined to consider manuscripts that have received prior publicity and dissemination in the form of reports.

Notwithstanding the controversy over their status, technical reports are becoming increasingly important as vehicles for the dissemination of technical information. In a study of references cited in various publications of the IEEE, Coile discovered a steady increase in the number of citations to report literature [20]. In 1973, the Technical Report Service of Bell Laboratories filled approximately 41,000 requests from its engineers for technical reports originating from outside the company [21].

The chief strength of the technical report seems to be the currency of the information it contains. As a vehicle for the dissemination of technical information, the technical report is much faster than a journal article, since the production of a report is not subject to the delays that are inherent in journal publishing. The role of the technical report relative to other forms of scientific and technical literature was neatly summarized by a journal editor responding to the COSATI Task Group survey mentioned earlier [22]:

> I think the technical reports-vs.-journal papers controversy is inherently a phony one. The technical report and the published paper each has its function, and to me the reasons seem obvious why neither can or should be expected to play the role of the other. . . . In the final analysis, I still believe the technical report incorporates a group of characteristics not found in any other single medium. It is a report of stewardship to the organization that funded the R & D on which it reports; it permits prompt dissemination of data on a highly flexible distribution basis; and, not being subject to externally imposed space restrictions, it can tell the total R & D project study, including exhaustive exposition and comprehensive tables and illustrations. In my opinion, this makes it neither better nor worse than the published paper as a medium of scientific communication. It means simply that the technical report can fulfill certain functions the published paper cannot, just as the published paper has valuable attributes not found in the technical report.

7.4 Security Classification

In view of the sensitive nature of the contents of some technical reports, their distribution is restricted to varying degrees by a system of security classification. Reports of research in aerospace, nuclear energy, and other subjects containing sensitive information of importance to national security are usually classified, at least initially, for a certain period of time. The quantity of classified report literature is virtually indeterminable. It is believed that there are about 20 million classified documents (including reproduced copies) within the Department of Defense. The National Archives is said to contain about 300 million pages of classified documents pertaining to the period 1946-1954. Each year millions of pages of classified documents are being added to this mass of classified literature [23].

Typical security designations are "top secret," "secret," "confidential," and "restricted circulation." Numerous other designators (e.g., "addressee only," "for U.S. Government use only") are used in government documentation to indicate various levels of access restriction to documents. Reports originating from industrial organizations and other nongovernment agencies may also be classified to protect the competitive advantage of industries or to satisfy contractual obligations when investigations of a sensitive nature are carried out. Government agencies such as the Department of Defense which issue security classified reports have established elaborate procedures for the controlled dissemination and handling of classified reports [24]. In general, individuals wishing to acquire classified reports must first of all establish the "need to know" the classified information, and must obtain security clearance from the appropriate issuing agency [25,26]. The Freedom of Information Act (P.L. 89-487), which became a law on July 4, 1967, was promulgated to maximize the disclosure of information to the general public, without prejudice to national security.

Certain types of classified reports are periodically reviewed for time-phased downgrading to a lower level of classification. A top secret document may, for example, be considered for reclassifying into a secret document after three years from the date of issue. The following is a consolidated index of declassified AD reports:

Declassified AD Documents Index. DDC TR-75/6. AD-A016 500/1GA. Springfield, Va.: National Technical Information Service, 1975. 120 p.

This index lists, in AD number order, over 46,000 technical reports originally accessioned by DDC as classified, but later declassified by automatic time-phased regrading action or by specific reclassification review. The index covers the last 16 years of classified document input into the DDC collection. An updated version of this index, published in 1977, is available as AD-A037 600 from NTIS.

The *Declassified Documents Reference System* (DDRS), published by Carrollton Press, Arlington, Va. (1975-), is a continuing retrospective collection of declassified documents combined with abstracts and subject index. The basic collection consists of full texts of 8032 declassified documents on microfiche. Abstracts of these documents and a subject index are available in hardcover volumes. The basic collection is supplemented annually by texts of documents on microfiche, and by abstracts and indexes in hardcover volumes.

7.5 Technical Report Numbers

One of the principal features of a technical report is the report number or designator that is assigned to the report for its identification, acquisition, and bibliographic control. Though the numbering of technical reports, as originally conceived, was intended to be a helpful device for facilitating the physical and bibliographic control of reports, the uncontrolled proliferation of reports and report numbering schemes has been a vexing problem to technical librarians and users of report literature. About 12,500 codes from nearly 4000 agencies were listed in the first edition of the *Dictionary of Report Series Codes* (New York: Special Libraries Association, 1962). This listing was admittedly incomplete. The proliferation of report numbering schemes may be attributed to increases in the numbers of reports and report-issuing agencies, and also to lack of adequate standardization and control measures to ensure some degree of uniformity in report numbering schemes.

7.5 Technical Report Numbers

One other factor that has further aggravated the report number problem is the assignment of several numbers to the same report. Several agencies, such as the research laboratory from where the report originates, the agency or agencies sponsoring the research project, and the clearinghouse that distributes the report, deal with a technical report at various stages of its production and distribution. Each of these agencies is likely to assign a new report number or modify an earlier designator appearing on the report. In addition to one or more report numbers and an accession number, a report also usually carries a grant number and a contract number. This multiplicity of numbers causes a great deal of confusion in the identification, acquisition, and treatment of technical reports. Auger has identified a Dutch standard (with the original number NEN-3005) which was given the number NLL-M-20984 with shelf location 5828.4F by the National Lending Library (now the British Library Lending Division). The same document was given the number N72-28275 (NLL-M-20984-(5828.4F):NEN-3005) by NASA for announcement in its *Scientific and Technical Aerospace Reports* [27].

When more than one agency sponsors a research project, each of the sponsoring agencies may assign a number to the resulting report. An extreme example of a document receiving seven different numbers has been noted in the *Directory of Engineering Document Sources* [28]:

Cosponsoring agency	Report number
Defense Supply Agency	DSAM-4130.1
Army	AR-708-8
Navy Supply Systems Command	NAVSUP-PUB-5006
Air Force	AFM-72-5
Marine Corps	MCO-P4423.17B
General Services Administration, Federal Supply Service	GSA-FSS-4130.1
Defense Atomic Support Agency	DASAI-4100.5A

Report numbers assigned at the point of origin usually contain an abbreviation or acronym indicating the identity of the performing agency. For example: USAMC-ITC-02-08-76-416 originated from the United States Army Materials Command, Texarkana, Texas, Intern Training Center. SAND-75-0079 was issued by Sandia Laboratories, Albuquerque, N. Mex. LMSC-HREC-TR-D496600 is from the Lockheed Missiles and Space Company, Huntsville, Ala., Research and Engineering Center.

Clearinghouses such as NTIS, DTIC, and the Scientific and Technical Information Facility of NASA also assign accession numbers to reports. NTIS continues to assign the PB numbers started by the erstwhile Publications Board in 1946.

Similarly DTIC assigns the AD numbers initiated by its predecessor, the Armed Services Technical Information Agency (ASTIA). Formerly, the abbreviation AD stood for "ASTIA Document"; the expansion currently preferred by the DTIC is "Accession Document." NASA assigns report numbers beginning with the acronym NASA to reports of NASA origin, and also an accession number beginning with the letter N to all documents announced in STAR.

Report numbers are usually made up of code designators indicating several of the following data elements:

Sponsoring agency

Issuing agency (performing agency)

Location of specific branch or department of the performing agency where the research was done

Distributing agency or clearinghouse

Subject matter

Type or form of report

Date of preparation or release

Individualizing identifier

Security classification code

It is obvious that not all report numbers contain designators for all of the above data elements. Most report numbers consist of three or four of the above elements in various permutations. Some typical report numbers are explained in Figures 7.1-7.3. The numbering schemes followed by various agencies range from fairly obvious codes (e.g., UCLA-ENG-7543 from the University of California, School of Engineering and Applied Sciences) to unintelligible and confusing concatenations of symbols. The compilers of the *Dictionary of Report Series Codes,* 2nd edition (New York: Special Libraries Association, 1973), have noted some examples of excessively long report numbers [29]:

AGARD-AC/82-D/4-PT-A SECT 1	(27 characters and spaces)
CERN/ECFA-66/WG2/US-SG2/TRe	(27 characters)
WALD/BL/RP/FVCOG-KPC-70	(23 characters)

The following two publications are very valuable in deciphering codes used in report numbers:

1. *Dictionary of Report Series Codes,* 2nd edition. Lois E. Godfrey and Helen F. Redman, eds. (New York: Special Libraries Association, 1973). This directory consists of two main indexes pairing corporate agencies and the report codes assigned by them. Identification of different types of codes

7.5 Technical Report Numbers

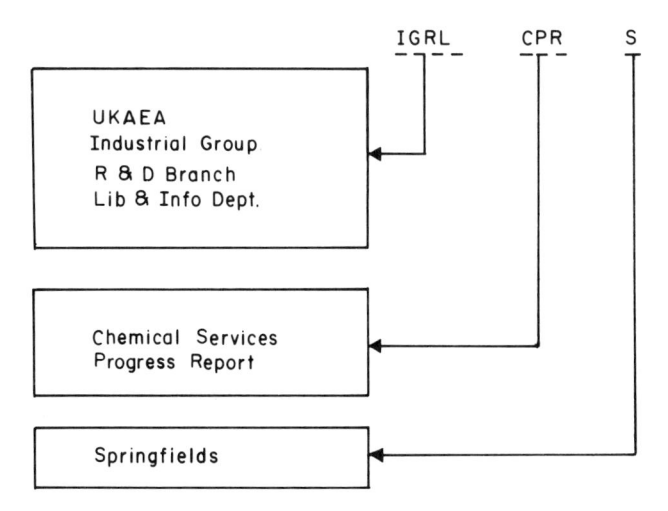

Figure 7.1 Expansion of report No. IGRL-CPR/S. (From Technical Literature, in *Encyclopedia of Library and Information Science*, v. 30.)

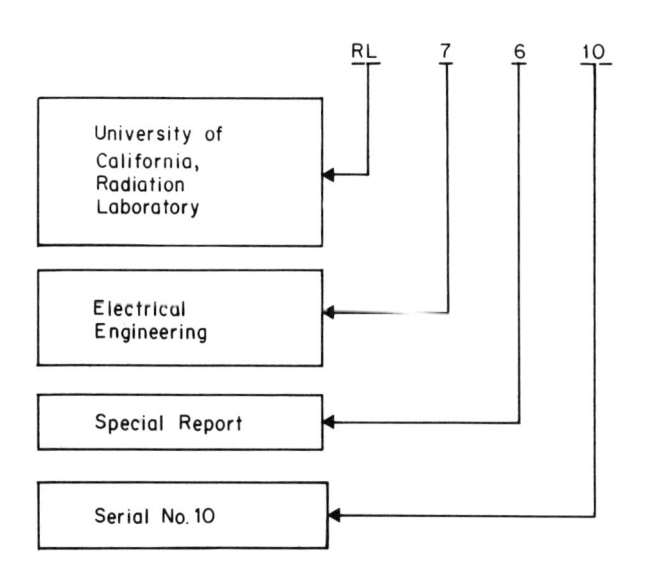

Figure 7.2 Expansion of report No. RL-7.6.10. (From Technical Literature, in *Encyclopedia of Library and Information Science*, v. 30.)

114   Technical Reports

```
                              NACA    MR   A   4   L   12   b
```

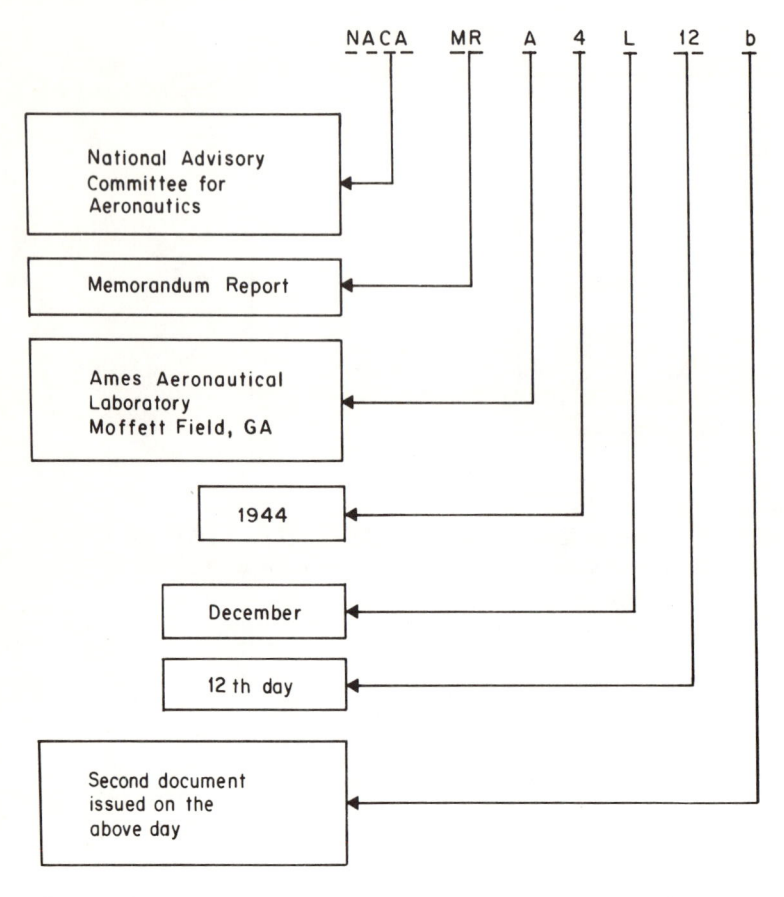

```
┌─────────────────────┐
│ National Advisory   │
│ Committee for       │←──────┘
│ Aeronautics         │
└─────────────────────┘

   ┌──────────────────────┐
   │ Memorandum Report    │←──────┘
   └──────────────────────┘

      ┌──────────────────────┐
      │ Ames Aeronautical    │
      │ Laboratory           │←──────┘
      │ Moffett Field, GA    │
      └──────────────────────┘

            ┌──────────┐
            │   1944   │←──────┘
            └──────────┘

               ┌──────────┐
               │ December │←──────┘
               └──────────┘

               ┌──────────┐
               │ 12 th day│←──────┘
               └──────────┘

┌──────────────────┐
│ Second document  │
│ issued on the    │←──────────────────────┘
│ above day        │
└──────────────────┘
```

Figure 7.3   Expansion of report No. NACA MR-A4L-12 b. (From Technical Literature, in *Encyclopedia of Library and Information Science,* v. 30.)

including codes of specific agencies is explained. The directory also contains a very informative introduction and a bibliography on report numbering.

2. *Directory of Engineering Document Sources,* 2nd edition. (New Port Beach, Calif.: Globe Engineering Documentation Services, Inc., 1974). This directory consists of a consolidated cross-index of document codes assigned by government agencies and industrial organizations to technical and management specifications, standards, and reports. The directory gives the meanings of codes as well as the addresses of organizations from which documents bearing the code numbers may be obtained.

7.5 Technical Report Numbers

The need for developing standards for technical report literature in general and report numbering in particular has been emphasized from time to time by technical librarians and information specialists [30,31]. COSATI has developed a set of guidelines for standardizing technical report format:

> U.S. Federal Council for Science and Technology. Committee on Scientific and Technical Information. Guidelines to Format Standards for Scientific and Technical Reports Prepared by or for the Federal Government. PB 180 600. December 1968.

COSATI has also developed guidelines for descriptive cataloguing of technical reports to promote uniformity in the bibliographic description of technical reports and to facilitate exchange of bibliographic data among federal agencies, libraries, and technical information centers. A revised edition of this standard is available from NTIS:

> Committee on Information Hang-ups. Washington, D.C. "Guidelines for Descriptive Cataloguing of Reports. A Revision of COSATI Standards for Descriptive Cataloguing of Government Scientific and Technical Reports." PB-277 951. AD-A050 900. March 1978.

The British Standards Institution has issued a specification entitled "Specification for the Presentation of Research and Development Reports" (BS 4811: 1972). The American National Standards Institute (ANSI) has also developed a standard for technical report format: "American National Standard for Scientific and Technical Reports: Format and Production," ANSI Z39.17-1971 (IEEE Std 268-1971). The ANSI standard for report numbering is based on the following principles: (1) Every report should have a unique identification number; (2) the number should be as brief as possible; (3) the number should designate the issuing organization, date of publication, and a serial number; and (4) the number should be usable in both machine and manual documentation systems [32]. The ANSI format for a report number is as follows:

A X X X/X X X -   X X   - X X X X (Local suffix)
   Group I   Group II   Group III

Group I: Issuing agency or corporate author or personal author. The first character is an uppercase letter; the remaining characters could be letters or numerals.

Group II: Last two digits of the year of publication.

Group III: A sequential number consisting of Arabic numerals only.

The local suffix in Group III is an optional field. A country code may be prefixed to the report number, if considered necessary.

7.6 Bibliographic Control of Technical Reports

Technical report literature has not received systematic coverage in the conventional abstracting and indexing services mainly because of (1) the large volume of journal articles and other "published" literature to be covered, (2) the difficulty of obtaining copies of technical reports, and (3) the uncertain status of technical reports. Major indexing and abstracting services occasionally cover technical reports, but the coverage is by no means comprehensive. This is not to say that the bibliographic control of report literature is inadequate. In fact, there are several secondary services that systematically index and abstract technical reports. In the United States, the three government agencies responsible for the production of a very large proportion of technical reports are: The Department of Defense (DOD), the National Aeronautics and Space Administration (NASA), and the Department of Energy (DOE). Each of these agencies has set up a separate technical information unit for the dissemination of technical information generated by it and for the bibliographic control of its report literature. In addition, at the national level, NTIS and the Superintendent of Documents have the responsibility for announcing and distributing reports and other documents of government origin. Similar national bibliographic and documentation centers exist in several other countries. In addition to national agencies, many international organizations such as the Advisory Group for Aerospace Research and Development (AGARD) and the International Atomic Energy Agency (IAEA) are also actively engaged in the production, bibliographic control, and distribution of technical reports. A few major national and international documentation centers and their information services and products pertaining to technical report literature will now be briefly described.

7.6.1 The U.S. Government Printing Office (GPO)

The GPO is a part of the Legislative Branch of the United States Government, and undertakes printing work in response to specific requests from the Congress and the Federal agencies. Publications printed by the GPO (about 20,000 current titles) are sold by the Superintendent of Documents, Washington, D.C. New titles and those in stock are announced in the *Monthly Catalog of United States Government Publications.* Recently a cumulative subject index to the *Monthly Catalog* covering the period 1900-1971 was published:

Cumulative Subject Index to the Monthly Catalog of the United States Government Publications, 1900-1971. Arlington, Va.: Carlton Press, 1975. 15 v.

Government publications printed by the Government Printing Office are listed in the *Monthly Catalog* under the name of the respective agencies. The GPO Monthly Catalog database is searchable online.

7.6 Bibliographic Control of Technical Reports

The following comprehensive bibliographic guides are helpful in locating government publications:

Laurence F. Schmeckebier and Roy B. Eastin. *Government Publications and Their Use,* 2nd edition. (Washington, D.C.: The Brookings Institution, 1969).

James Bennett Childs. "Government Publications (Documents)." *Encyclopedia of Library and Information Science.* (New York: Marcel Dekker, 1973). v. 10, pp. 36-140.

Joe Morehead. *Introduction to United States Public Documents.* (Littleton, Colo.: Libraries Unlimited, 1978).

7.6.2 National Technical Information Service (NTIS)

The history of NTIS can be traced back to 1945 when the Publications Board was established to handle the release of technical reports generated during World War II. A brief account of the series of developments culminating in the establishment of NTIS has been given in an earlier section. NTIS is one of the world's largest specialized information service organizations responsible for the bibliographic control and distribution of American and foreign technical reports and other specialty information products. Every day NTIS ships about 23,000 information products to customers all over the world. It receives copies of unclassified technical reports from about 300 other government agencies for announcement, distribution to depositories, and public sale. Its report collection consists of over a million titles; about 15% of these are of foreign origin. NTIS sells microfiche or paper copies of unclassified technical reports.

The principal announcement service for technical reports is the *Government Reports Announcements and Index* (GRAI), a biweekly abstracting and indexing service that has evolved through a series of changes from the *Bibliography of Scientific and Industrial Reports* started in 1946 by the erstwhile Office of Technical Services. An historical account of this announcement service is presented in Table 7.1. The coverage of GRAI is comprehensive; contents of about 70,000 technical reports are summarized each year. The report summaries are arranged under 22 subject categories and numerous sub-fields. These categories were endorsed by COSATI. The reports are indexed by subject, personal author, corporate author, contract number, and accession/report number. Annual cumulations of these five indexes are also published separately. A sample abstract entry from GRAI and the corresponding index entries are shown in Figure 7.4. GRAI database on magnetic tape is available for lease. The database can also be accessed online through database vendors.

A fast announcement service called *Weekly Government Abstracts* is published by NTIS in 26 different series covering a variety of areas ranging from

SAMPLE ENTRIES

MAIN ENTRY

Report entries are arranged by subject group and field. Within fields they are arranged alphanumerically by NTIS order number (accession number); alphabetic data precedes numeric.

Field 10 ENERGY CONVERSION
(NON-PROPULSIVE)
Group 10A Conversion Techniques

Order number ——— PB-254 315/5GA   PC A16/MF A01 ——— Price codes
Corporate author ——— Smithsonian Science Information Exchange, Inc., Washington, D.C.
Personal author ——— Information on International Research and Development Activities in the Field of Energy, David F. Hersey. May 76, 370p* ——— Report title
NSF/RA-760057, Grant NSF-AER74-20678

This directory is the product of a data collection effort undertaken by the Smithsonian Science Information Exchange (SSIE) on behalf of an interagency committee formed under the U.S. State Department to provide international cooperation in energy research and development. Included is information covering 1766 ongoing and recently completed energy research projects conducted in Canada, Italy, the Federal Republic of Germany, France, the Netherlands, the United Kingdom, and 25 other countries. In addition to the title and text of project summaries, the directory contains the following indexes: Subject Index, Investigator Index, Performing Organization Index, and Supporting Organization Index. ——— Abstract of report

INDEX ENTRIES

Index entries are arranged alphanumerically. Titles are included in all indexes except the Contract Number index.

SUBJECT

ENERGY
Information on International Research and Development Activities in the Field of Energy.
PB-254 315/5GI   10A

Entries are sequenced by major subject term and by NTIS order number.

CONTRACT GRANT NUMBER

NSF-AER74-20678
Smithsonian Science Information Exchange, Inc., Washington, D.C.
PB-254 315/5GI   10A

Entries are sequenced by contract or grant number, corporate author, and NTIS order number.

PERSONAL AUTHOR

HERSEY, DAVID F.
Information on International Research and Development Activities in the Field of Energy.
PB-254 315/5GI   10A

Entries are sequenced by personal author, report title, and NTIS order number.

ACCESSION/REPORT NUMBER

PB-254 315/5GI
Information on International Research and Development Activities in the Field of Energy.
PB-254 315/5GI   10A   PC A16/MF A0I

Entries are sequenced by NTIS order, original report, or monitor agency number. Price codes are given in this index.

CORPORATE AUTHOR

SMITHSONIAN SCIENCE INFORMATION EXCHANGE, INC., WASHINGTON, D.C.
Information on International Research and Development Activities in the Field of Energy.
(NSF/RA-760057)
PB-254 315/5GI   10A

Entries are sequenced by corporate author name, original report number, and NTIS order number. The monitor agency number is given following the report title.

Figure 7.4 Sample entries from *Government Reports Announcements and Index*, 1980.

7.6 Bibliographic Control of Technical Reports

administration and management, agriculture and food, and biomedical technology, through energy, health planning, physics, transportation, and urban and regional technology and development.

Microfiche copies of full texts of reports in any one or more of about 500 subject categories may be obtained on a standing order basis from NTIS. This biweekly standing order service is called Selected Research in Microfiche (SRIM). Automatic distribution of paper copies of reports is also now available.

NTIS undertakes custom searches in response to specific requests for report bibliographies. Online searches are made for relevant items in the GRAI database, and summaries of up to 100 items are made available. These bibliographies, compiled in response to specific requests, are subsequently made available for general sale as published searches. Lists of such published searches are issued by NTIS from time to time. Details of these and other services and materials (e.g., products of information analysis centers, press translations from the People's Republic of China, computer software, translations prepared by the Joint Publications Research Service, patent portfolios, etc.) may be obtained from the National Technical Information Service, 5285 Port Royal Road, Springfield, VA 22161.

7.6.3 Defense Technical Information Center (DTIC)

The Armed Forces Technical Information Agency (ASTIA) was established in 1951 under the operational control of the United States Air Force by merging two earlier agencies which were responsible for handling classified technical reports: (1) The Navy Research Section of the Library of Congress, established in 1946 and operated by the Office of Naval Research, and (2) the Central Air Documents Office at Dayton, Ohio, started in 1948 and operated by the U.S. Air Force. ASTIA gave accession numbers starting with the letters AD (ASTIA Document) to reports received from the DOD research facilities and their contractors. In 1963, ASTIA was renamed the Defense Documentation Center (DDC), and its operational control was transferred to the Defense Supply Agency of the DOD. In October 1979, DDC was renamed the Defense Technical Information Center (DTIC). The major functions of DTIC are

> To provide DOD research and development data and bibliographic services to an established user community
>
> To operate management information databases on the R & D projects of the DOD
>
> To develop new and improved information storage and retrieval systems

DTIC receives technical reports from various research facilities of the DOD and their contractors, and continues to assign the AD numbers initiated by

ASTIA, but the abbreviation AD now stands for Accession Document. Unclassified reports are sent to NTIS for announcement and distribution. Classified reports as well as some unclassified reports having distribution limitations are announced in a semimonthly publication entitled *Technical Abstracts Bulletin* (TAB). Until recently, TAB was a confidential document. The DTIC is now issuing TAB and TAB Indexes as unclassified biweekly publications. These are available to all registered user organizations. These publications do not contain any classified information, and carry the notation, "For Use Only by Registered Defense Technical Information Center Users." The earlier issues of TAB published as confidential documents will remain classified until they are automatically declassified according to the General Declassification Schedule.

DTIC has a report collection of over one million titles in almost all areas of science and technology and the social sciences. DTIC does not serve the general public directly. Its services are primarily directed to its own staff and DOD contractors. The genesis and functions of DDC are more fully described by Robert Rea [33].

7.6.4 National Aeronautics and Space Administration (NASA)

NASA was created in 1958 to supersede the National Advisory Committee for Aeronautics by the National Aeronautics and Space Act of 1958. The Act required that the aerospace activities of the United States should "contribute to the expansion of human knowledge of phenomena in atmosphere and space. The Administration shall provide for the widest practicable and appropriate dissemination of information concerning its activities and the results thereof." In furtherance of this national policy, the following statutory functions have been defined for NASA:

1. Conduct research for solving problems of flight within and outside earth's atmosphere
2. Develop, construct, test, and operate aeronautical and space vehicles
3. Conduct activities for space exploration with manned and unmanned vehicles
4. Arrange for the most effective utilization of the scientific and engineering resources of the United States with other nations engaged in aerospace activities for peaceful purposes
5. Provide the widest and most appropriate dissemination possible of information on NASA activities and their results

Six program offices, ten field centers, and the National Space Technology Laboratories constitute the principal components of NASA for planning, directing and managing its activities. A large part of NASA's R & D activity is conducted in its ten field centers and through contracts with research agencies in

7.6 Bibliographic Control of Technical Reports

academic institutions and industrial corporations. The results of NASA-sponsored R & D are recorded in approximately 10,000 technical reports each year. These reports are obtainable through NTIS. The following are the six major series of NASA reports:

1. *Technical Reports* (NASA TR-R-) contain scientific and technical information considered a significant and complete contribution to knowledge.
2. *Technical Notes* (NASA TN-D-) contain information of somewhat similar quality but of lesser scope than in the Technical Reports.
3. *Technical Memoranda* (NASA TM-X-) are working papers of limited scope and specialized utility.
4. *Technical Translations* (NASA TT-F-) are material originally published in a foreign language, but considered important enough to be translated and distributed in English.
5. *Contractor Reports* (NASA CR-) are reports of NASA-sponsored investigations performed by contractors and grantees.
6. *Special Publications* (NASA SP-) consist of a wide variety of publications including program summaries, conference proceedings, monographs, handbooks, bibliographies, historical accounts, and chronologies, compilations of charts and tables, and popular works. *The Book of Mars* (NASA SP-179), *Exploring Space with a Camera* (NASA SP-168), *Lunar Photographs from Apollo 8, 10 and 11* (NASA SP-246), and *This Island Earth* (NASA SP-250) are some examples of popular volumes in this series.

The NASA Office of Technology Utilization is especially concerned with the wide dissemination of NASA research results for application not only in aerospace industries, but also in manufacturing and processing industries, pollution control, housing and urban development, transportation, fire prevention and safety, crime prevention, etc. The Office of Technology Utilization issues three important series of publications especially addressed to persons and companies engaged in nonaerospace activities:

1. *NASA Tech Briefs* are one- or two-page descriptions of new techniques, concepts, innovations, and devices considered to be transferable across industrial or disciplinary lines for commercial exploitation. Several hundred Tech Briefs are issued each year in a number of categories including: electronics/electrical, physical sciences, materials/chemistry, life sciences, mechanics, machinery, equipment and tools, fabrication technology, and computer programs. Tech Briefs in any one or more categories can be selectively obtained on a subscription basis.
2. *Technology Utilization Reports* are more detailed descriptions of new technological developments of high promise. *Induction Heating Advances:*

Applications to 5800°F (NASA SP-5071), and *Earthquake Prediction from Laser Surveying* (NASA SP-5042) are representative examples.

3. *Technology Utilization Surveys* are comprehensive state-of-the-art surveys of new technological contributions by NASA researchers or NASA contractors. *Solid Lubricants* (NASA SP-5059) and *Air Pollution Monitoring Instrumentation* (NASA SP-5072) are representative examples.

NASA publishes a semimonthly abstracting journal entitled *Scientific and Technical Aerospace Reports* (STAR) which supersedes *NACA Research Abstracts* and *NASA Technical Publications Announcements.* Abstracts in STAR cover worldwide aerospace report literature, dissertations, theses, translations of reports, and NASA-owned patents and patent applications. The abstracts are grouped under 10 major subject divisions further divided into 74 specific subject categories and a general category. The abstracts are arranged in an unbroken series of accession numbers starting with the letter N and the last two digits of the year of accession, e.g., N72-10856. Individual papers in a composite work are often separately abstracted in STAR. Abstracts in STAR are indexed by subject, personal author, corporate source, contract number, and report/accession number. The indexes are cumulated semiannually and annually. The *NASA Thesaurus* (NASA SP-7050. Washington, D.C.: NASA, 1976) is helpful in using the subject index of STAR. Each issue of STAR also incorporates a separate section entitled "On-Going Research Projects" containing brief descriptions of aerospace-related research projects (not publications) in progress.

A complementary abstracting service entitled *International Aerospace Abstracts* (IAA) covering journal articles, books, conference papers, and translations of these is published semimonthly by the American Institute of Aeronautics and Astronautics (AIAA) under a NASA contract. Abstracts in IAA are arranged under the same 75 subject categories as those used in STAR, and are indexed by subject, personal author, contract number, meeting paper and report number, and accession number. IAA is released on the 1st and 15th of each month, and STAR is published on the 8th and 23rd each month. Thus, in coverage and timing, STAR and IAA are complementary abstracting services, covering between them worldwide aerospace literature. Each of these services can serve both current awareness and retrospective search functions. Copies of reports abstracted in STAR can be obtained from NTIS or other sources indicated in each abstract. Copies of documents abstracted in IAA are obtainable from the Technical Information Service, American Institute of Aeronautics and Astronautics, 555 West 57th Street, New York, NY 10019.

NASA operates a selective dissemination of information (SDI) service entitled Selected Current Aerospace Announcements (SCAN), especially designed for scientists and engineers employed by NASA and its contractors, NASA grantees, and consultants. This is a computer-generated biweekly alerting service covering

7.6 Bibliographic Control of Technical Reports 123

about 200 topics. Each subscriber receives *NASA/SCAN Notification* in technical areas of his interest, and can use the *Notifications* to order complete copies of desired documents.

A number of Industrial Application Centers (formerly known as Regional Dissemination Centers) have been established by NASA in universities and nonprofit institutions to promote wider dissemination of information acquired by NASA to industrial firms, research organizations and other interested users. Technical information specialists in these Centers provide SDI service, retrospective search service, and distribute NASA publications to users.

Mention must be made of the computerized database of NASA, consisting of bibliographic details of material indexed in STAR and IAA since 1962. The NASA information facility conducts searches in response to specific requests from users. The database is made available to a number of institutions in the United States and Europe to satisfy local bibliographic information needs. Remote users in the United States and some European countries can access the database through the NASA/RECON service. More detailed information on NASA's technical information activities and publications may be obtained from the NASA Scientific and Technical Information Facility, P.O. Box 8757, B.W.I. Airport, MD 21240.

7.6.5 U.S. Department of Energy (DOE)

During the Second World War, large numbers of technical reports were produced. Most of these were classified documents because of the sensitive nature of information relating to weaponry in the context of the war. In 1942, the Metallurgical Laboratory of the University of Chicago started a centralized indexing service for nuclear science and technology literature under the code name of "Manhattan District." This project eventually led to the establishment of the Technical Information Center of the U.S. Atomic Energy Commission (USAEC). The USAEC itself was established under the Atomic Energy Act of 1954, and the Technical Information Center was assigned the responsibility of declassifying and releasing the technical reports. The Atomic Energy Act of 1954 (Section 141) required that: "The dissemination of scientific and technical information relating to atomic energy should be permitted and encouraged so as to provide that free interchange of ideas and criticism which is essential to progress and public understanding and to enlarge the fund of technical information."

In 1946 USAEC started publishing the *Abstracts of Declassified Documents*. Its title changed to *Nuclear Science Abstracts* (NSA) in 1948. NSA covered USAEC reports, journal articles, books, patents, dissertations, conference papers and proceedings, bibliographies, translations and reports on nuclear science and technology and related areas from all over the world. NSA was a semimonthly publication with quarterly, annual, and quinquennial cumulations. Abstracts were arranged in broad subject categories, and were indexed by subject, personal author, corporate author, and report number.

Under an agreement with IAEA, USAEC was sending bibliographic details of literature covered in NSA in machine-readable form to the International Nuclear Information System (INIS) for inclusion in its *Atomindex*. Thus, for a time, there was substantial duplication of coverage in NSA and *Atomindex*; the NSA was therefore discontinued in June 1976. However, it still remains a valuable bibliographic tool for retrospective searching of the world's output of nuclear scientific and technical literature for the period 1946-1976.

In view of the urgency and importance of energy conservation as a national effort in the United States, a number of organizational changes have taken place in rapid succession in the last few years. In 1975, the USAEC was superseded by the Energy Research and Development Administration (ERDA) which was established by the Energy Reorganization Act of 1974 to reorganize and consolidate federal R & D activities relating to enhancement and efficient utilization of energy resources with due regard to the protection of environmental quality and public health and safety. Besides the military and production activities and the nuclear R & D activities of the former USAEC, the functions of a number of other federal agencies relating to energy and environment were also transferred to ERDA by a series of congressional acts. In January 1976, the Technical Information Center of ERDA, located in Oak Ridge, Tennessee, began publishing a semimonthly abstracting journal entitled *ERDA Research Abstracts*. This journal covered technical reports, journal articles, conference papers and proceedings, books, patents and theses on energy systems, conservation, environmental protection, biology, medicine, and related subjects. The abstracts were indexed by personal author, subject, corporate source, and report number. The indexes were cumulated semiannually and annually. The title of *ERDA Research Abstracts* has since changed to *Energy Research Abstracts*, and this is being issued by the Department of Energy.

On October 1, 1977, the Department of Energy (DOE) was established by combining the ERDA, the Federal Energy Administration, the Federal Power Commission, parts of the U.S. Bureau of Mines, and other smaller government agencies. The DOE is a large department with over 20,000 employees and a budget in excess of $10 billion. A subdivision of DOE called the Energy Information Administration was created at the same time to collect, process, evaluate, and disseminate energy information. In the meantime, the DOE Technical Information Center continues to function at Oak Ridge, Tennessee, as the collector, publisher, and disseminator of DOE's scientific and technical literature.

DOE's energy database covers six basic areas: energy conservation, fossil fuel, environment and safety, nuclear energy, solar and geothermal energy, and national security. About 14,000 records are added to this database each year, and the database is accessible online to qualified users [34].

7.6 Bibliographic Control of Technical Reports

7.6.6 Environmental Protection Agency (EPA)

The EPA was created as an independent agency within the Executive Branch of the U.S. Government in 1970 to protect and enhance the environment to the fullest extent possible under the laws enacted by the U.S. Congress. Its mission is to control and abate pollution in the areas of air, water, solid waste, pesticides, noise, and radiation. The agency works in cooperation with federal, state, and local governments, as well as with industries, universities, and citizen groups. It carries out an integrated program of research, monitoring, standardization, and enforcement activities. Designed to serve as "the public's advocate for a livable environment," the EPA is specifically charged with publishing its written comments on environmental impact statements, especially when a proposal is determined to be unsatisfactory from the standpoint of public health or welfare or environmental quality.

Numerous series of technical reports are issued by the various organizational units of EPA. For example, EPA's Office of Research and Development issues the following series of reports:

Environmental Health Effects of Research

Environmental Protection Technology

Ecological Research

Environmental Monitoring

Socioeconomic Environmental Studies

Scientific and Technical Assessment Reports

Interagency Energy-Environment Research and Development

Special Reports

Miscellaneous Reports

The EPA library system consists of 28 agency-wide libraries located in all the ten regional offices, the major laboratories, and specialized information centers in the Program Offices. A complete collection of EPA technical reports on microfiche is maintained in each library.

The *EPA Cumulative Bibliography 1970-1976* (PB-265 920) published in 1976 by NTIS, provides a cumulative listing, with abstracts, of all reports entered into the NTIS collection through 1976 by the EPA and its predecessor agencies. Access points to this report bibliography are by report title, subject (keyword), corporate or personal author, contract number, and accession/report number. Beginning in March 1977, a quarterly abstract bulletin entitled *EPA Publications Bibliography* has been issued by NTIS. This abstracting service covers EPA technical reports and journal articles entered into the NTIS collection. The following indexes are provided: title index, subject index, sponsoring

EPA office index, corporate author index, personal author index, contract number index, and accession/report number index. The indexes are cumulated annually. EPA technical reports are distributed by NTIS.

Information on foreign environmental pollution control technology, laws, and regulations, obtained by EPA through an exchange program from some 60 environmental organizations in other countries, is abstracted and announced in the *Summaries of Foreign Environmental Reports* (monthly) distributed by NTIS.

EPA has a number of nonbibliographic databases containing data on chemicals, toxic substances and hazardous materials, pesticides, air quality, water quality, and other topics. Information on these nonbibliographic databases can be obtained from any of the regional offices of EPA or from the Office of Public Affairs (A-107), United States Environmental Protection Agency, Washington, DC 20460.

7.6.7 The RAND Corporation

The RAND Corporation, located in Santa Monica, California, is an independent nonprofit organization engaged in R & D largely sponsored by government and private agencies in the physical, social, and biological sciences. A large number of publications are issued in applied engineering and technological fields such as aerodynamics, communications systems, computing technology, pollution control, and transportation. The major publication series are: Reports (R series), RAND Memoranda (RM series), Papers (P series), and Books. The RM series was discontinued in 1971. Unclassified RAND publications including technical reports are abstracted and indexed by author and subject in *Selected RAND Abstracts,* a quarterly service published by the Corporation. Copies of RAND reports may be obtained from the Corporation. A subscription service is also available. Unclassified RAND reports are also distributed by NTIS. The Corporation has a technical report collection of over 250,000 reports.

7.6.8 The British Library Lending Division (BLLD)

The BLLD is a depository for reports and other literature on atomic energy and related topics. It receives all unclassified reports from NTIS, NASA, DOE, RAND Corporation, and other similar sources in over 90 countries. Technical reports held in the BLLD are announced in the *BLLD Announcement Bulletin: A Guide to British Reports, Translations and Theses,* published monthly. Copies listed in the Bulletin may be acquired from the BLLD, Boston Spa, England.

7.6.9 United Kingdom Atomic Energy Authority (UKAEA)

UKAEA was created as a public corporation by the Atomic Energy Authority Act of 1954 to be responsible for R & D in all aspects of nuclear energy and for

7.6 Bibliographic Control of Technical Reports

dissemination of its results. Unclassified reports of UKAEA are regularly announced in the following lists:

UKAEA List of Publications Available to the Public, monthly, compiled by the Atomic Energy Research Establishment (AERE) at Harwell

Government Publications, published at irregular intervals by HMSO

BLLD Announcement Bulletin, monthly published by the British Library Lending Division, Boston Spa

R & D Abstracts, published semimonthly by the Technology Reports Center

Details of UKAEA publications are furnished in *Guide to UKAEA Documents,* 5th edition, edited by J. R. Smith (London: HMSO, 1973). Unclassified UKAEA reports are available from HMSO, BLLD, and the Technology Reports Center.

7.6.10 International Atomic Energy Agency (IAEA)

IAEA was established in July 1957 as an autonomous intergovernmental organization under the auspices of the United Nations to promote the peaceful uses of atomic energy throughout the world and to foster the exchange of scientific and technical information on the peaceful applications of atomic energy. IAEA is a large international organization of over one hundred member states. A specialized information system called the International Nuclear Information System (INIS) was established in 1970 after a four-year feasibility study, to disseminate worldwide information on nuclear science and technology [35-37]. INIS is a cooperative venture in which IAEA and its member states participate in the centralized bibliographic control of nuclear literature based on decentralized input from participating member states. Over 44 member states and 12 international and regional organizations systematically scan nuclear literature in their respective regions and provide bibliographic input to INIS, which then processes the input records, creates a machine-readable master file, and provides several services:

1. *INIS Atomindex,* a monthly abstracting journal
2. Abstracts on microfiche in one of the four official languages of INIS (English, French, Russian, Spanish), available on an annual subscription basis
3. INIS Atomindex database on magnetic tape, with monthly updates

INIS also makes available on microfiche full texts of nonconventional literature such as technical reports, theses, and conference papers. The *INIS Thesaurus* (Vienna: International Atomic Energy Agency, 1975) is useful in providing subject access to literature abstracted and indexed in *INIS Atomindex.*

7.6.11   European Atomic Energy Community, Brussels (EURATOM)

EURATOM was established in 1958 to coordinate and supplement national programs for the promotion of nuclear industries in member countries. Nine European countries are members of EURATOM. EURATOM's Joint Research Center consists of establishments in Ispra (Italy), Mol (Belgium), Petten (The Netherlands), and Karlsruhe (Federal Republic of Germany). Technical reports issued by EURATOM are listed in *Euro-Abstracts* and also in *INIS Atomindex*.

7.6.12   Advisory Group for Aerospace Research
         and Development (AGARD)

AGARD was established in 1952 as an agency of the North Atlantic Treaty Organization (NATO) to bring together the leading personalities of the NATO nations in the field of aerospace science and technology to promote:

Exchange of scientific and technical information

Advances in the aerospace sciences

Cooperation among member nations in aerospace R & D

Advice and assistance to the North Atlantic Military Committee and other NATO agencies and member countries in the field of aerospace science and technology

The following are some of the important report series of AGARD:

AGARD-AG-	AGARDographs
AGARD-AR-	Advisory Reports
AGARD-BUL-	Bulletins
AGARD-CP-	Conference Proceedings
AGARD-LS-	Lecture Series
AGARD-R-	Technical Reports

AGARD reports are not stocked in or sold by its headquarters in Nevilly Sur Seine, France. The national distribution centers in member countries have the responsibility of distributing AGARD reports. The national distribution center in the United States is NASA; but NASA, unlike the national distribution centers in other countries, does not supply copies of AGARD reports. In the United States, AGARD reports are announced in STAR and GRAI and are distributed by NTIS. Complete retrospective listings of AGARD publications have been published. These are:

AGARD Index of Publications, 1952-1970.

AGARD Index of Publications, 1971-1973.

Since 1974, annual supplements to these cumulative lists have been published.

7.6.13 European Space Agency (ESA)

ESA was established with its headquarters in Paris, France, in May 1957 to promote cooperation among European countries in space research and technology and the application of this know-how to peaceful purposes. It replaces the European Space Research Organization (ESRO) and the European Organization for the Development and Construction of Space Vehicle Launchers (ELDO). ESA issues the following series of technical reports, very similar to those of NASA described earlier:

ST/TR-	*Scientific/Technical Reports:* formal documents considered to be of permanent scientific value.
SR/TN-	*Scientific/Technical Notes:* less formal and narrower in scope than SR/TR- reports, but still considered useful additions to technical literature
SM/TM-	*Scientific/Technical Memoranda:* informal documents for rapid dissemination of results of research on specific subjects
SP-	*Special Publications*
CR-	*Contractors Reports,* originating from ESA contractors
PSS-	*Procedures, Standards and Specifications*

Unclassified ESA reports are announced in STAR.

The organizations mentioned above are representative of numerous other national and international organizations actively engaged in the generation and dissemination of technical literature in atomic energy, aerospace, and other scientific and technical disciplines. The OECD Nuclear Energy Commission in Paris, the European Organization for Nuclear Research (CERN) in Geneva, and the Joint Institute for Nuclear Research in Dubna near Moscow are examples of other international organizations concerned with nuclear energy information. A fairly comprehensive listing of the world's major atomic energy report series has been prepared by J. P. Chillag [38]. Brief descriptions of the activities and publications of international organizations may be found in the *Yearbook of International Organizations* (Brussels, Belgium: Union of International Associations), and other similar reference works.

References

1. C. P. Auger, ed., *Use of Report Literature* (Hamden, Conn.: Shoe String Press, 1975), p. 6.
2. Auger, *Use of Report Literature* (Ref. 1), pp. 14-15.

3. U.S. Federal Council for Science and Technology. Committee on Scientific and Technical Information. Task Group on the Role of the Technical Report. The role of the technical report in scientific and technological communication. PB 180 944. (1968) pp. 25-26.
4. Neil Brearley, The role of technical reports in scientific and technical communication, *IEEE Transactions on Professional Communication*, PC-16(3) 117-119 (September 1973).
5. Edward Rosen, Copernicus published as he perished, *Nature*, 241(5390): 433-434 (February 16, 1973).
6. Auger, *Use of Report Literature* (Ref. 1), p. 9.
7. J. E. Tallman, History and importance of technical report literature, *Sci-Tech News*, 15:44-46 (Summer 1961).
8. J. E. Tallman, History and importance of technical report literature, *Sci-Tech News*, 15:164-165+ (Winter 1962).
9. J. E. Tallman, History and importance of technical report literature, *Sci-Tech News*, 16:13 (Spring 1962).
10. N. T. G. Boylan, A history of the dissemination of PB reports, *Journal of Library History*, 3:156-161 (April 1968).
11. N. T. G. Boylan, Technical reports, identification and acquisition, *RQ*, 10:18-21 (Fall 1970).
12. National Academy of Sciences, National Academy of Engineering. *Scientific and Technical Communication: A Pressing National Problem and Recommendations for its Solution* (Washington, D.C.: National Academy of Sciences, 1969), p. 116.
13. Do techical reports belong to literature? *Nature*, 236:275 (April 7, 1972).
14. Science, government and information. The responsibilities of the technical community and the government in the transfer of information. A report of the President's Science Advisory Committee (Washington, D.C.: The White House, 1963).
15. *Scientific and Technical Communication* (Ref. 12), pp. 116-117.
16. J. P. Chillag, Problems with reports, particularly microfiche reports, *ASLIB Proceedings*, 22(5):201-216 (May 1970).
17. Science, government and information (Ref. 14), p. 11.
18. Brearley, The role of technical reports, (Ref. 4).
19. Dwight E. Gray and Staffan Rosenborg, Do technical reports become published papers? *Physics Today*, 10(6):18-21 (June 1957).
20. Russell C. Coile, Information sources for electrical and electronics engineers, *IEEE Transactions on Engineering Writing and Speech*, EWS-12(3):71-78 (October 1969).
21. Eileen W. English, Hits and misses: Securing report literature, *Special Libraries*, 66:237-240 (May-June 1975).
22. COSATI, The role of the technical report (Ref. 3), p. 108.
23. Irving Klempner, The concept of "national security" and its effects on information transfer, *Special Libraries*, 64(7):263-269 (1973).
24. U.S. Department of Defense. *Industrial Security Manual for Safeguarding Classified Information*. Attachment to DD Form 441, DOD 5220.22-M (Washington, D.C.: U.S. Government Printing Office, 1966).

References

25. Virginia Sternberg, Accountability, *Encyclopedia of Library and Information Science* (New York: Marcel Dekker, 1968), v. 1, 55-60.
26. Mary Vasilakis, Classified Material (Security), *Encyclopedia of Library and Information Science* (New York: Marcel Dekker, 1971), v. 5, pp. 174-185.
27. Auger, *Use of Report Literature* (Ref. 1), pp. 33-34.
28. D. P. Simonton, *Directory of Engineering Document Sources* (New Port Beach, Calif.: Global Engineering and Documentation Services, Inc., 1972), Preface.
29. L. E. Godfrey and H. F. Redman, *Dictionary of Report Series Codes*, 2nd edition (New York: Special Libraries Association, 1973), p. 9.
30. Mortimer Taube, Memorandum for a conference on bibliographical control of government scientific and technical reports, *Special Libraries*, 39(5): 54-60 (May-June 1948).
31. Helen F. Redman, Report number chaos, *Special Libraries*, 53(10):574-578 (December 1962).
32. American National Standards Institute. Committee Z39, Standard for technical report numbering, *Special Libraries*, 63(11):541-542 (November 1972).
33. Robert H. Rea, Defense Documentation Center, *Drexel Library Quarterly*, 10(1-2):21-38 (January-April 1974).
34. Ted Albert, Information programs in the new DOE, *Bulletin of the American Society for Information Science*, 4(4):34 (April 1978).
35. M. Middleton, INIS: International Nuclear Information System, *Australian Library Journal*, 23(4):136-140 (May 1974).
36. D. S. R. Murty, International Nuclear Information System (INIS), *Annals of Library Science and Documentation*, 18(1):22-30 (March 1971).
37. C. W. Pelzer, The International Nuclear Information System, *ASLIB Proceedings*. 24:38-55 (January 1972).
38. J. P. Chillag, Nuclear Energy, in *Use of Report Literature* (Ref. 1), pp. 182-183.

8

STANDARDS AND SPECIFICATIONS

8.1 Introduction

Standards are fundamental to many aspects of modern life including science, technology, industry, commerce, health, and education. Standards and specifications are documents that stipulate or recommend: (1) minimum levels of performance and quality of goods and services, and (2) optimal conditions and procedures for operations in science, industry, and commerce, including production, evaluation, distribution, and utilization of materials, products, and services. Standards are established by general agreement among representatives of consumers, designers, manufacturers, distributors, and other concerned groups. Standards and specifications are important components of technical literature. Thousands of standards and specifications are used by scientists, engineers, technologists, designers, manufacturers, and consumers of virtually every type of material, product, or service ranging from common items of everyday use such as paper clips, toys, food, and drugs to extremely complicated equipment and components used in nuclear reactors and space vehicles.

In scientific research standards are essential to ensure reproducibility of research and accuracy and reliability of the results of research. In industrial and commercial practice, standards are essential (1) to prevent avoidable wastage of resources and manpower, (2) to enhance safety, speed, and productivity, (3) to ensure uniformity, reliability, and excellence of product quality, and (4) to achieve overall efficiency and economy. Before the industrial revolution of the 19th century, individual craftsmen were producing consumer goods in small-scale cottage industries, largely through manual operations. They had to work with nonstandard and widely divergent methods, materials, and tools, and the resulting products were of unpredictable and nonuniform quality. Modern industry is based on team effort requiring the concerted application of the knowledge and skills of specialists in various scientific and technological disciplines and techniques. In mass production process, a large number of components are separately produced or acquired, and these are then assembled

8.1 Introduction

into complex finished products by a number of technicians, each performing a single operation in succession using sophisticated tools and techniques. Thus, manual production and assembly of parts are largely replaced by machine-aided or automated assembly of interchangeable components separately manufactured or acquired from ancillary industries. The industrial revolution spurred the development of standards in the 19th century. With the advent of computer-aided engineering design and computer-controlled production processes in the 20th century, standards and specifications have become even more crucial, especially in science and industry.

Standards for measurement and quality control have been in existence from ancient times, but the early standards were quite crude. The Egyptian cubit, for example, was the distance from the tip of the middle finger of an "average" man to his elbow. The inch was originally the length of three grains of barley placed end to end. The first attempt to establish a scientific standard for the measurement of length was made in 1496 when the distance between two marks on a bronze bar was decreed to be the Imperial yard. This standard bronze bar is still preserved in England. The metric system of weights and measures began in France during the 1790s. The French government passed a law in 1840 forbidding the use of other systems of weights and measures. Most countries have now adopted the metric system of weights and measures in which the kilogram and the meter are respectively the standard units of mass and length. Standards for the optimum composition of chemicals and drugs were formulated and published in the first edition of the *British Pharmacopoeia* (BP) in 1613. Numerous revised editions of BP have since been published. The Food and Drug Act in England requires that chemicals used in the manufacture of foods and drugs conform to BP requirements. Similar standards and specifications for foods, drugs, and chemicals exist in other countries.

Standardization of interchangeable parts for machines and equipment began in the United States in 1798 when Eli Whitney, inventor of the cotton gin, developed the idea of using machine-made components in the manufacture of guns. He received a contract from the federal government for the manufacture of 10,000 rifles for the United States Army, and set up a musket factory in Hamden, Conn. The principle of interchangeability of parts was adopted in clock manufacturing in 1802 by Eli Terry. Standards for interchangeable threaded parts (e.g., screws, nuts, and bolts) was developed in 1841 by Sir Joseph Whitworth in England. But acceptance of the Whitworth standard thread was slow. The need for standardization to facilitate interchangeability of threaded components was dramatically brought home during the great Baltimore fire in February 1904. A large number of fire engine companies went to Baltimore by special trains from as far away as Wilmington, Chester, Philadelphia, New York, Altoona, Harrisburg, and Annapolis to contain the fire had to helplessly stand back and watch the devastating fire consumer over 1500 buildings along with telegraph, telephone and power supply facilities in an area of more than 70 city blocks in down town Baltimore. Reason: These firefighting companies that arrived from different cities had hoses with nonstandard threads that did not match with one another or with the street hydrants in Baltimore [1].

Standardization of iron and steel structural components was proposed by Sir John Barry, a British civil engineer, and his recommendations led to the establishment of the Engineering Standards Committee in 1901 to standardize steel structural members. The Committee soon expanded its scope to other engineering materials and components, and eventually became the British Standards Institution.

8.2 Specifications

The terms "standard," "specification," and "standard specification" are often used interchangeably. Although both standards and specifications stipulate acceptable levels of dimensions, quality, performance, or other attributes of materials, products, and processes, the scope of applicability of a standard is usually much larger than that of a specification. For example, a state government agency awarding a contract for the construction of a school building may specify to its contractor the minimum level of illumination required in the various parts of the proposed building. Such a statement is incorporated in a "specification." On the other hand, the recommendations for minimum levels of brightness in school buildings all over the state, or the entire country, may be developed by representatives of academic institutions, government agencies, technical societies, and industrial organizations, acting in concert. Such a recommendation, when approved by an appropriate standards body (e.g., the American National Standards Institute in the United States), becomes a "standard." According to the American Naitonal Standards Institute (ANSI), a specification is "a concise statement of the requirement for a material, process, method, procedure or service, including, whenever possible, the exact procedure by which it can be determined that the conditions are met within the tolerances specified in the statement; a specification does not have to cover specifically recurring subjects or subjects of wide use, or even existing objects." A standard is defined by ANSI as follows: "A standard is a specification accepted by recognized authority as the most practical and appropriate current solution of a recurring problem."

A specification may therefore be thought of as a purchase document that that contains the description of the technical features of a product, material, process, or service that are required to meet the specific needs of a purchaser or an industry [2]. A standard consists of similar descriptions of technical features, formulated by agreement, authority, or custom, but applicable to a broader range of situations, at a corporate, national, or international level. Standards are frequently amended or revised to keep up with changes in technology or accepted current practice. New materials, methods, and tools representing improvements over existing ones (or occasioned by changes in consumer preferences) are codified into new standards which then modify or replace earlier standards.

8.3 Types of Standards and Specifications

The term "standards and specifications" covers a variety of documents including standards, specifications, codes of practice, recommendations, guidelines, nomenclature and terminology, and so on. Although several distinct types of standards and specifications may be recognized on the basis of their purpose or formulating agency, many standards and specifications are composite in nature and possess the characteristics of more than one type. Various kinds of specifications, including dimensional specifications, "technique" specifications, production specifications, performance specifications, packaging specifications, etc., have been described by Reeves [3]. Based on their purpose, standards may be categorized into the following major types [4,5]:

1. *Dimensional Standards:* These specify standard dimensions to achieve interchangeability in assembly of components in manufacturing industries and other applications, and to facilitate replacement of worn-out or damaged parts; an example is ISO 1020-1969: "Dimensions of Daylight Loading Spools for Double 8 mm Motion Picture Film."

2. *Materials Standards:* Materials standards specify the composition, quality, chemical or mechanical properties of materials such as alloys, fuels, paints, etc.; an example is ANSI/ASTM D 1655-77: "Standard Specification for Aviation Turbine Fuels."

3. *Performance Standards:* A performance standard specifies the minimum performance or quality of a product or component so that it can fulfill its intended functions at acceptable levels of efficiency and safety; an example is BS 4579:Part 2:1973: "Specifications for the Performance of Compression Joints in Electric Cable and Wire Connectors. Part 2: Nickel, Iron and Plated Copper Conductors."

4. *Standards of Test Methods:* These recommend conditions, procedures, and tools for testing, evaluating and comparing the quality or performance of materials and products; an example is ANSI B38.1-1970: "Methods of Testing for Household Refrigerators, Combination Refrigerator-Freezers, and Household Freezers."

5. *Codes of Practice:* These specifiy procedures for installation, operation, maintenance, and other industrial operations to achieve safety and uniformity in

such operations; an example is CP 1021:1973: "Code of Practice for Cathodic Protection" (British Standards Institution).

6. *Standards of Terminology and Graphic Symbols:* Terminology and symbols used in technical communication, drawings, and flowcharts are also standardized to enable engineers and technologists to communicate unambiguously; examples are IEEE Std 100-1972: *IEEE Standard Dictionary of Electrical and Electronics Terms* (716 pages) and ANSI Y14.2-1973: "Engineering Drawing and Related Documentation Practices: Line Conventions and Lettering." Standard nomenclature for chemical compounds recommended by the International Union of Pure and Applied Chemistry (IUPAC) and similar agencies may also be considered in this category.

7. *Documentation Standards:* These are standards for layout, production, reproduction, distribution, classification, indexing, and bibliographic description of documents. Committee Z39 of ANSI and Technical Committee 46 of the International Organization for Standardization are concerned with developing such standards. Typical examples of this kind of standard are ANSI/ASTM E250-76: "Standard Recommended Practice for Use of CODEN," and ANSI Z39.18-1974: "American National Standard Guidelines for Format and Production of Scientific and Technical Reports." British Standard 1000 and its parts are various editions of the Universal Decimal Classification schedule.

Besides the technical standards described above, which are mainly useful in applied R & D and industrial practice, there are also fundamental standards dealing with the measurement of length, mass, time, temperature, various forms of energy, force, and other quantifiable fundamental entities that are basic to all scientific and technical endeavors.

Hemenway has identified two principal types of product standards, based on their function: (1) standards for uniformity, and (2) standards of quality [6]. Standards for uniformity are created to achieve simplicity by reducing needless variety in product sizes, speeds, and other attributes. The second function of standards for uniformity is to facilitate interchangeability of products and components manufactured by different companies. The quality of the product is not the dominant consideration in formulating standards for uniformity. Standards for uniformity are helpful in specifying dimensions and other features of products, e.g., bed sizes (king, queen, double, twin), electric lamp sockets, screw threads, etc. Standards for railroad track gauge also enforce uniformity. There are various track gauges in use in different countries: 5 ft 6 in. in South America, 4 ft 8½ in. in the United States, and 3 ft 6 in. in South Africa. The tracks on the original Great Western Railroad were 7 ft apart. Three different track gauges (broad, meter, and narrow gauges) are presently in use in India. In any given railroad system, uniformity of track gauge is extremely important for safety of passengers and interchangeability of tracks, locomotive engines, and rolling stock.

8.4 Sources of Standards and Specifications

Quality standards usually specify the minimum levels of product quality or performance, and are helpful in assessing the quality of products. The United States Department of Agriculture requires that commercial peanut butter must contain at least 90% peanuts by weight, or that orange drink must contain at least 10% orange juice; these are examples of product quality standards.

Though uniformity and quality are two important and distinct attributes of products, the distinction between a quality stanard and a standard for uniformity gets blurred in many cases; standardization of one of these attributes is very often likely to enhance the other also.

8.4 Sources of Standards and Specifications

Standards and specifications are fomulated by: (1) companies, (2) technical associations and professional societies, (3) government agencies, (4) national standards bodies, and (5) international standards agencies.

8.4.1 Corporate Sources

Most large designing and manufacturing companies have their own standards department staffed and supervised by standards engineers who develop new standards or modify published standards for internal use. Sometimes company standards developed for internal use in the company form the basis for national standards for industry-wide use. Company standards are usually confidential documents, not intended for publication or public use; they are primarily intended to streamline design and production procedure, and to achieve uniformity, safety, quality control, and economy within the company.

8.4.2 Technical Associations and Professional Societies: ASTM

A large number of technical associations such as the Technical Association of the Pulp and Paper Industry (TAPPI) and the American Petroleum Institute (API), trade associations such as the National Electrical Manufacturers' Association, and professional societies such as ASME are actively engaged in the formulation of standards applicable to the industries which they serve. For example, the *ASME Boiler and Pressure Vessel Code* (New York: American Society for Mechanical Engineers, Boiler and Pressure Vessel Committee, 1977) is a well-known and important set of standards. The *Code* provides rules for the construction of boilers, pressure vessels, and nuclear components, including requirements for material, design, fabrication, examination, inspection, and sampling. The *Code* is continually updated, and consists of 11 sections available in bound or loose-leaf format:

I. Power Boilers
II. Material Specifications

III. Nuclear Power Plant Components
IV. Heating Boilers
V. Non-Destructive Examination
VI. Recommended Rules for Care and Operation of Heating Boilers
VII. Recommended Rules for Care of Power Boilers
VIII. Pressure Vessels
IX. Welding and Brazing Qualifications
X. Fiberglass-Reinforced Plastic Pressure Vessels
XI. Rules for In-Service Inspection of Nuclear Power Plant Components

Standards made by technical associations and professional societies are often approved and adopted as national standards by the national standards bodies (e.g., ANSI in the United States and BSI in the United Kingdom). For example, ANSI Z11.299-1971 "Measurement of Liquid Hydrocarbons by Turbine Meter System" was originally a standard prepared and published as API Std 2534 in 1970 by the American Petroleum Institute; it was approved as an American National Standard and reissued in 1971 by ANSI with a new number.

Though a large number of engineering societies and technical associations create standards, the bulk of the standardization activity is carried out by only a few organizations. A statistical distribution similar to Bradford's law of scattering of scientific literature seems to be applicable to the contribution of societies and associations to standardization effort. Hemenway has pointed out that in 1964, over one half of all industry-wide voluntary standards were written by only three societies in the United States: the American Society for Testing and Materials (ASTM), the SAE, and the Aerospace Industries Association. Twenty percent of the standards were prepared by a further 15 organizations. The remaining 25-30% of the standards were developed by a very large number of societies and associations, each contributing one or a few standards [7].

Information on the activities and publications of various trade associations, professional societies, and other similar organizations may be obtained directly from the organization concerned. Brief descriptions of these organizations may also be found in directories, e.g., the *Encyclopedia of Associations* (Detroit, Mich.: Gale Research Company). A quarterly supplement entitled *New Associations and Projects* updates the *Encyclopedia* between successive editions. Another extremely valuable publication that gives descriptive information on numerous sources of standards and specifications is the guide entitled *Standards and Specifications Information Sources* (Management Information Guide No. 6) compiled by Erasmus J. Struglia (Detroit, Mich.: Gale Research Company, 1965). This guide also describes general sources of standards information, directories, bibliographies, indexes, and periodicals pertaining to standardization.

The ASTM is the largest society solely dedicated to standardization. It was founded in 1898 for "the development of standards on characteristics and

8.4 Sources of Standards and Specifications

performance of materials, products, systems, and services; and the promotion of related knowledge." ASTM operates through 132 main technical committees. It is the world's largest source of voluntary concensus standards, with a membership of over 27,500 of whom about 15,000 serve as technical experts on committees. The committees are composed of technical experts who are ASTM members, representatives of manufacturers and consumers, and general interest participants. All the currently approved ASTM standards are printed in the annual *Book of ASTM Standards*. The 1979 edition of the *Book* consists of over 44,000 pages in 48 volumes (or parts), and includes texts of over 5800 ASTM standards of various kinds. The last volume is an index. In 1949, the *Book* had a total of 8300 pages in six volumes. Besides standards formally approved and adopted by ASTM, the *Book* contains tentatives, tentative revisions, proposals, test methods, definitions, recommended practices, classifications, and specifications. These are various types of standards and specifications issued by ASTM, and are explained in the foreword in the first volume of the *Book*. ASTM also publishes numerous technical reports, papers, books, an annual catalog of publications, and the following serials:

ASTM Standardization News (formerly, *Materials Research and Standards*), monthly

ASTM Journal of Testing and Evaluation (formerly, *Journal of Materials*, quarterly), bimonthly

Journal of Forensic Sciences, quarterly

ASTM Proceedings, annual, incorporating the annual reports of ASTM committees

Copies of standards and other publications of ASTM can be obtained from the American Society for Testing and Materials, 1916 Race Street, Philadelphia, PA 19103.

8.4.3 Government Agencies: NBS

Government agencies are among the largest producers and users of standards and specifications in most countries. The United States Government is the world's largest bulk purchaser of virtually every kind of commodity and service, and, predictably, it is also one of the largest producers and users of standards and specifications. The number of standards and specifications in the military field alone is estimated to be well over 200,000 [8,9]. The total number of standards and specifications generated through other agencies is practically indeterminable. However, not all these standards and specifications are actually produced by the government agencies. The DOD, for example, endorses and adopts for its own use standards produced or approved by ANSI or other organizations. Best known is the MIL (or military) series of specifications dating back to 1944. The MIL series consists of all approved specifications of the United States Army, Navy, and Air Force. The Defense Standardization Act of 1952 provided for the coordination of standardization effort throughout the United States armed forces. An earlier series of specifications known as the JAN series (Joint Army Navy series), which was in use when some of the air forces of the United States were a part of the U.S. Army (USAAF), is now merged into the MIL series.

The designator for a military specification consists of three parts: The initials MIL (or JAN for older specifications), an initial letter of the title, and a serial number, as illustrated in the following examples:

MIL-P-46210 "*P*lastic Molding and Extrusion Material, Polysulfone"

MIL-L-21260 "*L*ubricating Oil"

Military standards are similarly designated, except that the second component consists of the letters STD instead of the first letter of the title:

MIL-STD-698A "Quality Standard for Aircraft Pneumatic Tire and Inner Tubes"

MIL-STD-1256A "Rubber-coated Parts for Machine Guns, 7.62mm, M60"

The *Department of Defense Index of Specifications and Standards* (DODISS) is an important source of information on unclassified federal, military, and departmental specifications and standards as well as those industry standards that are adopted by the DOD. DODISS is issued annually (with bimonthly supplements) and is available from the Superintendent of Documents, U.S. Government Printing Office, Washington, DC 20402. New military standards and specifications and amendments are regularly announced in a feature entitled "Status of Technical Manual Specifications and Standards" in *Technical Communication* (Journal of the Society of Technical Communication), published quarterly. The following are additional sources of information on military standards and specifications:

Naval Ordnance Systems Command. *Index of Ordnance Specifications and Weapons Specifications.* Issued annually, with mid-year supplements. Available from: Commanding Officer, Naval Ordnance Station, Central Technical Documents Office, Louisville, KY 40214. Attn: Code 80121.

Government Specifications and Standards for Plastics Covering Defense Engineering Materials and Applications. Revised. Norman E. Beach. (Plastics Technical Evaluation Center, Picatinny Arsenal, Dover, N.J.) Available from NTIS as AD-697 181.

Copies of unclassified military specifications and standards listed in the DODISS may be obtained from the DOD Single Stock Point (DOD-SSP) located

8.4 Sources of Standards and Specifications 141

in the Naval Publications and Forms Center (NPFC), 5801 Tabor Avenue, Philadelphia, PA 19120. Besides military standards and specifications, the DOD-SSP also distributes industry standards, federal standards and specifications, military handbooks, and other government standards documents. A booklet entitled *Department of Defense Single Stock Point for Specifications and Standards: A Guide for Private Industry* (available from the Naval Publications and Forms Center, 5801 Tabor Avenue, Philadelphia, PA 19120) describes the DOD-SSP and provides information on the acquisition of military standards and specifications, which are required by those doing business with the DOD.

Federal standards and specifications are administered by the General Services Administration. Federal specifications are designated by numbers with three components indicating: (1) the Federal procurement group to which the item belongs, (2) the first letter of the specification title, and (3) a serial number; for example,

L-P-349C   "*P*lastic Molding and Extrusion Material, Cellulose Acetate Butyrate"

In this designator, the first letter L represents the Federal procurement group Cellulose Products and Synthetic Resins; the letter P is the first letter of the title of the specification; and the third component is a serial number. Similarly, Federal standards have designators consisting of three components:

FED-STD-601   "Rubber: Sampling and Testing"

Federal standards and specifications are indexed in the following publication:

U.S. General Services Administration. *Index of Federal Specifications and Standards*, published annually since 1952 by the General Services Administration, with monthly supplements

A key government agency dedicated to standardization and research at a national level is the National Bureau of Standards (NBS). NBS was established by an Act of Congress on March 3, 1901 with the broad objective of strengthening the nation's science and technology and facilitating their application for public benefit. Rexmond Cochrane has written a comprehensive history of the NBS including an account of the developments that led to its inception at the turn of the century [1]. The Bureau consists of several field centers and five organizational units: (1) the Institute for Basic Standards, (2) the Institute for Materials Research, (3) the Institute for Applied Technology, (4) the Institute for Computer Sciences and Technology, and (5) the Office for Information Programs. Each of the Institutes consists of several divisions and offices. NBS conducts research to provide:

A basis for the nation's physical measurement system

Scientific and technological services for industry and government

A technical basis for equity in trade

Technical services to promote public safety

The Bureau also assists industries in developing and publishing voluntary product standards. The publication program of NBS is quite extensive and includes a wide variety of publications relating to standardization:

The Journal of Research of the National Bureau of Standards (bimonthly).

Dimensions/NBS (formerly, *Technical News Bulletin*) published monthly to disseminate latest research results, with emphasis on work done at NBS.

Monographs: These are major contributions to technical literature based on the Bureau's research and standardization activities.

Handbooks: These are recommended codes of practice developed in cooperation with industries, professional societies, and regulatory bodies.

Special Publications including conference proceedings, bibliographies, annual reports, etc.

Applied Mathematics Series: Mathematical tables and manuals.

Building Science Series: Test methods, performance criteria, and the durability and safety characteristics of building elements and systems.

Technical Notes: Research reports similar to monographs, but of more restricted scope.

Voluntary Product Standards developed in cooperation with industry according to procedures prescribed by the U.S. Department of Commerce.

Consumer Information Series: Nontechnical presentation of NBS research results on consumer products.

In 1976, its 75th anniversary year, NBS published a total of nearly 110,000 pages. About 79% of this material was published in the Bureau's own publications; the rest appeared in non-NBS journals, books, and proceedings. A 725-page catalog of NBS publications for 1976 (with abstracts, author index, and keyword index) was issued in June 1977 [10].

NBS publications are distributed by the Superintendent of Documents, U.S. Government Printing Office, Washington, DC 20402. Federal Information Processing Standards issued by NBS pursuant to the Federal Property and Administrative Services Act of 1949, and Interagency Reports on NBS research work sponsored by outside agencies, are distributed by NTIS. Several current awareness services (e.g., Cryogenic Data Center Current Awareness Service) and recurring literature surveys (e.g., a quarterly literature survey on superconducting devices and materials) are also issued by NBS.

8.4 Sources of Standards and Specifications

Two specialized activities within the NBS are: (1) the National Standard Reference Data System (NSRDS), and (2) Standards Information Service (NBS-SIS). The aims and activities of NSRDS are described in detail in Chapter 14, Handbooks and Data Compilations.

The Standards Information Service at the NBS was established in 1965 by the U.S. Department of Commerce at the suggestion of the Panel on Engineering and Commodity Standards of the Commerce Technical Advisory Board (better known as the LaQue Panel, after its chairman, Francis L. LaQue) [11]. NBS-SIS maintains an extensive reference collection of: (1) over 240,000 standards issued by various agencies (including government agencies) in the United States, by international standards bodies, and by national standards agencies in some 61 countries, and (2) over 600 technical reference books and 100 periodicals relating to standards. Besides this reference collection, NBS-SIS has a complete Visual Search Microfilm File (VSMF) of all federal standards and specifications, military specifications, standards, handbooks, and drawings. The VSMF collection also includes standards of the International Organization for Standardization (ISO) and the International Electrotechnical Commission (IEC), and some 6000 American vendor catalogs and federal construction regulations. NBS-SIS does not sell or distribute standards, but acts as a reference and referral center. It answers over 6000 inquiries each year, and conducts custom online bibliographic searches from a computerized database through the NBS Terminal Oriented Data Analysis and Retrieval System (TODARS). The reference collection is open to the public.

NBS-SIS has compiled and published the following indexes to standards and specifications:

Index of International Standards. NBS Special Publication 390, 1974.

An Index of U.S. Voluntary Engineering Standards. NBS Special Publication 329, 1971.

World Index of Plastics Standards. NBS Special Publication 352, 1971.

An Index of State Specifications and Standards. NBS Special Publication 375, 1973.

The above indexes are available from NTIS. Supplements to the *Index of U.S. Voluntary Engineering Standards* are sold by the Superintendent of Documents, U.S. Government Printing Office, Washington, DC 20402. An *Index of U.S. Standardization Activities* prepared by NBS-SIS is also available from the Superintendent of Documents.

8.4.4 National Standards Organizations: BSI, ANSI

Most industrially advanced countries and many developing countries have a national standards organization to prepare, approve, and publish standards and to coordinate standardization efforts at a national level. There are now over 60 national standards organizations in all. Standards are usually prepared by technical committees consisting of representatives of industry, professional societies, technical and trade associations, consumer groups, and government agencies. Requests for new standards may come from any of these groups, or the standards organization may initiate work on a new standard. The procedure for developing a new national standard is fairly involved and time-consuming. The responsibility for preparing a new standard is assigned to a technical committee. The committee then carries out the necessary research and prepares a draft standard. The draft is circulated to interested parties and comments are invited. The draft is revised in the light of comments received, and a final standard is prepared and proposed for adoption. A similar procedure is followed for revising or amending an existing standard. The national standards organization then announces the new standard, revision, or amendment, it its publications and catalogs, and prints the approved standards for distribution. In most countries, national standards are not mandatory; they are intended for voluntary adoption by producers of goods and services and other interested groups.

The British Standards Institution (BSI), with its headquarters in London, is the oldest national standards organization. It was founded by the Institution of Civil Engineers in 1901 as a single committee of eight members, then known as the Engineering Standards Committee. The scope of the Committee's activities expanded to include all branches of engineering and technology, a Royal Charter was granted in 1929, and the Committee became known as the British Engineering Standards Association. The present name was adopted in 1931. The BSI is now a very large organization: There are 4200 technical committees with a membership of approximately 22,000 drawn from industry, research institutions, and consumer groups. BSI has its own technical staff of over 200. The Institution is financed by membership fees, grants, sale of publications, and fees received for certification and testing of products. Its library has a complete collection of British standards for reference, and a large lending collection of overseas and international standards.

There are currently about 7000 British standards, and approximately 600 new standards are made each year. British standards are listed and indexed in the *British Standards Institution Yearbook.* New standards and revisions are announced in the *BSI News* published monthly.

BSI operates a certification marking system under which it permits manufacturers of goods to mark their products with the BSI's monogram (popularly known as the "Kite mark"). The presence of the Kite mark on a product is an indication to the purchaser that the product has been manufactured and tested according to the appropriate British standards. BSI periodically scrutinizes the quality of products to ensure that products marked with the Kite mark do indeed satisfy the quality requirements set by the standards. BSI claims that each year £500 million worth of products are sold with the Kite mark. Through

8.4 Sources of Standards and Specifications

a government-supported service entitled Technical Help to Exporters (THE), BSI also renders technical assistance to British industries in exporting their products.

In the United States, the American National Standards Institute (ANSI) is the national agency for coordinating and promoting standardization activity at a national level. ANSI's predecessor, the American Engineering Standards Committee, was established in 1918 by ASTM and four other engineering societies: the American Society of Mining and Metallurgical Engineers, the American Institute of Electrical Engineers, the ASME, and the ASCE. In 1928, the name was changed to American Standards Association (ASA). (The speed of photographic films is still designated by ASA numbers). For a short time in the mid-1960s the organization was known by the name United States of America Standards Institute (USASI). It was thought that this name might create an erroneous impression that the Institute was a government agency, and the present name was adopted in October 1969.

ANSI is a federation with a membership of over 160 technical and trade associations and professional societies, about 1000 companies, and a few consumer associations. The major functions of ANSI are:

To coordinate national standardizing activity by reducing duplication, overlapping, and variations in standards.

To facilitate the development of new standards and the revision or amendment of existing standards.

To approve standards developed by other agencies (e.g., professional societies, industries, etc.) as American National Standards.

To represent United States interests in international standards bodies. ANSI is the U.S. member of the International Organization for Standardization (ISO), the IEC, the Pan-American Standards Commission, and other international agencies.

To serve as a clearinghouse for information on American, foreign, and international standards.

ANSI operates through technical committees, and its standards depict the consensus of representatives of various groups as to the best current practice. Standards developed or approved by ANSI are not mandatory injunctions; they are recommendations for voluntary adoption by manufacturers, sellers, and purchasers of goods and services. ANSI standards are listed in the annual *ANSI Catalog*. New standards, revisions, and amendments are announced in a periodical supplement entitled *Listing of New and Revised American National Standards*. International standards of ISO and IEC are also listed in the annual *ANSI Catalog* and its supplements. Copies of ANSI standards as well as international standards are obtainable from the American National Standards Institute, 1430 Broadway, New York, NY 10018.

Beginning from January 1977, ANSI has adopted a new numbering system for standards. For standards developed by an industry, professional society, or other sponsor, the designation given by the sponsor is retained, and acronyms for the sponsor's name and ANSI are prefixed. For example, ANSI/ASTM C735-76 "Test for Acid Resistance of Ceramic Decorations on Returnable Beer and Beverage Glass Containers," was originally developed by ASTM and designated as C735-76, and was later approved as an American National Standard by ANSI. If there is no sponsor designation for a standard prepared by a sponsor, ANSI assigns an appropriate designator. For standards prepared by ANSI committees, the appropriate committee designation is used. Thus, ANSI Z39.5-1969 "Abbreviation of Titles of Periodicals" is a standard developed by the ANSI committee Z39 which is concerned with standardization in the fields of librarianship, documentation, and publishing. The earlier practice of assigning new designators to standards prepared by outside sponsors and approved by ANSI would result in the same standard receiving two or three different numbers, as indicated in the following example. ANSI K61.1-1972 "Safety Requirements for the Storage and Handling of Anhydrous Ammonia" replaced two standards: Standard CGA-G-2.1-1972 prepared by the Compressed Gas Association, Inc., New York, and Standard TFI-M-1-1972 prepared by the Fertilizer Institute, Washington, D.C. All these three designators are printed on the ANSI standard, and the ANSI designator itself does not indicate the sponsorship of the standard.

Instances of a standard issued by one agency being adopted by another agency, with or without a new designator, are fairly common. Cross-references to standards adopted by two or more ogranizations can be found in:

Standards Cross-Reference List, 2nd edition. Compiled by LaDonna Tompson, Kathy Beckman, and Patricia Ricci. Minneapolis, Minn.: MIS Systems Corporation, 1977.

8.4.5 International Standards Agencies

The need for standardization at an international level is becoming increasingly acute in view of increased international cooperation and multinational involvement in such areas as scientific research, technological development, industrial production, space exploration, national defense-related activities under treaties such as NATO, and cultural interchange among nations. These factors emphasize the need for (1) disseminating national standards information at an international level, and (2) coordinating standardization effort at an international level, at least in some vital sectors, to facilitate international communication and exchange of ideas, products, and technological know-how. Manufacturing

8.4 Sources of Standards and Specifications

industries exporting their products are required to conform to, or at least be familiar with, the standards and specifications affecting their products in the target countries where their products are likely to be marketed. For instance, electrical goods manufactured in the United States for export to England should be designed to operate at 220 volts and not at 110 volts. J. E. Holmstrom is reported to have said that "differences in the design of screw threads as between British and American practice added not less than $100 million to the cost of the Second World War" [12]. Many national standards organizations (including ANSI) provide listings and distribute copies of foreign and international standards. As noted earlier, BSI has set up a special service known as Technical Help to Exporters (THE) to provide information and advice on foreign standards and to translate foreign standards of importance to British exporters.

The International Electrotechnical Commission (IEC), established in 1906, was the first international standards organization to be concerned with the coordination and unification of national standards. IEC is composed of nearly 40 national committees, and its recommendations, representing international consensus, are intended to serve as a basis for national standards in member countries. For example, the British Standard BS 5142:July 1974 "Specification for Standard Cells" is a substantial reproduction of an earlier IEC standard IEC 428:1973. IEC standards are listed in its annual *Catalogue of Publications,* and also in the catalogs of many national standards organizations (including ANSI and BSI).

The International Organization for Standardization (ISO) was established in 1946 to achieve "international agreement on industrial and commercial standards and thus facilitate international trade as well as interchange of scientific and technological data relevant to standards." ISO has 64 national members (including ANSI and BSI), and its work is accomplished through more than 100 technical committees. Each committee covers a specific subject area, and its membership consists of representatives from national organizations and other concerned groups. ISO TC 46 is concerned with standardization in the field of library science and documentation. Numerical and alphabetical lists of ISO standards are published in its annual catalog of publications, and new standards and revisions are announced in the monthly *ISO Bulletin.* ISO standards are also announced in the catalogs of many national standards organizations. In the United States, copies of ISO standards may be obtained from ANSI.

Other international standards organizations and their activities have been described by Struglia [13] and Houghton [14]. A computer program for generating a unified internatiol catalog of all standards has been developed at the ISO Information Center in Geneva. This program is available to other standards institutes wishing to produce a computerized listing of their own standards [15].

References

1. Rexmond C. Cochrane, *Measures for Progress. A History of the National Bureau of Standards* (Washington, D.C.: National Bureau of Standards, 1966), pp. 84-86.
2. A. S. Tayal, Acquisition and updating of standards and specifications in technical libraries, *UNESCO Bulletin for Libraries,* 25:198-204 (July 1971).
3. S. K. Reeves, Specifications, standards and allied publications for U.K. military aircraft, *ASLIB Proceedings,* 22(9):432-448 (September 1970).
4. Bernard Houghton, *Technical Information Sources,* 2nd edition (Hamden, Conn.: Linnet Books, Shoe String Press, 1972), pp. 67-68.
5. Denis J. Grogan, *Science and Technology: An Introduction to the Literature,* 3rd edition (London: Clive Bingley, 1976), pp. 275-276.
6. David Hemenway, *Industrywide Voluntary Product Standards* (Cambridge, Mass.: Ballinger Publishing Co., 1975), pp. 8-9.
7. Hemenway (Ref. 6), p. 81.
8. Grogan, *Science and Technology* (Ref. 5), p. 279.
9. Houghton, *Technical Information Sources* (Ref. 4), p. 72.
10. *Publications of the National Bureau of Standards. 1976 Catalog.* SP 305, Supplement 8 (Washington, D.C.: U.S. Department of Commerce, National Bureau of Standards, 1977).
11. William J. Slattery, Standards Information Service (NBS), *Information Hotline,* 8(9):30-31 (October 1976).
12. Grogan, *Science and Technology* (Ref. 5), p. 274.
13. Erasmus J. Struglia, *Standards and Specifications Information Sources.* Management Information Guide 6 (Detroit, Mich.: Gale Research Company, 1965).
14. Houghton, *Technical Information Sources* (Ref. 4), pp. 79-81.
15. Barteld E. Kuiper, Towards a world catalog of standards, *UNESCO Bulletin for Libraries,* 27:155-159+ (May 1973).

9

HOUSE JOURNALS

9.1 Introduction

House journals are serial publications published by industrial corporations, business houses, public utility companies, and other similar organizations for circulation to their own employees or to their customers, shareholders, and libraries, or the scientific and technical community at large. The terms "house magazine" and "house organ" are also sometimes used. House journals largely or mainly meant for circulation to the company's own employees may be termed internal house journals; those that are largely addressed to outside constituents such as customers or prospective customers, shareholders, or the general public may be called external house journals. There are some composite house journals, circulated to company employees as well as outside users.

The main purpose of an external house journal is to promote the organization, its products, and its services. Internal house journals are intended to promote employee relations.

The origin of house journals has been traced back approximately 2000 years to the court letters of the Han dynasty in China in 200 B.C. These court letters were the forerunners of the modern official gazettes of national or local governments. *Travelers Record*, started by the Travelers Insurance Companies at Hartford, Conn. on March 1, 1865, is said to be the first external house journal. Its current title is *Protection*. *NCR Factory News*, started in 1887 by John H. Patterson, President of the National Cash Register Company, is said to be the first internal house journal. The term "house journal" was first used in 1891 in connection with this journal [1].

Widely divergent estimates have been made of the number of house journals. Grogan believes that about 2000 house journals are published in the United Kingdom and about 10,000 in the United States [2]. According to Walker, approximately, 17,000 house journals are published in the United States, about 10,000 in Europe, and 6000 in Japan [1]. If the definition of house journal is extended to include the promotional publications of all kinds of publishers

149

(businesses, industries, colleges and universities, labor unions, fraternities, clubs, churches, and so on), then the number of house journals published in the United States alone might be closer to 50,000.

9.2 Internal House Journals

An internal house journal serves the function of a newspaper within an organization. It carries personnel news, benefit plans, expansion plans, "suggestion box" features, calendar of events, and management views and news on a wide variety of topics of interest to employees of the organization. Large corporations are likely to have several internal house journals. Large multinational corporations usually publish several different language editions of their house journals.

The internal house journal is supported and published by the management, and is distributed to employees free of charge. It is predominantly a medium of the management, and rarely projects the employee viewpoint. A few underground employee newspapers have appeared recently to air employee grievances and viewpoints and to counterbalance the bland journalism of the company press. Joann Lublin has cited several examples of such employee house journals [3]:

The Stranded Oiler, published by and for the employees of the Standard Oil Company of California

The Met Lifer of the employees of the San Francisco offices of the Metropolitan Life Insurance Company

AT&T Express of the employees of the Pacific Telephone and Telegraph Company (the West Coast affiliate of the American Telephone and Telegraph Company)

Brookhaven Free Press, circulated monthly to about 3000 employees to the Brookhaven National Laboratory in Upton, New York

These employee house journals are produced and supported by the employees themselves, and have elicited varying degrees of response from the management, ranging from nonchalance to genuine concern and positive action to alleviate employee grievances. Employee house journals tend to be produced sporadically and to contain material of uncertain quality. Many of them are short-lived. The *GE Resistor*, circulted to 900 employees of the General Electric Company in the Philadelphia area, ceased publication after eight months.

9.3 External House Journals

As mentioned earlier, the external house journal is primarily a public relations tool, designed to promote company policies, plans, activities and products to

9.3 External House Journals

outside constituents. Depending on the contents and level of treatment, two types of external house journals can be identified: (1) "popular" (or nontechnical) house journals, and (2) technical house journals.

9.3.1 Popular House Journals

Popular house journals are aimed at nontechnical readers and contain descriptions of company products and services and general articles on current issues and developments relating to the industry. The articles and news items are usually supported by very attractive color photographs and other graphic material. *Age of Tomorrow,* published (in English) bimonthly by Hitachi Limited, Tokyo, Japan, is a typical example. This house journal frequently contains articles depicting the history, culture, and mythology of Japan. *Petroleum Today*, published quarterly by the American Petroleum Institute, and *Philips Technical Review,* published since 1936 by N. V. Philips Gloeilampenfabrieken, Eindhoven, Netherlands, are other examples of popular house journals. *Philips Technical Review* is published in three separate language editions (English, German, and Dutch) of identical contents. This journal is devoted to the investigations, processes, and products of the laboratories and plants which form part of or cooperate with enterprises of the Philips groups of companies. It covers a wide range of subjects and the articles are intended for both the subject specialist and the nonspecialist reader with some technical or scientific training. Annual and quinquennial indexes to *Philips Technical Review* are also published.

9.3.2 Technical House Journals

Technical house journals are aimed at the technical specialist, and contain scholarly research and review articles. These are often indistinguishable from regular research journals published by professional societies, academic institutions, commercial publishing houses, and government agencies. The oldest technical house journal is *Compressed Air,* published since 1896 by Ingersoll-Rand Company. The *IBM Journal of Research and Development, Bell System Technical Journal,* and *Endeavour* are examples of well-known technical journals.

In a survey conducted jointly by the Graduate School of Library Science, Drexel University, and the Graduate Library School in Antwerp, Belgium, Drott et al. studied various characteristics of 266 technical house journals published in the United States, the United Kingdom, and France [4]. For the purpose of this study [4], a technical house journal was defined as

> A journal that publishes at least one article per year which would aid a person with the equivalent of a four-year college degree in engineering or science in performing his duties in research, development, or production.

This definition excludes material of a business intelligence nature and also eliminates the type of product announcement and specification data which could only lead an engineer to contact the company selling the product. We have also excluded material dealing with marketing, finance, and management as well as material aimed solely at a lay audience.

The subjects covered in technical house journals encompass almost all the major branches of engineering and technology, such as chemical engineering, agriculture, construction, electronics, medicine, and metallurgy. Substantive articles in technical house journals include reports of research carried out at the company and elsewhere, research reviews, state-of-the-art reviews, and tutorial articles. Besides substantive articles, technical house journals also usually contain a variety of other material such as descriptions of company products, editorials, book reviews, calendar of conferences, product advertisements and reader service cards, lists of papers and patents of company technical staff, and not infrequently abstracts of substantive articles in one or more foreign languages. Several technical house journals in the sample studied by Drott and others had annual indexes and cumulated indexes.

Several large multinational corporations publish more than one technical house journal. *Philips Journal of Research* published bimonthly by Philips Research Laboratories, Eindhoven, Netherlands, contains scholarly papers on research carried out in the various Philips laboratories. *Philips Telecommunication Review* is published by Philips' Telecommunicatie Industrie B.V., Hilversum, Netherlands, in cooperation with the Philips group of companies and associates in 17 countries. This journal is dedicated to radio communication, radio relay, television, radar, navaigational aids, switching and transmission in telephony and telegraphy, data processing and conveyance, and traffic systems. Summaries of articles are given in French, German, and Spanish. The journal is distributed free of charge to those seriously interested.

The following four technical house journals are published by various divisions of the Exxon Corporation:

1. *Exxon Air World* (formerly, *Esso Air World*). Exxon International Company, a division of Exxon Corporation, New York, N.Y. 1947-, quarterly.

2. *Exxon Aviation News Digest* (formerly, *Esso Aviation News Digest*). Exxon International Company, a division of Exxon Corporation, New York, N.Y. 1947-, weekly.

3. *Exxon Chemicals Magazine.* Exxon Chemical Company, Houston, Tex. 1968-, quarterly.

4. *Exxon Oilways* (variant title: *Oilways*). Exxon Company USA, a division of Exxon Corporation, Houston, Tex. 1934-, bimonthly.

Endeavour is an interesting example of a periodical that started out as a technical house journal of an industrial firm and eventually became a regular scientific journal published by a large commercial publishing house. *Endeavour,* "A Review of the Progress of Science and Technology in the Service of Mankind" (subtitle), was founded in 1942 by the Imperial Chemical Industries, Limited, London. For some time the company published five separate language editions: English, French, German, Italian, and Spanish. Two cumulated indexes were also published covering the periods 1942-1951 (volumes 1 through 10), and 1952-1961 (volumes 11 through 20). From January 1977, the publication of *Endeavour* was entrusted to Pergamon Press.

Drott and coworkers found that only about 4% of the articles in technical house journals were again published in other regular journals. About 7% of the journals studied had subscription charges for at least some categories of users. Press runs varied greatly, ranging from 140 to 300,000.

9.4 Bibliographic Control of House Journals

House journals display, to an extreme degree, all the bibliographic vagaries of other kinds of journals. These are:

Change of title without notice or comment

Use of same title for several publications

Irregular publication schedule

Absence of date, volume number, or issue number

A high mortality rate

Drott and coworkers found that coverage of house journals in abstracting and indexing services was very low, but the delay in coverage was commendably short.

House journals are listed in standard lists of periodicals such as *Ulrich's International Periodicals Directory*. Drott and others found that coverage of British house journals was surprisingly low in the *Union List of Serials* and *New Serial Titles*. Also, entries in the "House Organs" section of *Standard Periodical Directory* were badly out of date; some titles listed in this directly had ceased publication as much as forty years ago. In the article on house journals in *Progress in Library Science,* Haberer has given complete and accurate description of over 220 house journals; an additional list of nearly 100 house journals is also appended to this article [5]. The *British Union Catalog of Periodicals* (BUCOP) and the *World List of Scientific Periodicals* also include technical house journals. Additional listings of house journals can be found in the following publications:

1. *Gebbie House Magazine Directory.* Burlington, Ida.: (National Research Bureau, Inc. 1946-, annual).
2. Special Libraries Association. New York Chapter. Technical Services Group. *Technical House Organs: A Directory and Union List of Titles in New York Chapter Libraries.* (New York: Special Libraries Association, 1971).
3. *Standard Periodical Directory,* 6th edition, 1979-80. (New York: Oxbridge Communications, Inc. 1978).

References

1. Albert Walker, House Journals, *Encyclopedia of Library and Information Science* (New York: Marcel Dekker, 1974), v. 11, pp. 61-74.
2. Denis J. Grogan, *Science and Technology: An Introduction to the Literature,* 3rd edition (London: Clive Bingley, 1976), p. 298.
3. Joann S. Lublin, Underground papers in coporations tell it like it is—or perhaps like it isn't, *Wall Street Journal* (November 3, 1971), p. 16.
4. M. Carl Drott, Toni C. Bearman, and Belver C. Griffith, The hidden literature: The scientific journals of industry, *ASLIB Proceedings,* 27(9):376-384 (September 1975).
5. Isabel H. Haberer, House journals, *Progress in Library Science* (London: Butterworths, 1967), v. 1, pp. 1-96.

10
TRADE CATALOGS

10.1 Introduction

The term "trade catalog" is generally used to denote a type of literature produced by manufacturers and distributors of virtually every kind of material, product, or service ranging from pets, books, drugs, and chemicals to extremely complicated equipment and components used in research and industry. The basic purpose of trade catalogs is twofold: (1) to provide information on the various attributes of a product, process, material, or service; and (2) to stimulate sales of the product, process, material, or service. The earliest trade catalogs were book lists distributed by booksellers. A catalog of books issued in 1564 by Georg Willer, an Augsburg bookseller, is believed to be the first such catalog. This was a subject list of 256 books. The first trade catalogs of books in England were those published by Andrew Munsell (d. 1596). Benjamin Franklin issued a catalog of books in 1744 in which he offered "those who live remote the same justice as though present . . . provided they send him the necessary cash purchase price as listed."

According to Romaine, the first American drug catalog was issued in 1760 by John Tweedy of Newport. The title page of this catalog read [1]:

> A Catalogue of Druggs, and of Chymical and Galenical Medicines; sold by John Tweedy at his shop in New Port, Rhode-Island. And for him in New-York, at the Sign of the Unicorn and Mortar.

The second drug catalog of John Day & Co., printed in 1771, had a similar title page:

> A Catalogue of Druggs, Chymical and Galenical Preparations, Shop Furniture, Patent Medicines, and Surgeon's Instruments, sold by John Day & Co., Druggists and Chymists, in Second-Street, Philadelphia. Printed by John Dunlap in Market-Street, Philadelphia: M,DCC,LXXI. 8vo. 33pp.

During the 1780s metal manufacturers of Birmingham and Sheffield were sending out elaborately illustrated catalogs to their agents in France, Italy, the

155

Americas, and also perhaps to India and Russia. In many of these early British trade catalogs, the manufacturers' names were omitted so that the customers were forced to order the goods through agents. In contrast to this, American trade catalogs almost always contain the manufacturer's name so that goods could be ordered directly from the manufacturer. This opportunity for direct mail order has probably been an important factor in the profusion of trade catalogs in the United States [2].

Retrospective collections of trade catalogs and bibliographies of such collections are becoming increasingly important for the newly emerging disciplines of industrial archaeology, business history, and history of technology. An excellent annotated guide to early American trade catalogs is Lawrence B. Romaine's *A Guide to American Trade Catalogs, 1744-1900* (New York: R. R. Bowker, 1960). Emphasizing the need for collecting and preserving trade catalogs, Romaine suggested that business historians should rather preserve "ten good, well considered, thoroughly illustrated, American trade catalogs in place of four tons of manuscript ledgers, possibly two tons of correspondence, and a scattering of leaflets, circulars, sales managers' instructions and charts of business development" [3]. Columbia University, the Center for Research Libraries in Chicago, and the Smithsonian Institution library have large collections of trade catalogs. Some public libraries, notably the Enoch Pratt Free Library in Baltimore, Maryland, and the Toledo Public Library in Ohio also have fairly large collections of trade catalogs.

Catalogs of merchandise such as those issued by Sears & Roebuck Co. or Montgomery Ward are of interest to general consumers. Specialized catalogs of chemicals and drugs and industrial products, processes, and materials are directed toward specialized groups of users such as chemists, medical practitioners, and engineers. Product catalogs are particularly important to engineers and technologists engaged in the design, development, production, and marketing of industrial products. In manufacturing industries, designers would prefer to use existing components, equipment, materials or subassemblies to achieve economy, compatibility, interchangeability, and marketability of their products. Specialized product catalogs (e.g., *Chemical Engineering Catalog* published by the Reinhold Publishing Company) enable engineers and technologists to:

Ascertain whether a specialty device, component, material, or process suitable for a specific purpose is available on the market

Compare the quality and economy of similar items available from various manufacturers to determine the most appropriate product for the purpose on hand

Obtain information on the manufacturer or supplier of a desired item.

10.2 Characteristics and Types of Trade Catalogs

In a survey of 1800 engineers, Davis found that those engaged in design and development were the most frequent users (90.5%) of manufacturers' catalogs. Catalogs were also used by 77.27% of consulting engineers and 64.84% of engineers engaged in facilities planning [4]. As sources of product and company information, trade catalogs are also useful to managers, technical and sales personnel, and buyers.

10.2 Characteristics and Types of Trade Catalogs

A cursory glance through the *Thomas Register of American Manufacturers* is enough to reveal the vast variety of products described in trade catalogs. Apart from this tremendous variety in the products themselves, there is considerable diversity in the format, size, and source of catalogs, as well as in the nature and amount of information provided therein. Trade catalogs range from a small advertisement in a periodical or a piece of paper briefly announcing a single product to elegantly bound multivolume compendia or frequently updated loose-leaf services describing in great detail thousands of products of a large number of manufacturers, often with photographs, drawings, and even bibliographic references to literature. Some product advertisements are so indistinguishable from a magazine article that editors of some magazines (e.g., the *Reader's Digest*) find it necessary to label them as "Advertisement" to distinguish them from feature articles. The following are some of the general characteristics of trade catalogs:

1. Trade catalogs contain application-oriented descriptive information rather than discussions of theoretical principles. However, literature describing medicines and complex scientific instruments frequently includes a brief discussion of relevant background research, supported by charts, diagrams, equations, and literature references.

2. Trade catalogs are primary sources in which information about products or processes can appear prior to its publication in technical journals or other forms of literature. In fact, much of the information about specific commercial products depicted in trade catalogs is not likely to be published at all in other forms of literature.

3. Much of the information contained in trade catalogs loses currency very quickly, as new products and processes are constantly being developed and improvements made in existing ones.

4. The amount of information provided in trade catalogs varies from a very brief announcement to a very elaborate description of the product or process. In general the following types of data are likely to be included in product catalogs:

Historical overview of the manufacturing company

Research background leading to the development of the product

Product description (dimensions, capacities, materials of construction, shape, size, color, and other physical, chemical, or engineering properties)

Applications

Operating characteristics (ranges of speed, rates of input and output, modes of operation, power requirements, etc.)

Installation instructions (foundation, fixtures, wiring and assembly instructions, instructions for unpacking, safety precautions, etc.)

Operation and maintenance procedures

Mode of acquisition, terms of licensing or lease, shipping and insurance data, delivery terms, availability of spares and auxiliaries, etc.

Illustrations, including photographs, sectional or cutoff views, exploded views, details of parts or subassemblies, layout plans, circuit diagrams, etc.

List of customers, recommendations or testimonials from satisfied customers (sometimes with photographs), excerpts from reviews

Names, addresses, and telephone numbers of regional offices, agents, local distributors, service centers

Trade catalogs are often undated. Price information is usually not included, and has to be requested by interested customers.

5. Trade catalogs are almost always supplied free. Even if they are priced, manufacturers' representatives usually give them free of charge to prospective customers. Some technical journals (e.g., *Chemical Engineering*) with an abundance of product news and advertisements are supplied free to qualified professional engineers.

6. Because of their dual function of information and persuasion, trade catalogs come in a great variety of sizes, formats, and colors. ANSI and BSI have formulated several standards specifying the format and contents of trade catalogs:

BS 1311:1955 "Manufacturers' Trade and Technical Literature." This British Standard gives the range of recommended paper sizes and also the type of information a trade catalog should contain.

BS 4462:1969 "Guide for the Preparation of Technical Sales Literature for Measuring Instruments and Process Control Equipment"

BS 4940:1973 "Recommendations for the Presentation of Technical Information about Products and Services in the Construction Industry"

ASA Z39.6-1966 "American Standard Specification for Trade Catalogs"

10.2 Characteristics and Types of Trade Catalogs

These standards are rarely followed by publishers of trade catalogs. The design of trade catalogs seems to be governed by the psychology and economics of advertising and selling rather than by considerations of their use and preservation.

In spite of the bewildering variety in the physical characteristics of trade catalogs, a few basic types may be identified:

Advertisements and announcements in technical journals and trade magazines

Special issues and supplements of journals

Manufacturers' catalogs and data sheets

Product descriptions supplied in trade expositions, convention exhibits, etc.

Directories of industries, products, and companies

Trade catalog services

Each of these major types of trade catalogs will be briefly discussed below. Catalogs of books (trade bibliographies) and instruction manuals supplied with equipment and machinery are not usually considered to be trade catalogs; these are not discussed in this chapter.

10.2.1 Advertisements and Announcements in Periodicals

Most technical journals and trade magazines carry product advertisements as well as product news and notes compiled by staff writers. Almost always each issue has an advertisers index. Some technical journals (e.g., *Production Equipment Digest, Chemical Processing*, and *Product Engineering*) consist almost entirely of new product announcements and advertisements. Another feature that is becoming increasingly common in journals containing product announcements and advertisements is the inclusion of one or more "reader service cards" (RSC) in each issue. The RSC is a simple mechanism for the reader of the journal to obtain more detailed information about a product or service advertised or announced in the journal from the manufacturer or distributor. This mechanism is sometimes known by other names, e.g., product information service or reader inquiry service. Trade names such as Inform-O-Gram and UTILICARDS are also used in some journals. The RSC (Figure 10.1) contains a block of consecutive numbers keyed to the announcements and advertisements in the journal issue, and space for the user's return address. The user simply circles the appropriate numbers, writes his name and address on the card, and then mails it to the journal publisher. Almost invariably, RSCs are "business reply cards" with postage prepaid by the journal publisher. Upon receipt of the RSC from the user, the journal publisher transmits the user's request for product information to the appropriate advertisers, and the advertiser then sends the desired

BUSINESS REPLY CARD
FIRST CLASS   Permit #27346   Philadelphia, Pa.

POSTAGE WILL BE PAID BY ADDRESSEE

analytical chemistry
P.O. BOX #7826
PHILADELPHIA, PA 19101

JUNE 1980   VALID THROUGH OCTOBER 1980

TO VALIDATE THIS CARD, PLEASE CHECK ONE ENTRY FOR EACH CATEGORY BELOW:

ADVERTISED PRODUCTS:

1	2	3	4	5	6
7	8	9	10	11	12	13	14	15	16	17
18	19	20	21	22	23	24	25	26	27	28
29	30	31	32	33	34	35	36	37	38	39
40	41	42	43	44	45	46	47	48	49	50
51	52	53	54	55	56	57	58	59	60	61
62	63	64	65	66	67	68	69	70	71	72
73	74	75	76	77	78	79	80	81	82	83
84	85	86	87	88	89	90	91	92	93	94
95	96	97	98	99	100	101	102	103	104	105
106	107	108	109	110	111	112	113	114	115	116
117	118	119	120	121	122	123	124	125	126	127
128	129	130	131	132	133	134	135	136	137	138
139	140	141	142	143	144	145	146	147	148	149
150	151	152	153	154	155	156	157	158	159	160
161	162	163	164	165	166	167	168	169	170	171
172	173	174	175	176	177	178	179	180	181	182
183	184	185	186	187	188	189	190	191	192	193
194	195	196	197	198	199	200	201	202	203	204
205	206	207	208	209	210	211	212	213	214	215
216	217	218	219	220	221	222	223	224	225	226
227	228	229	230	231	232	233	234	235	236	237
238	239	240	241	242	243	244	245	246	247	248
249	250	251	252	253	254	255	256	257	258	259

NEW PRODUCTS:

401	402	403	404	405	406	407
408	409	410	411	412	413	414	415	416	417	418
419	420	421	422	423	424	425	426	427	428	429
430	431	432	433	434	435	436	437	438	439	440
441	442	443	444	445	446	447	448	449	450	451
452	453	454	455	456	457	458	459	460	461	462
463	464	465	466	467	468	469	470	471	472	473
474	475	476	477	478	479	480	481	482	483	484
485	486	487	488	489	490	491	492	493	494	495

READER SURVEY:

301	302	303	304	305	306	307
308	309	310	311	312	313	314	315	316	317	318
319	320	321	322	323	324	325	326	327	328	329
330	331	332	333	334	335	336	337	338	339	340
341	342	343	344	345	346	347	348	349	350	351

Intensity of product need:
- [ ] 1. Have salesman call
- [ ] 2. Need within 6 mos.
- [ ] 3. Future project

Primary field of work:
- [ ] A. Energy
- [ ] B. Environmental
- [ ] C. Medical/Biological
- [ ] D. Drug/Cosmetic
- [ ] E. Forensic/Narcotic
- [ ] F. Textile/Fiber
- [ ] G. Metals
- [ ] H. Pulp/Paper/Wood
- [ ] I. Soaps/Cleaners
- [ ] J. Paint/Coating/Ink
- [ ] K. Electrical/Electronic
- [ ] L. Instrument Dev./Des
- [ ] M. Plastic/Polymer/Rub
- [ ] N. Agricultural/Food
- [ ] O. Inorganic Chemicals
- [ ] P. Organic Chemicals

Primary area of employment:
INDUSTRIAL
- [ ] A. Research/Development
- [ ] B. Quality/Process Control

MEDICAL/HOSPITAL
- [ ] C. Research/Development
- [ ] D. Clinical/Diagnostic

GOVERNMENT
- [ ] E. Research/Development
- [ ] F. Regulate/Investigate

COLLEGE/UNIVERSITY
- [ ] G. Research/Development
- [ ] H. Teaching

INDEPENDENT/CONSULTING
- [ ] I. Research/Development
- [ ] J. Analysis/Testing

This copy of Analytical is:
- [ ] 1. Personally addressed to me in my name.
- [ ] 2. Addressed to other person or to my firm.

NAME: _____
TITLE: _____
FIRM: _____
STREET: _____
CITY: _____
STATE: _____ ZIP: _____
PHONE: (___ ___ ___) ___ ___ ___ - ___ ___ ___ ___

Figure 10.1 Reader Service Card from *Analytical Chemistry*. (Reprinted with permission from *Analytical Chemistry*, 52(8). Copyright 1980 American Chemical Society.)

10.2 Characteristics and Types of Trade Catalogs

information directly to the user. Most advertisers add the inquirer's name to their mailing list, and continue to mail promotional literature periodically.

In an investigation reported by Bottle and Emery, acquisition of product information through RSC was found to be a very slow process [5]. Response times for receipt of the literature requested from advertisers approximated a normal distribution with an overall mean of 5.8 weeks from the date of mailing the RSC to the journal publisher. At the end of ten weeks, one quarter of the items requested had not been received. The investigators concluded that "if one sends off a RSC one is unlikely to receive more than half the items requested until 5½ weeks later."

Some journal publishers periodically send a bunch of postage-paid product inquiry cards to journal subscribers. Each card contains a description of one product and the address of its manufacturer or supplier. The user has to simply write his own address on the card and mail it. This variation brings the needed trade literature to the user faster than the RSC, as the request goes directly from the user to the advertiser.

10.2.2 Special Issues and Supplements of Journals

Many technical journals publish an annual special issue or supplement usually called "Buyers' Guide" or "Directory Issue." Annual buyers' guides issued by *Chemical Engineering, Chemical Week, Electronics, Hydrocarbon Processing, Machinery*, and *Nuclear News* are typical examples. The special issues are independent publications containing product and company information and advertisements. The *Nuclear News Buyers' Guide* is an annual special issue of the monthly magazine *Nuclear News* published by the American Nuclear Society, Inc., LaGrange Park, IL 60525. The 1978 *Buyers' Guide* contains: (1) An "Industry Report 1977-78" (an annual survey of inportant developments in the nuclear industry); (2) a world list of nuclear power plants; (3) a directory of nuclear products, materials, and services; and (4) a directory of suppliers located in the United States and 22 other countries. Each year in November, the journal *Science* (published by the American Association for the Advancement of Science) issues a special directory of scientific instruments. A listing of special issues and supplements can be found in the *Guide to Special Issues and Indexes of Periodicals*, 2nd edition, by Charlotte M. Devers, Doris B. Katz, and Mary M. Regan (New York: Special Libraries Association, 1976).

The American Chemical Society's annual *Lab Guide* is a directory of laboratory supply houses, manufacturing companies, instruments, equipment, chemicals, supplies, analytical and research services, trade names, and new books in analytical chemistry. The beginnings of the *Lab Guide* go back to 1939 when the October 15 issue of *Industrial and Engineering Chemistry, Analytical Edition*, included contributed articles on apparatus and instruments, and indexes of products and companies. In 1940 and 1941, the entire October 15 issues were devoted to articles on "American Apparatus, Instruments and Instrumentation" and to a directory of instruments and related apparatus. This feature was discontinued during the war years, and a "Buyers' Guide" was included in the Reviews issue of *Analytical Chemistry* during the years 1955-1965. Since 1966, the *Lab Guide* has been published as a separate publication.

In most libraries these special issues and supplements are treated as reference works of the directory type and are shelved along with other directories in the reference collection; these special issues are not usually bound with the other regular issues of the journal.

10.2.3 Manufacturers' Catalogs and Data Sheets

These may range from a single sheet of paper with a technical description of one product to a bound volume containing detailed descriptions and technical data on numerous products of a company, or a loose-leaf service kept up-to-date by periodical supplements. Catalogs of this category contain substantial technical data about products and their applications; some catalogs (e.g., *Alcoa Aluminum Handbook* published by the Aluminum Company of America, Pittsburgh) have attained the status of a reference work. These are supplied directly by the manufacturers or their agents to prospective customers. Engineers seem to prefer this kind of trade catalog for extended use; the other kinds of catalogs are more useful for current awareness or for generating requests from users for more detailed information on specific products.

10.2.4 Trade Fair Catalogs

Manufacturers exhibit their products and distribute catalogs at conventions and conferences organized by trade associations or professional societies. Trade catalogs are also issued at international trade fairs and expositions such as the famous Leipzig and Frankfurt Fairs which have been held from the middle of the 16th century. Besides product and company data, these catalogs are also likely to contain general information concerning the convention city and the sponsoring organizations.

10.2.5 Directories

These are independent publications containing data on a number of companies and their products, usually in one branch of engineering or technology (e.g., chemical engineering, aviation). *World Aviation Directory*, published semi-annually by Ziff-Davis Publishing Company, New York, is an international directory of air carriers, manufacturers of aircraft and related equipment,

10.2 Characteristics and Types of Trade Catalogs

government agencies, and other organizations concerned with aviation in some 160 countries. Books and other publications on aviation and aerospace technology are also listed. The "Buyers' Guide" section has a company index, a product cross index, and a trade name index.

The *Directory of Chemical Producers, USA*, published by Stanford Research Institute, is kept up-to-date by quarterly supplements. The publishers of *Chemical Engineering Catalog* (New York: Reinhold Publishing Company, 1916-) take pride in the fact that their directory is "not a thirteenth edition of a monthly magazine . . . *Chemical Engineering Catalog* is a book of reference for engineers, buyers and others seeking information on chemicals, equipment, machinery and supplies." The Reinhold Publishing Company also publishes an annual *Chemical Materials Catalog*. A more complete discussion of directories can be found in Chapter 13, Directories and Yearbooks.

10.2.6 Trade Catalog Services

In view of the growing importance of trade catalog collections in certain libraries, a number of commercial catalog services have started providing copies of manufacturers' catalogs on a continuing basis; this subscription service sometimes also includes an inquiry service. The Thomas Publishing Company, New York, publishers of the well-known *Thomas Register of American Manufacturers* (annual), supplies microfilm copies of catalogs of most of the manufacturers listed in the annual *Register*. Sweet's Catalog Service (offered by Sweet's Industrial Systems, Palo Alto, California) makes available trade catalogs and similar related material from industrial suppliers in microfilm cartridges. The catalogs are indexed by product name, trade name, and company name. Subscribers can keep their trade catalog collections current through a bimonthly updating service.

Another trade catalog service on microfilm is the Visual Search Microfilm Files (VSMF) offered by Information Handling Services, Inc., Englewood, Colorado. Besides separate files of trade catalogs in different subjects (e.g., design engineering, integrated circuits, marine engineering), files of military standards and specifications, international standards, industry standards, and documents of federal regulatory agencies are also available in VSMF. Catalogs of over 10,000 manufacturers from the United States and England are available on microfilm cassettes. New cassettes are supplied periodically. Subscribers can print hard copies of desired catalogs on a microfilm reader-printer.

A number of such catalog services have been described and compared by Wall [6]. Wall has also described an SDI service for trade catalogs offered by Indata Limited, in England. Interest profiles of subscribers are matched every week with new additions to a computerized database of trade catalogs, and subscribers receive weekly lists of new items as well as data cards containing specifications of the products. Acquisition and subject indexing of the catalogs are done in collaboration with the British Scientific Instruments Research Association.

New Technology Index published bimonthly since 1976 by Technology Clearinghouse, Inc., Wilmington, Delaware, is an indexing service that provides information on new technical products and services. Products and services are indexed by basic function as well as generic classification.

10.3 Acquisition and Control of Trade Catalogs

Traditionally, trade catalogs have not been systematically acquired and organized in libraries with the same zeal that is accorded to other types of technical literature. Certain characteristics of trade catalogs render any such systematic acquisition and organization very tedious and also perhaps unnecessary. Trade catalogs are seen as ephemeral material that becomes obsolete very quickly; they are also thought of as expendable material because of the ease with which they can be obtained by the user free of charge. Most engineers who regularly use trade catalogs do not find it difficult to acquire and maintain their own personal collection of trade catalogs. Conventional abstracting and indexing services do not cover trade catalogs. The librarian's traditional predilection for the more enduring and scholarly material is also perhaps another contributing factor. Most librarians prefer to acquire and maintain a collection of directories of various sorts, and leave the acquisition of catalogs to the individual user. The acquisition of trade catalogs is relatively simple. Many manufacturers routinely mail their catalogs and promotional material to technical librarians, purchasing agents, and information officers in companies. Reader service cards can be used to acquire additional material free of charge. Where such a facility does not exist, a request on official letterhead to the manufacturer is usually adequate for acquiring the desired catalogs.

The organization of trade catalogs is also relatively simple. The catalogs themselves may be filed in vertical files or pamphlet boxes alphabetically by manufacturer's name. A card index may be maintained to provide access by product name. A simple coordinate index of the Uniterm type is adequate for this purpose. A more difficult problem is the constant weeding that is necessary to keep the collection current. Trade catalogs usually do not contain date of publication, and it is difficult to determine their currency. Without constant attention and weeding, a collection of current trade catalogs will soon become a retrospective collection of archival material, more useful for historical purposes than for obtaining current and accurate information on companies and their products. In libraries where there is a constant demand for trade catalogs, it is easier to subscribe to one of the commercial catalog services mentioned earlier. Subscription to a commercial catalog service will ensure the availability of up-to-date information on companies, materials, products, and processes.

One additional feature of trade catalogs is the extensive use of trade names to identify specific products. Many directories of companies and products include a trade name index. Where such an index is not available, the following two publications will be very helpful for obtaining explanations of product trade names and the names and addresses of their manufacturers:

1. *Trade Names Dictionary*, edited by Ellen T. Crowley (Detroit, Mich.: Gale Research Company, 1976). This is a "guide to trade names, brand names, product names, coined names, model names, and design names, with addresses of their manufacturers, importers, marketers or distributors."
2. *Chemical Synonyms and Trade Names: A Dictionary and Commercial Handbook*, by William Gardner. 8th edition revised and enlarged by Edward I. Cooke and Richard W. I. Cooke (Oxford: Technical Press, 1978). This handbook contains over 35,000 definitions of trade names; where available, names of manufacturers are also given.

References

1. Lawrence B. Romaine, *A Guide to American Trade Catalogs, 1744-1900* (New York: R. R. Bowker, 1960), Introduction.
2. A. Hyatt-Mayor, Foreword in *A Guide to American Trade Catalogs* (Ref. 1), p. vii.
3. Lawrence B. Romaine, American trade catalogs vs. manuscript records, *Library Resources and Technical Services*, 4(1):63–65 (Winter 1960).
4. R. A. Davis, How engineers use literature, *Chemical Engineering Progress*, 61(3):30–34 (March 1965).
5. Robert T. Bottle and Betty L. Emery, Information transfer by reader service cards. A response time analysis, *Special Libraries*, 62(11):469-474 (November 1971).
6. R. A. Wall, Trade literature problems, *Engineer*, 225:453-454 (March 15, 1968); 225:489-491 (March 22, 1968).

11

BIOGRAPHICAL LITERATURE

11.1 Introduction

Literature on the lives of scientists and their views and accomplishments constitutes an important part of the literature of science. Advances in science are often facilitated by direct personal communication within the scientific community. Scientists require biographical information about other scientists for a variety of reasons:

- To establish contact with those engaged in similar scholarly or research pursuits for exchange of information
- To request or disseminate reprints, preprints, and other publications
- To identify experts who can review manuscripts, books, and grant proposals
- To identify experts for committee assignments, conference panels, and other professional activities
- To write biographical sketches

Biographical information sought may range from the address or telephone number of a known person to exhaustive biodata including education, research interests, previous and current employment data, society memberships, publications, and contributions to science.

11.2 General Biographical Works

General biographical publications such as *Dictionary of American Biography, Dictionary of National Biography,* and *Chambers's Biographical Dictionary* provide biographical data on scientists also. Besides these general works, very many specialized biographical works are available for obtaining biographical data on scientists and engineers. *World Who's Who in Science: A Biographical Dictionary of Notable Scientists from Antiquity to Present*, edited by Allen G.

11.3 Specialized Biographical Works

Debus (Chicago: Marquis Who's Who, Inc., 1968), is an example of general biographical publications that include information on scientists and engineers irrespective of their specialization, national origin, or other factors. This work contains about 30,000 entries; approximately one-half of them are biographical sketches of historical figures. Information on living scientists is largely assembled from questionnaires. Though the scope is international, American and Western European scientists are covered more comprehensively than those of other countries.

The *McGraw-Hill Encyclopedia of World Biography* (New York: McGraw-Hill, 1973) is especially useful for finding biographical information even when the name of the biographee is not known. Designed to meet the needs of school, college, and public libraries, this 12-volume encyclopedia has about 5000 biographical articles and 200 historical maps. The average length of the articles is 800 words, and each article has at least one illustration. The 12th volume contains an index and 17 "study guides" which are topical outlines used in planning the *Encyclopedia*. Each study guide includes the names of biographees. Thus, if one wanted to collect biographical information on European mathematicians of the 17th century, the names of biographees (Pascal, Leibniz, Descartes, and others) may first be ascertained from the study guide on science in volume 12, and the biographical information on each one of these biographies may then be obtained from the encyclopedia.

11.3 Specialized Biographical Works

The scope of specialized biographical works in science and technology is often limited by the national origin, subject specialization, or professional society affiliation of the biographees. Again, some are retrospective works, listing only deceased scientists, and others are current biographies that include only living personalities. The following are typical of specialized biographical works with restrictions of time, national origin, or subject specialization of biographees:

Dictionary of Scientific Biography (New York: Charles Scribners, 1970-), sponsored by the American Council of Learned Societies, includes biographies of only deceased scientists from all periods of history. So far 13 volumes have been published, and the work is still in progress.

American Men and Women of Science, 14th edition (New York: R. R. Bowker, 1979) has approximately 130,000 entries in 8 volumes. Geographic and discipline indexes make up the 8th volume. The first edition, issued in 1906, had 4000 entries in a single volume.

Current Bibliographic Directory of the Arts and Sciences (CBD) is an international directory of current authors in all branches of science and the arts whose works are indexed in the various publications of the Institute for

Scientific Information (ISI). Names of authors come from nearly 5800 journals and about 2500 books indexed by ISI each year. CBD has three sections: (1) author section, (2) geographic section, and (3) organization section. (A sample of each is shown in Figure 11.1.) A special feature of CBD is that each entry in the author section has abbreviated citations of an author's publications.

Each year CBD lists over 350,000 authors; slightly more than one-half of these are from outside the United States. A statistical data section contains tables ranking countries, states, and cities according to the output of scholarly publications. Beginning in 1979 CBD has replaced an earlier publication known as *Who Is Publishing in Science* (WIPIS). The 1978 edition of WIPIS listed 339,078 published authors from 237 countries. The data reported in WIPIS have been used for a number of studies in the sociology of science [1-5].

Biographical Notes Upon Botanists (Detroit, Mich.: Gale Research Co., 1965) contains some 44,000 biographical entries in three volumes taken from the annotated card file maintained in the New York Botanical Gardens Library. *Who's Who in Science in Europe: A Reference Guide to European Scientists* (Guernsey, British Isles: Francis Hodgson, 1977) has over 40,000 entries in four volumes. *Who's Who in Soviet Science and Technology*, 2nd edition, compiled by Ina Telberg (New York: Telberg Book Co., 1964) is said to be the first biographical directory of Soviet scientists in English; it contains biographical data on some 1000 living Soviet scientists. For scientists born before the 1917 Revolution, two dates of birth are given: The first date is according to the Julian calendar in use when the person was born, and the second date is its equivalent in the Gregorian calendar in use today. *Soviet Men of Science, Academicians and Corresponding Members of the Academy of Sciences of the USSR* (Princeton, N.J.: D. Van Nostrand, 1963) compiled by John Turkevich, with financial support from the National Science Foundation, gives biographical information on scientists; engineers and technologists are not listed.

11.4 Biographical Serials

Typical of biographical serials are *Biographical Memoirs of the Fellows of the Royal Society* (London: The Royal Society, 1955-), and *Biographical Memoirs* of the National Academy of Sciences of the USA. The former is a continuation of *Obituary Notices of the Fellows of the Royal Society* (London: The Royal Society, 1932-1954). The *Biographical Memoirs* of the National Academy of Sciences is a continuing series of volumes published annually beginning from 1877. These volumes are designed to provide "a record of the lives and works of the most distinguished leaders of American science as witnessed and interpreted by their colleagues and peers." Each biographical essay is written by

11.4 Biographical Serials

(a)

ROBINSON JW
BELL TEL LABS INC
MURRAY HILL, NJ, 07974
  REV SCI INS 49(2):205-207 78
ROBINSON JW
LOUSIANA STATE UNIV, DEPT CHEM
BATON ROUGE, LA, 70803
  AM LABORAT 10(10):41 73
  ANALYT CHIM 100(SEPT):301-312 78
  ATMOS ENVIR 12(4):957 78
  ATMOS ENVIR 12(5):1247-1248 78
  SPECT LETT 11(2):73-88 78
  SPECT LETT 11(9):715-724 78
ROBINSON JWL
CTR HOSP UNIV VAUDOIS, DEPT
CHIRURG EXPTL, SERV CHIRURG A
CH-1011 LAUSANNE, SWITZERLAND
  DIGESTION 16(1-2):217-218 77
  GASTRO CL B 1(11):950-951 77
  GASTRO CL B 2(3):279-286 78
ROBINSON K
AUSTRALIAN NATL UNIV, RES SCH
CHEM
CANBERRA 2600, ACT, AUSTRALIA
  PHYS LETT A 64(5):467-469 78
ROBINSON K
INST COMMONWEALTH STUDIES
LONDON, ENGLAND
  J COM C POL 15(2):197-198 77
  J COM C POL 16(1):1C3-106 78
  J COM C POL 16(2):119-135 78
ROBINSON K
NATL HOSP, DEPT CLIN
NEUROPHYSIOL
LONDON WC1N 3BG, ENGLAND
  J NEUR SCI 36(1):147-156 78
ROBINSON K
TUFTS UNIV, SCH DENT MED

(b)

LOUISBURG COLL
  NORTH CAROLINA LOUISBURG
LOUISE CHILD CARE CTR
  PENNSYLVANIA PITTSBURGH
LOUISIANA AGR EXPT STN
  LOUISIANA BATON ROUGE
LOUISIANA LAND EXPLORAT CO
  LOUISIANA NEW ORLEANS
LOUISIANA OFF FAMILY SERV
  LOUISIANA NEW ORLEANS
LOUISIANA PACIFIC CORP
  CALIFORNIA OROVILLE
LOUISIANA STATE EMPLOYEES RETIREMENT SYST
  LOUISIANA BATON ROUGE
LOUISIANA STATE PLANNING OFF
  LOUISIANA BATON ROUGE
LOUISIANA STATE UNIV
  COSTA RICA SAN JOSE
  LOUISIANA ALEXANDRIA
  LOUISIANA BATON ROUGE
  LOUISIANA CHASE
  LOUISIANA CROWLEY
  LOUISIANA EUNICE
  LOUISIANA HOMER
  LOUISIANA MONROE
  LOUISIANA NEW ORLEANS
  LOUISIANA SHREVEPORT
LOUISIANA STATE UNIV HOSP
  LOUISIANA SHREVEPORT
LOUISIANA STATE UNIV OBSERV
  LOUISIANA BATON ROUGE
LOUISIANA TECH UNIV
  LOUISIANA RUSTON
LOUISIANA WILDLIFE & FISHERIES COMMISS
  LOUISIANA LAKE CHARLES
LOUISVILLE GEN HOSP
  KENTUCKY LOUISVILLE

(c)

BATON ROUGE
LOUISIANA STATE UNIV
  DUNCAN JA
  DUNIGAN EP
  EATON HC
  EFFERSON JN
  ELLIS FW
  EPPS EA
  FALK WW
  FARBER SC
  FARMER RC
  FELDMAN AD
  FELPS WS
  FERRELL RE
  FINDLEY AM
  FISCHER JI
  FIVIZZANI AJ
  FLEETWOOD SC
  FORESTER JS
  FOSTER LL
  FOWLER JF
  FRANKE DE
  FRENCH WL
  FULTON RW
  GALTON DM
  GAMBRELL RP
  GANDOUR RD
  GARAY R
  GARTON D
  GIBBS DE
  GILDERSLEEVE RP
  GILMAN DF
  GODKE RA
  GOOD BK
  GOSSELINK JG
  GOYERT JC
  GRAVES GR

Figure 11.1 Sample entries from ISI's *Current Bibliographic Directory of the Arts and Sciences*, 1978. (a) Author section, (b) organization section, and (c) geographic section.

an individual familiar with the discipline and the scientific career of the biographee, and includes a portrait and a chronological bio-bibliography. Both these biographical memoirs include biographies of only deceased members of the respective societies.

Biographical sketches of living persons who have achieved prominence and recognition recently may be found in the serial *Current Biography* (New York: H. W. Wilson, 1940-). The aim of *Current Biography* is to provide "brief, objective, accurate, and well-documented biographical articles about living leaders in all fields of human accomplishment the world over." The articles in *Current Biography* are two to three pages long, and contain portraits and bibliographical references. At the end of each year, articles in the monthly issues of *Current Biography* are cumulated in one sequence, revised, and printed in a single volume known as *Current Biography Yearbook*. Three decinnial indexes for the years 1940-1950, 1951-1960, and 1961-1970 as well as a cumulated index for the years 1940-1970 have been published.

11.5 Collective Biographies

Collective biographical works are monographs or anthologies containing separate biographical sketches of a number of biographees. The selection of biographees may be governed by their scientific discipline, time period, or national origin. The following are examples of collective biographies:

Early Seventeenth Century Scientists, edited by R. Harré (Oxford: Pergamon Press, 1965.)

Makers of Science: Mathematics, Physics, Astronomy by Ivor B. Hart. (London: Oxford University Press, 1923.)

The Golden Age of Science: Thirty Portraits of the Giants of 19th Century Science by Their Scientific Contemporaries, edited by Bessie Zaban Jones. (New York: Simon and Schuster, in cooperation with the Smithsonian Institution, 1966.)

The Laureates. Jewish Winners of the Nobel Prize by Tina Levitan. (New York: Twayne Publishers, 1960.) This is a collected biography of 40 Jewish intellectuals who received the Nobel Prize between 1905 and 1959.

Almost always the entries in collective biographical works are arranged alphabetically by biographee's name. Other arrangements such as chronological, geographical, and subject arrangement are relatively rare. Entries in Isaac Asimov's *Biographical Encyclopedia of Science and Technology* (Garden City, N.Y.: Doubleday, 1964) are arranged chronologically. Collective biographical works often contain subject and geographical indexes to supplement the

11.6 Biographical Monographs and Autobiographies

alphabetical arrangement of biographical sketches. *American Men and Women of Science*, 13th edition (New York: R. R. Bowker, 1976) has discipline and geographic indexes in a separate volume.

Biographical sketches usually contain portraits and biobibliographies of biographees. The *Dictionary of Scientific Biography* lists important original works of the biographee and secondary works about the biographee.

11.6 Biographical Monographs and Autobiographies

Prominent scientists and engineers are often the subject of biographical monographs and autobiographies. *Coulomb and the Evolution of Physics and Engineering in Eighteenth Century France* by C. Stewart Gilmore (Princeton University Press, 1971) is a fine example of a biographical monograph. It contains an extensive bibliography of works by and about Coulomb. *Louis Pasteur* by S. J. Holmes (New York: Dover, 1961), *A Portrait of Isaac Newton* by Frank E. Manuel (Harvard University Press, 1968), and *Ramanujan, the Man and the Mathematician* by S. R. Ranganathan (Bombay: Asia Publishing House, 1967) are other representative examples.

Personal correspondence and papers of scientists are a valuable source of information on their lives, professional contributions, and views. The *Posthumous Works of Robert Hooke*, first published by the Royal Society of London in 1705, is an excellent example of biographical works of this genre. This book was republished in 1969 by Johnson Reprint Corporation as Number 73 in *Sources of Science*, a series that includes many similar classic biographical works on Newton, Faraday, William Harvey, Kepler, and other famous scientists. The following are some additional examples:

The Born-Einstein Letters. Correspondence Between Albert Einstein and Max and Hedwig Born from 1916 to 1955 with Commentaries by Max Born. Translated by Irene Born. (New York: Walker and Company, 1971.)

Partners in Science, Letters of James Watt and Joseph Black, edited with an introduction and notes by Eric Robinson and Douglas McKie. (London: Constable, 1970.)

The Papers of Joseph Henry. Volume 1: *December 1797-October 1832, the Albany Years*; Volume 2: *November 1832-December 1835, the Princeton Years.* (Washington, D.C.: The Smithsonian Institution Press, 1972, 1975.) These volumes contain Henry's personal and professional correspondence, lecture notes, minutes of meetings, and other documents.

A Scientific Autobiography of Joseph Priestly 1733-1804. Selected Scientific Correspondence, edited with commentary by Robert E. Schofield. (Cambridge, Mass.: The MIT Press, 1966.)

Max Born's *My Life and My Views* (New York: Charles Scribner's Sons, 1968), and Julian Huxley's *Memories* (New York: Harper & Row, 1970) are typical autobiographies of prominent scientists.

11.7 Other Sources of Biographical Information

Biographical information on scientists may be found in festschrift volumes. *Stephen Timoshenko: 60th Anniversary Volume* (New York: Macmillan, 1938) consists of "contributions to the mechanics of solids dedicated to Stephen Timoshenko by his friends on the occasion of his sixtieth birthday anniversary." General encyclopedias (e.g., *Encyclopedia Americana* and *Encyclopaedia Britannica*) contain biographical articles, often with portraits and bibliographies. Some biographical entries may also be found in dictionaries and encyclopedias of science such as *Hackh's Chemical Dictionary*, 4th edition (New York: McGraw-Hill, 1969), and the *Harper Encyclopedia of Science*.

Primary journals are important sources of biographical information. Besides overtly biographical articles, they may contain brief biographical sketches of contributors, obituaries, and "personalia" columns in each issue. Newspapers and news digest services (e.g., *Facts on File, Asian Recorder, Keesing's Contemporary Archives*), and membership directories of professional societies are other sources of biographical information.

11.8 Bibliographic Control of Biographical Literature

Biographical works can be identified through the standard guides to literature (e.g., Walford, Sheehy, and *American Reference Books Annual*). *Biographical Dictionaries and Related Works* by Robert B. Slocum (Detroit, Mich.: Gale Research Co., 1967) is an extensive bibliography of biographical works. The subtitle indicates the scope of this work: "An international bibliography of collective biographies, bio-bibliographies, collections of epitaphs, selected genealogical works, dictionaries of anonyms and pseudonyms, historical and specialized dictionaries, biographical materials in government manuals, bibliographies of biography, biographical indexes, and selected portrait catalogs." About 4800 biographical and related works are indexed by author, title, and subject. Supplements were issued in 1972 and 1978.

Several retrospective and continuing indexes are available for tracing biographical information. *Index to Scientists of the World from Ancient to Modern Times: Biographies and Portraits*, compiled by Norma O. Ireland (Boston: F. W. Faxon Co., 1962) is an index to biographical information on nearly 7500 scientists in some 338 collected works in the English language. Another very

Selected List of Biographical Works 173

useful collective index is the *Biographical Dictionaries Master Index* (Detroit, Mich.: Gale Research Co., 1975) in three volumes; this is a guide to nearly 750,000 biographical sketches in over 50 current who's whos and other works of collective biography, including *American Men and Women of Science*. The emphasis in the works indexed, and hence in the Master Index, is largely on living persons who are prominent in the United States scene. Two supplements indexing an additional 300,000 biographical documents are being issued.

A recurring bibliography of biographic material is the *Biography Index* (New York: H. W. Wilson Co., 1947-). This cumulative index to biographical material in books and magazines is published monthly, with annual and triennial cumulations. It covers over 1500 periodicals, books, and incidental biographical material such as prefaces and chapters in nonbiographical books.

Biographical material can also be traced through abstracting and indexing services and indexes to newspapers (e.g., *The New York Times Index*). A selected list of biographical works is appended to this chapter.

Selected List of Biographical Works

Indexes to Biographies

1. *Biographical Dictionaries and Related Works*. Robert B. Slocum (Detroit, Mich.: Gale Research Co., 1967), 1056p.
   A bibliography of 4829 biographical works representing material from 108 countries or regions.
2. *Biographical Dictionaries and Related Works: Supplement*. Robert B. Slocum (Detroit, Mich.: Gale Research Co., 1972), 852p.
   An annotated bibliography of 3342 biographical works.
3. *Biographical Dictionaries Master Index, 1975-76* (Detroit, Mich.: Gale Research Co., 1975), 3 v. Biennial editions and supplements.
   A guide to nearly 750,000 biographical sketches in about 50 collective biographical works.
4. *Biography Index* (New York: H. W. Wilson Co., 1946-), quarterly; annual and three-year cumulations.
   A guide to biographic material appearing in about 2200 periodicals indexed in the Wilson indexes.
5. *Chicorel Index to Biographies* (New York: Chicorel Library Publishing Corporation, 1974).
   Contains over 21,000 index entries to biographies and autobiographies in all branches of knowledge.
6. *Current Biography Cumulated Index, 1940-1970* (New York: H. W. Wilson Co., 1973), 113p.
7. *Index to Scientists of the World from Ancient to Modern Times: Biographies and Portraits*, compiled by Norma O. Ireland (Boston, Mass.: F. W. Faxon Co., 1962).

Index to biographical data on nearly 7500 scientists in some 338 collective biographies.
8. *Marquis Who's Who Publications Index to All Books, 1977* (Chicago, Ill.: Marquis, 1978-), annual.
An annual index to biographical sketches in current editions of 14 Marquis Who's Who volumes; 280,000 sketches indexed in 1978.

General Biographical Works

9. *Chamber's Biographical Dictionary*. J. O. Thorne, ed. (New York: St. Martin's Press, 1969), 1432p.
Over 15,000 entries. Original edition published in 1897; successive editions issued in 1929 and 1946.
10. *Current Bibliographic Directory of the Arts and Sciences* (Philadelphia, Penn.: Institute for Scientific Information, 1979-), annual.
Directory of authors currently publishing in 5800 journals and 2500 books indexed in the various publications of ISI. For each author, abbreviated citations of current publications are given. About 350,000 entries each year.
11. *Current Biography* (New York: H. W. Wilson Co., 1940-), monthly except December.
Biographical articles on prominent contemporary personalities in all fields. Source references are given at the end of each article. Earlier articles are often revised and updated. Brief obituaries are given for persons whose biographies have previously appeared in *Current Biography*. See also *Current Biography Cumulated Index, 1940-1970* and *Current Biography Yearbook*.
12. *Current Biography Yearbook* (New York: H. W. Wilson Co., 1940-).
Annual cumulation of the monthly *Current Biography*.
13. *Dictionary of National Biography*. Sir Leslie Stephen and Sir Sidney Lee, eds. (London: Smith, Elder, 1908-); supplements.
14. *Dictionary of American Biography* (New York: Charles Scribner's Sons, 1928-1937), 20 v. and index. Reprinted in 1943.
Sponsored by the American Council of Learned Societies. Supplements.
15. *A Concise Dictionary of American Biography* (New York: Charles Scribner's Sons, 1964), 1273p.
16. *International Who's Who* (London: Europa Publications, 1935-), annual.
17. *McGraw-Hill Encyclopedia of World Biography* (New York: McGraw-Hill, 1973), 12 v.
About 5000 biographical sketches. Twelfth volume contains list of contributors and consultants, study guides, and an index. Most articles have at least one illustration.
18. *New Century Cyclopedia of Names*. Clarence L. Barnhart and William D. Halsey, eds. (New York: Appleton, 1954), 3 v.
19. *Speakers and Lecturers: How to Find Them* (Detroit, Mich.: Gale Research Co., 1979), 464p.

Selected List of Biographical Works 175

Alphabetical listing of speakers and lecture bureaus including academic and governmental agencies, professional societies, and trade associations. Indexes.

20. *Webster's Biographical Dictionary: A Dictionary of Names of Noteworthy Persons, with Pronunciations and Concise Biographies* (Springfield, Mass.: Merriam-Webster, 1972), 1697p.
21. *Who's Who* (London: Black; New York: St. Martin's Press, 1949-), annual.
22. *Who Was Who* (London: Black, 1929-).
A companion to *Who's Who*, containing biographies of deceased persons.
23. *Who's Who in America: A Biographical Dictionary of Notable Living Men and Women* (Chicago: Marquis, 1899-).
The 40th edition (1978) contains more than 72,000 entries in two volumes. Special features: "Thoughts on My Life" written by biographees included in selected sketches, and a necrology listing deceased biographees from the 39th edition.
24. *Who Was Who in America* (Chicago: Marquis, 1899-), biennial.
A companion to *Who's Who in America*.

Specialized Biographies of Scientists and Engineers

25. *American Men and Women of Science.* Jaques Cattell Press, ed. 14th edition (New York: R. R. Bowker, 1979), 8 v.
Seven biographee volumes and a geographic and discipline index volume. Profiles of 130,000 scientists.
26. *American Men and Women of Science: Agricultural, Animal and Veterinary Sciences.* Jaques Cattell Press, ed. (New York: R. R. Bowker, 1974). 832p.
27. *American Men and Women of Science: Biology 1977.* Jaques Cattell Press, ed. (New York: R. R. Bowker, 1977), 1134p.
Biographical sketches of 20,133 currently active American and Canadian biological scientists. Disciplinary and geographic indexes.
28. *American Men and Women of Science: Chemistry 1977.* Jaques Cattell Press, ed. (New York: R. R. Bowker, 1977), 1672p.
Biographical sketches of 29,689 currently active American and Canadian scientists. Disciplinary and geographic indexes.
29. *American Men and Women of Science: Consultants 1977.* Jaques Cattell Press, ed. (New York: R. R. Bowker, 1977), 1100p.
15,447 biographical sketches, with disciplinary and geographic indexes.
30. *American Men and Women of Science: Physics, Astronomy, Mathematics, Statistics and Computer Sciences, 1977* (New York: R. R. Bowker, 1977), 1294p.
24,896 biographical sketches with disciplinary and geographic indexes.
31. *Asimov's Biographical Encyclopedia of Science and Technology.* Isaac Asimov (Garden City, N.Y.: Doubleday, 1972), 805p.
Lives and achievements of 1195 great scientists from ancient times to the present, chronologically arranged.

32. *A Biographical Dictionary of American Civil Engineers* (New York: American Society of Civil Engineers, 1962), 163p. (ASCE Historical Publications No. 2).
33. *Biographical Dictionary of American Science: The Seventeenth Through Nineteenth Centuries.* Clark A. Elliott (Westport, Conn.: Greenwood Press, 1979), 360p.
    A guide to the lives and scientific contributions of nearly 900 Americans born between 1606 and 1867.
34. *Biographical Dictionary of Botanists Represented in the Hunt Institute Portrait Collection.* Hunt Botanical Library, Carnegie-Mellon University, Pittsburgh, Penn. (Boston, Mass.: G. K. Hall, 1972), 451p.
    The portrait collection consists of more than 17,000 woodcuts, engravings, lithographs, and some drawings and paintings of over 11,000 botanists from all countries and all times.
35. *A Biographical Dictionary of Scientists.* 2nd edition. William I. Trevor (New York: Wiley Interscience, 1975).
    Brief biographical sketches of deceased scientists and technologists.
36. *Biographical Memoirs* (Washington, D.C.: National Academy of Sciences, 1877-).
    Continuing series of volumes containing the biographies of deceased members of the Academy and bibliographies of their published scientific contributions.
37. *Biographical Memoirs of the Fellows of the Royal Society* (London: The Royal Society, 1955-), annual.
    Continues *Obituary Notices of the Fellows of the Royal Society*, 1932/35-1954 (v. 1-9).
38. *Biographical Notes upon Botanists.* John H. Barnhard (Detroit, Mich.: Gale Research Co., 1965), 3 v. 1658p.
    Annotated file containing biographical details of some 44,000 botanists, maintained in the New York Botanic Gardens Library.
39. *Consultants and Consulting Organizations Directory.* 4th edition (Detroit, Mich.: Gale Research Co., 1979).
    Over 5500 entries describing consulting organizations and consultants. Subject index. Periodically updated by an interedition supplement called *New Consultants*.
40. *Dictionary of British Scientists, 1966-67.* 3rd edition (New York: R. R. Bowker, 1966), 2 v.
41. *Dictionary of Scientific Biography.* Charles C. Gillispie, editor-in-chief (New York: Scribner, 1970-1980), 16 v.
42. *Directory of Medical Specialists.* 18th edition, 1977-78 (Chicago: Marquis, 1977), 2 v.
    About 190,000 biographical sketches of specialists certified by the 22 boards that constitute the American Board of Medical Specialties.

Selected List of Biographical Works

43. *Engineers of Distinction: A Who's Who in Engineering.* Jean Gregory, ed. 2nd edition (New York: Engineers Joint Council, 1973), 401p.
44. *McGraw-Hill Modern Scientists and Engineers.* Mel Bolden and Richard A. Roth (New York: McGraw-Hill, 1980), 3 v.
45. *National Faculty Directory, 1979* (Detroit, Mich.: Gale Research Co., 1979), 2 v. 2668 p.
    Lists about 480,000 educators in junior colleges, colleges, and universities in the United States and in selected Canadian institutions.
46. *Nobel Lecturers, Including Presentation Speeches and Laureates' Biographies. Physics, 1901-1962.* Edited by the Nobel Foundation (Amsterdam: Elsevier, 1964-67), 3 v.
47. *Prominent Scientists of Continental Europe.* John Turkevich and Ludmilla Turkevich (New York: American Elsevier, 1968).
48. *Soviet Men of Science: Academicians and Corresponding Members of the Academy of Sciences of the USSR.* John Turkevich (Princeton, N.J.: Van Nostrand, 1963).
49. *Who's Who in Atoms* (Guernsey, British Isles: Francis Hodgson, 1959-), irregular.
    The 1976 edition lists over 10,000 personalities in the nuclear field from all over the world.
50. *Who's Who in Aviation* (New York: Harwood & Charles, 1973-), irregular.
51. *Who's Who in Consulting*, 2nd edition. Paul Wasserman and Janice McLean, editors (Detroit, Mich.: Gale Research Co., 1973), 1011p.
52. *Who's Who in Ecology* (New York: Special Reports, Inc., 1973-).
53. *Who's Who in Engineering* (New York: Lewis Historical Publishing Co., 1922/23-1964), 9 v.
54. *World Who's Who in Science: A Biographical Dictionary of Notable Scientists from Antiquity to Present.* Allen G. Debus, ed. (Chicago: Marquis Who's Who, Inc., 1968).
    Approximately 30,000 biographical sketches.
55. *Who's Who in Science in Europe* (Guernsey, British Isles: Francis Hodgson, 1967-), irregular.
    The 1977 edition contains about 40,000 biographical entries.
56. *Who's Who in Soviet Science and Technology*, 2nd edition. Ina Telberg, compiler. Revised and enlarged by Antonio Dimitriev and V. G. Telberg (New York: Telberg Book Co., 1964), 301p.
57. *Who's Who of British Engineers*, 2nd edition (Athens, Ohio: Ohio University Press, 1968), 784p.
58. *Who's Who of British Scientists, 1969/70-1971/72* (London: Longmans, 1970-71), 2 v.

References

1. Derek J. De Solla Price, Measuring the size of science, *Proceedings of the Israel Academy of Sciences and Humanities*, 4:98–111 (1969).
2. H. Inhaber, Scientific cities, *Research Policy*, 3:182–200 (1974).
3. Derek J. de Solla Price and S. Gursey, Some statistical results for the number of authors in the states of the United States and the nations of the world, in *Who is Publishing in Science* (Philadelphia, Penn.: Institute for Scientific Information, 1977), pp. 26–34.
4. L. J. Carter, Research Triangle Park succeeds beyond its promoters' expectations, *Science*, 200(4349):1469–70 (30 June 1978).
5. Eugene G. Kovach, Country trends in scientific productivity, in *Who is Publishing in Science* (Philadelphia, Penn.: Institute for Scientific Information, 1978), pp. 33–40.

12

DICTIONARIES AND THESAURI

12.1 Dictionaries

Dictionaries are among the most commonly used reference books. A dictionary consists of an alphabetical list of words with their meaning, etymology, pronunciation, and usage, sometimes with graphic illustrations. Some dictionaries contain biographical entries and names of places also. The terminology of science is extremely specialized, and each scientific discipline has its own terminology that researchers and practitioners use for communication. Of the few thousand specialized scientific and technical dictionaries that exist, some are bilingual or multilingual dictionaries, especially useful in translation work; others are monolingual. The scope and level of the dictionaries also vary considerably. Some contain terms from all branches of science and technology (e.g., *The McGraw-Hill Dictionary of Scientific and Technical Terms*, 2nd edition, 1978, with nearly 108,000 entries and 3000 illustrations), and others are limited to a narrow branch of science or technology. The following is an example of a very specialized multilingual dictionary:

> Herzka, A. *Elsevier's Lexicon of Pressurized Packaging (Aerosols)*. English, French, Italian, Spanish, Rumanian, German, Dutch, Norwegian, Swedish, Danish, Russian, Czech, Servo-Croation, Slovenian, Bulgarian, Hungarian, Finnish, Greek, Hebrew, Arabic, Japanese (Amsterdam: Elsevier Publishing Co., 1964).

Some dictionaries go far beyond merely furnishing definitions or meanings of terms; they contain short articles, and resemble an encyclopedia. *James and James Mathematics Dictionary*, 4th edition (D. Van Nostrand, 1976), is an example of this type: "Although this is by no means a mere word dictionary, neither is it an encyclopedia. It is a correlated condensation of mathematical concepts, designed for time-saving reference work. Nevertheless, a general reader can come to an understanding of concepts in which he has not been schooled by looking up the unfamiliar terms in the definition at hand and following this procedure down to familiar concepts" (from the preface to the third edition, 1968). This dictionary describes 8000 terms, concepts, and formulas, and has a multilingual index. Appendices contain: denominate numbers, differentiation formulas, tables of integrals, and a list of symbols and abbreviations arranged by subject.

Technical literature abounds in a variety of abbreviations, acronyms, and contractions that may bewilder even the specialist reader. Acronyms such as UNESCO and RADAR have become so common that some authors forget that these are acronyms and write them in lowercase letters. Numerous specialized dictionaries of abbreviations and acronyms have appeared recently. These range from short journal articles containing lists of abbreviations in a specialized field to multivolume dictionaries listing hundreds of thousands of items. A bibliography of over 60 such specialized dictionaries can be seen in an article entitled "Abbreviations" by Virginia Sternberg in the *Encyclopedia of Library and Information Science* (New York: Marcel Dekker, 1968), v. 1, pp. 1-12.

Acronyms, Initialisms and Abbreviations Dictionary, 6th edition, by Ellen T. Crowley and others (Detroit, Mich.: Gale Research Co., 1978) is a three-volume set consisting of the following parts:

Vol. 1: *Acronyms, Initialisms and Abbreviations Dictionary*

Vol. 2: *New Acronyms, Initialisms and Abbreviations* (annual supplement to Vol. 1)

Vol. 3: *Reverse Acronyms, Initialisms and Abbreviations Dictionary*

This edition lists some 178,000 entries. The first edition published in 1960, and simply entitled *Acronyms Dictionary*, listed a mere 12,000 terms. In the second edition published a few years later, the number of entries had increased nearly fourfold. Later editions were entitled *Acronyms and Initialisms Dictionary*. The fifth and sixth editions have been called *Acronyms, Initialisms and Abbreviations Dictionary*. The changing title reflects the expanding coverage of this publication. Since 1968, annual supplements have been issued between editions to keep pace with the proliferating acronyms and abbreviations.

Lists of abbreviations may be found in dictionaries and encyclopedias as well as in handbooks such as the *Handbook of Chemistry and Physics* (Chemical Rubber Company, Cleveland, Ohio), and *Machinery's Handbook* (Industrial Press, New York). A few selected dictionaries of acronyms, abbreviations, eponyms, trade names, signs, and symbols are listed at the end of this chapter.

Glossaries of terms are sometimes issued as standards by standards organizations. "American National Standard Industrial Engineering Terminology: Biomechanics" (ANSI Z94.1-1972) is one of a series of standard terminologies approved by ANSI. Other standards in this series include terminologies in cost engineering, data processing and systems design, engineering economy, materials

12.3 Bibliographies of Dictionaries

processing, applied mathematics, production planning and control, and similar subjects. The following are typical of glossaries issued by professional associations, government agencies, and international organizations:

A Glossary of Petroleum Terms, 4th edition. Peter Hepple, ed. (London: Institute of Petroleum, 1967).

IEEE Standard Dictionary of Electrical and Electronic Terms, 2nd edition. IEEE Std 100-1977 (New York: IEEE, 1977). This glossary has an extensive list of abbreviations and acronyms on pp. 797-882.

Glossary of Oceanographic Terms, 2nd edition. B. B. Baker, Jr., W. R. Deebel, and R. D. Geisenderfer (Washington, D.C.: United States Naval Oceanographic Office, 1966).

Glossary of Terms and Definitions in the Field of Friction, Wear and Lubrication (Tribology). (Paris: Organization for Economic Cooperation and Development, Research Group on Wear of Engineering Materials, 1969).

12.2 Thesauri

Thesauri are different from dictionaries in that they do not contain meaning or definition of words, except to a very limited extent in the form of scope notes and synonyms. Thesauri are controlled vocabularies that display relationships among terms in a scientific or technical discipline to facilitate indexing and retrieval of documents. The relationships most commonly displayed in thesauri are: Hierarchical (e.g., generic-specific) relationship, and collateral (e.g., synonymy and nonspecific) relationship. A few sample entries from the *Thesaurus of Engineering and Scientific Terms* are shown in Figure 12.1. This is a general thesaurus containing terms from all branches of science and engineering for use as a vocabulary reference in indexing and retrieving technical information. The *Thesaurus of Metallurgical Terms* (2nd edition, 1976) is a vocabulary listing "for use in indexing, storage and retrieval of technical information in metallurgy" published by the American Society for Metals.

12.3 Bibliographies of Dictionaries

The large number of specialized scientific and technical dictionaries has necessitated bibliographies of dictionaries. The following are examples of such bibliographies:

Bibliography for Interlingual Scientific and Technical Glossaries, 5th edition (Paris: UNESCO, 1969).

Metallic textiles
USE Fire resistant textiles
Metalliferous mineral deposits
0807
UF †Chromate mineral deposits
†Molybdenum mineral deposits
†Tungsten mineral deposits
†Vanadium mineral deposits
BT Mineral deposits
NT Aluminum ore deposits
Copper ore deposits
Iron ore deposits
Lead ore deposits
Manganese ore deposits
Mercury ore deposits
Molybdenum ore deposits
Nickel ore deposits
Precious metal ore deposits
Tin ore deposits
Titanium ore deposits
Tungsten ore deposits
Uranium ore deposits
Zinc ore deposits
RT Chromite ore deposits
Geochemical prospecting
Geological surveys
—Metalliferous minerals
—Mines (excavations)
—Nonmetalliferous mineral deposits
Ore sampling
Metalliferous minerals 0807
UF Ores (metal sources)
NT —Aluminum ores
Bauxite
Beryl
Cassiterite
Chalcopyrite
Chromites
Cinnabar
—Copper ores
Cryolite
Galena
Hematite
Ilmenite
—Iron ores
—Lead ores

—Spraying
Substrates
—Vapor deposition
Metallographic structures 1106
Excludes structural elements, shapes and forms
UF Metallurgical structures
RT —Microstructure
Metallographs 1106
RT Metallography
Optical microscopes
—Photographs
Photomicrography
Metallography 1106
UF †Electron metallography
RT Abrasion
—Anisotropy
—Crystal lattices
Crystallography
Debye-Scherer method
Inclusions
Isotropy
Metallographs
Microporosity
Microradiography
—Microscopes
—Microscopy
—Microstructure
—Orientation
Photomacrographs
Photomicrography
Physical metallurgy
—Polishing
—Radiography
Replicas
Thermionic emission microscopy
X ray diffraction
Metalloid alloys 1106
Alloys in which a metalloid is a significant minor constituent
NT Aluminum boron hardeners
Aluminum copper silicon alloys
Aluminum magnesium silicon alloys
Aluminum silicon alloys
Aluminum titanium boron hardeners

RT Assaying
Identifying
—Metals
Neutron activation analysis
—Sampling
Metallurgical coke
USE Coke
and Metallurgical fuels
Metallurgical converters 1309
BT Furnaces
Metallurgical furnaces
NT Acid converters
Air blown converters
Basic converters
Bessemer converters
—Bottom blown converters
Copper converters
Kaldo converters
Oxygen blown converters
Side blown converters
—Steel converters
Top blown converters
Metallurgical engineering 110
For specific descriptors related to t broad subject consult the Subject Category Index
Metallurgical fuels 2104
UF †Metallurgical coke
RT Blast furnace gas
Charcoal
—Coke
Heating fuels
Metallurgical furnaces 1309
BT Furnaces
NT Acid converters
Air blown converters
Basic converters
Bessemer converters
Blast furnaces
—Bottom blown converters
Copper converters
Furnace cupolas
Kaldo converters
—Metallurgical converters
Openhearth furnaces
Oxygen blown converters

Figure 12.1 Sample entry from *Thesaurus of Engineering and Scientific Terms* (American Association of Engineering Societies, New York, 1969; reproduced by permission).

Foreign Language and English Dictionaries in the Physical Sciences and Engineering. A Selective Bibliography, 1952-1963. NBS Miscellaneous Publication No. 258. (Washington, D.C.: National Bureau of Standards, 1964).

A Bibliography of Scientific, Technical, and Specialized Dictionaries: Polyglot, Bilingual, and Unilingual. Charles W. Rdchenbach, and Eugene R.

Selected List of Dictionaries

Garnett (Washington, D.C.: Catholic University of America Press, 1969). This is a list of 1257 items indexed by language, subject, and compiler.

International Bibliography of Specialized Dictionaries, 6th revised edition (v. 4: *Handbook of International Documentation and Information*). (New York: K. G. Saur Publishing, Inc., 1978). This is a listing of about 9000 dictionaries in the fields of science, technology, and economics published in countries throughout the world. Entries are arranged under ten major categories (e.g., Agriculture, Transportation) and 120 subdivisions.

Dictionaries, Encyclopedias and Other Word-Related Books, 2nd edition. (Detroit, Mich.: Gale Research Co., 1979), 1335p. This is a classified guide to some 25,000 LC catalog card reproductions in two volumes, with a keyword index in each volume. General and specialized encyclopedias, English and polyglot dictionaries are described.

Specialized dictionaries in specific disciplines and technologies can be identified through appropriate guides to literature described in Chapter 18, "Bibliographic Control of Scientific and Technical Literature."

A short list of thesauri arranged under subjects may be found in F. W. Lancaster's *V abulary Control for Information Retrieval* (Washington, D.C.: Information Resources Press, 1972), pp. 227-228. Another source of information on standardized vocabularies is the *Bibliography of Standardized Vocabularies* (v. 2: *Infoterm Series*) compiled by Helmut Felber, Magdalena Krommer-Benz, and Adrian Manu. This is a revised version of an earlier bibliography published by UNESCO in 1955, and contains information on some 8000 standardized vocabularies including those published by standards associations throughout the world.

Selected List of Dictionaries

General Science and Technology

1. *Chambers's Dictionary of Science and Technology*. T. C. Collocott (New York: Barnes and Noble, 1972), 1328p.
   Gives British spellings and meanings, with notes on American usage.
2. *Compton's Illustrated Science Dictionary*. Charles A. Ford, editor-in-chief (Chicago: F. E. Compton, Division of Encyclopaedia Britannica, Inc., 1963), 632p.
3. *Dictionary of Inventions and Discoveries*. Ernest F. Carter. 2nd edition (London: Frederick Muller, 1969), 204p.
4. *McGraw-Hill Dictionary of Scientific and Technical Terms*. 2nd edition (New York: McGraw-Hill, 1978).
   About 108,000 definitions.
5. *The Penguin Dictionary of Science*. E. B. Uvarov and D. R. Chapman (New York: Schocken Books, 1972), 443p.

Dictionaries of Abbreviations and Acronyms

6. *ANSI Y1.1-1972 American National Standard Abbreviations for Use on Drawings and in Text* (New York: American National Standards Institute, 1972).
7. *Abbreviations Dictionary.* New international 5th edition. (New York: Elsevier-North Holland, Inc., 1978), 654p.
   Over 160,000 entries of abbreviations, acronyms, anonyms, apellations, contractions, eponyms, geographical equivalents, historical and mythological characters, initials and nicknames, short forms and slang shortcuts, and signs and symbols. Appendices cover such items as airlines, astronomical constellations, birthstones, capital cities, diacritical and punctuation marks, Greek and Russian alphabet.
8. *Acronyms, Initialisms and Abbreviations Dictionary.* Ellen T. Crowley. 6th edition (Detroit, Mich.: Gale Research Co., 1978), 3 v.
   - v. 1: *Acronyms, Initialisms and Abbreviations Dictionary.*
   - v. 2: *New Acronyms, Initialisms and Abbreviations* (annual supplement).
   - v. 3: *Reverse Acronyms, Initialisms and Abbreviations Dictionary.*

   Contains some 178,000 entries in all.
9. *Anglo-American and German Abbreviations in Science and Technology* (New York: R. R. Bowker; Munich: Verlag Dokumentation, 1977-1978), 3 parts.
   - Part 1: A-E, 607p.
   - Part 2: F-O, 831p.
   - Part 3: R-Z, 750p.

   Lists some 150,000 acronyms and abbreviations drawn from 800 scientific and technical periodicals. Over two-thirds of the items listed are Anglo-American.
10. *Cassell's Dictionary of Abbreviations.* J. W. Gurnett and C. H. J. Kyte (London: Cassell & Co., 1966).
    About 21,000 abbreviations are listed.
11. *A Dictionary of Acronyms and Abbreviations: Some Abbreviations in Management, Technology, and Information Science.* 2nd edition. Eric Pugh (Hamden, Conn.: Archon Books, 1970), 389p.
12. *A Second Dictionary of Acronyms and Abbreviations: More Abbreviations in Management, Technology and Information Science.* Eric Pugh (Hamden, Conn.: Archon Books, 1974), 410p.
13. *Dictionary of Russian Technical and Scientific Abbreviations.* Henryk Zalucki (Amsterdam/New York: Elsevier Publishing Co., 1968), 387p.
    About 7300 entries arranged according to the Russian alphabet.
14. U.S. Library of Congress. Aerospace Technology Division. *Glossary of Russian Abbreviations and Acronyms* (Washington, D.C.: Superintendent of Documents, Government Printing Office, 1967), 806p.
    Contains approximately 23,600 entires.

Selected List of Dictionaries 185

15. *Space-Age Acronyms, Abbreviations and Designations.* 2nd edition. Reta C. Moser (New York: IFI/Plenum Press, 1969), 534p.
16. *World Guide to Abbreviations (Internationales Wortenbuch der Abkurzungen von Organizationen).* 2nd edition. Paul Spillner (New York: R. R. Bowker; Munich: Verlag Dokumentation, 1970-1973), 3 v. 1295p.

    v. 1: A-H, 1970.
    v. 2: I-R, 1972.
    v. 3: S-Z, 1973.

    Lists some 50,000 Roman alphabet abbreviations used by governmental, commercial, cultural, religious, and other institutions in 120 countries. Bibliography.
17. *A World Guide to Abbreviations of Organizations.* 5th edition. F. A. Buttress (London: Leonard Hill, 1974), 473p.
    About 18,000 entries, of which over 5000 are for abbreviations of organizations in Continental Europe. Russian organizations are not listed.

Dictionaries of Eponyms and Trade Names

18. *Chemical Synonyms and Trade Names.* 8th edition. Revised and enlarged by Edward I. Cooke (Cleveland, Ohio: Chemical Rubber Company Press, 1979).
19. *A Dictionary of Eponyms.* Cyril Leslie Beeching (New York: K. G. Saur, 1979), 140p.
    Over 200 eponyms are listed.
20. *A Dictionary of Named Effects and Laws in Chemistry, Physics, and Mathematics.* 3rd edition. Denis W. G. Ballentyne and D. R. Lovett (London: Chapman and Hall, 1970), 355p.
21. *Engineering Eponyms.* 2nd edition. Charles P. Auger (New York: Nicholas Publishing Co., 1976), 122p.
    An annotated bibliography of some named elements, principles, and machines in mechanical engineering (subtitle).
22. *Eponyms Dictionaries Index.* James Ruffner, Jennifer Berger, and Georgia Schoenung (Detroit, Mich.: Gale Research Co., 1977), 730p.
    A reference guide to persons, both real and imaginary, and the terms derived from their names, with references to sources of further information. Over 30,000 entries.
23. *Trade Names Dictionary.* Ellen T. Crowley (Detroit, Mich.: Gale Research Co., 1976), 2 v.
    A guide to trade names, brand names, product names, coined names, model names, and design names, with addresses of their manufacturers, importers, marketers, or distributors. Approximately 106,000 entries. Periodically supplemented by *New Trade Names.*

Dictionaries of Signs and Symbols

The following dictionaries of abbreviations, signs, and symbols are published by the Odyssey Press, New York:

24. *Dictionary of Aeronautical and Aerospace Technology Abbreviations, Signs and Symbols.*
25. *Dictionary of Architectural Abbreviations, Signs and Symbols.*
26. *Dictionary of Civil Engineering Abbreviations, Signs and Symbols.*
27. *Dictionary of Computer & Control Systems Abbreviations, Signs and Symbols.*
28. *Dictionary of Electrical Abbreviations, Signs and Symbols.*
29. *Dictionary of Electronics Abbreviations, Signs and Symbols.*
30. *Dictionary of Industrial Engineering Abbreviations, Signs and Symbols.*
31. *Dictionary of Mechanical Engineering Abbreviations, Signs and Symbols.*
32. *Dictionary of Nuclear Abbreviations, Signs and Symbols.*
33. *Dictionary of Physics and Mathematics Abbreviations, Signs and Symbols.*
34. *Encyclopedia of Engineering Signs and Symbols.*
35. *Encyclopedia of Letter Symbols for Science and Engineering.*

Standard specifications for signs and symbols approved by ANSI (e.g., ANSI/ASTM E135-78 *Standard Definitions of Terms and Symbols Relating to Emission Spectroscopy*) can be traced through ANSI's annual *Catalog*.

Mathematics

36. *Chinese-English Glossary of the Mathematical Sciences.* John De Francis (Providence, R.I.: American Mathematical Society, 1963), 277p.
37. *Dictionary of Mathematics.* T. Alaric Millington and William Millington (New York: Barnes and Noble, 1971), 259p.
38. *The International Dictionary of Applied Mathematics.* W. F. Freiberger and others (Princeton, N.J.: D. Van Nostrand, 1960), 1173p.
39. *James and James Mathematics Dictionary.* 4th edition (New York: Van Nostrand Reinhold, 1976), 518p.
   Describes 8000 terms, concepts, and formulas. Multilingual index.
   Appendices contain: Denominate numbers, differentiation formulas, tables of integrals, and a list of symbols and abbreviations arranged by subject.
40. *The New Mathematics Dictionary and Handbook.* Robert W. Marks (New York: Grosset and Dunlap, 1965), 186p.
   "A thesaurus of historical, biographical, and tabular data and a manual of mathematical procedures and operations, as well as a working dictionary whose scope includes the most advanced areas of the new mathematics." (from preface).
41. *Romanian-English Dictionary and Grammar for the Mathematical Sciences.* Sydney H. Gould and P. E. Obreanu (Providence, R.I.: American Mathematical Society, 1967), 51p.

Selected List of Dictionaries 187

42. *Russian-English Dictionary of the Mathematical Sciences.* A. J. Lohwater and Sydney H. Gould (Providence, R.I.: American Mathematical Society, 1961), 267p.
43. *Russian-English Vocabulary, with a Grammatical Sketch, to be Used in Reading Mathematical Papers* (New York: American Mathematical Society, 1972), 66p.

Computer Science

44. *Dictionary for Computer Languages.* Hans Breuer. APIC Studies in Data Processing, No. 6, Automatic Programming Information Center, Brighton College of Technology, London (New York: Academic Press, 1966).
45. *A Dictionary of Computers.* Anthony Chandor and others (Harmondsworth: Penguin, 1970), 407p.
46. *A Dictionary of Microcomputing.* Philip E. Burton (New York: Garland STPM Press, 1976), 190p.
    Over 900 definitions, with many examples, sketches, diagrams, and tables.
47. *Elsevier's Dictionary of Computers, Automatic Control and Data Processing in Six Languages: English/American, French, Spanish, Italian, Dutch, and German.* 2nd edition. W. E. Clason (Amsterdam: Elsevier, 1973).
48. *Funk and Wagnalls Dictionary of Data Processing Terms.* Harold A. Rodgers (New York: Funk and Wagnall, 1970), 151p.
49. *Standard Dictionary of Computers and Information Processing.* 2nd edition. Martin H. Weik (Rochelle Park, N.J.: Hayden Book Co., 1977), 350p. Explanation of over 12,500 terms, with many examples and illustrations. Bibliography.

Physics and Astronomy

50. *Chinese-English and English-Chinese Astronomical Dictionary.* Hong-Yee Chiu (New York: Consultants Bureau, 1966), 176p.
51. *Concise Dictionary of Physics and Related Subjects.* James Thewlis (Oxford: Pergamon Press, 1973), 366p.
52. *Dictionary of Astronomical Terms.* Ake Wallenquist (Garden City, N.Y.: Natural History Press, 1966), 265p. Translated from Swedish by Sune Engelbrektson.
53. *Dictionary of Pure and Applied Physics.* Louis De Vries and W. E. Clason (Amsterdam: Elsevier, 1963-64), 2 v.
    v. 1: German-English, 1963, 367p.
    v. 2: English-German, 1964, 341p.
54. *Dictionary of Semi-Conductor Physics and Electronics: English-German, German-English.* Werner Bindmann (Oxford: Pergamon Press, 1965), 615p.
55. *Glossary of Astronomy and Astrophysics.* Jeanne Hopkins (Chicago: University of Chicago Press, 1976), 169p.

Chemistry, Chemical Engineering, and Technology

56. *The Condensed Chemical Dictionary.* 8th edition. Gessner G. Hawley (New York: Van Nostrand Reinhold, 1971), 971p.
57. *Dictionary of Chemistry and Chemical Technology in Six Languages: English, German, Spanish, French, Portuguese, Russian.* 2nd edition (Oxford: Pergamon Press, 1966), 1325p.
58. *Dictionary of Drying.* C. W. Hall (New York: Marcel Dekker, 1979).
59. *Dictionary of Fuel Technology.* Alan Gilpin (London: Butterworths, 1969), 275p.
60. *Dictionary of Organic Compounds: The Constitution and Physical, Chemical, and other Properties of the Principal Carbon Compounds and their Derivatives.* 4th edition (New York: Oxford University Press, 1965). 6 v., annual supplements.
61. *Dictionary of Rubber.* Kurt F. Heinisch (New York: John Wiley, 1974), 545p.
   Includes a list of 443 producers and marketing organizations.
62. *Dictionary of Rubber Technology.* Alexander S. Craig (London: Butterworths, 1969), 228p.
63. *Elastomeric Materials. The International Plastics Selector. Desk Top Data Bank* (San Diego, Calif.: The International Plastics Selector, Inc., 1977), 875p.
64. *Elsevier's Dictionary of Chemical Engineering.* W. E. Clason (Amsterdam: Elsevier, 1968), 2 v.

   v. 1: Chemical Engineering Laboratory and Equipment.

   v. 2: Chemical Engineering Processes and Products.

   In six languages: English/American, French, Spanish, Italian, Dutch, and German.
65. *Elsevier's Dictionary of Industrial Chemistry.* A. F. Dorian (Amsterdam: Elsevier, 1964), 2 v.
   In six languages: English/American, French, Spanish, Italian, Dutch, and German.
66. *Glossary of Chemical Terms.* Clifford Hampel and Gessner G. Hawley (New York: Van Nostrand Reinhold, 1976), 288p.
   Approximately 20,000 concise definitions of terms used in chemistry and chemical industries.

Geology, Mining, and Metallurgy

67. *A Dictionary of Alloys.* Eric N. Simons (London: Muller, 1969), 191p.
68. *Dictionary of Applied Geology, Mining and Civil Engineering.* Archibald Nelson and K. D. Nelson (New York: Philosophical Library, 1967), 421p.
69. *Dictionary of Ferrous Metals.* Eric N. Simons (London: Muller, 1970), 244p.
70. *Dictionary of Metallurgy.* Donald Birchon (New York: Philosophical Library, 1965), 409p.

Selected List of Dictionaries

71. *Elsevier's Dictionary of Metallurgy in Six Languages: English/American, French, Spanish, Italian, Dutch, and German.* W. E. Clason (Amsterdam: Elsevier Publishing Co., 1967), 634p.
72. *Glossary of Geology.* Margaret Gary, Robert McAfee, and Carol L. Wolf (Washington, D.C.: American Geological Society, 1972), 805p.
73. *Glossary of Mining Geology in English, Spanish, French, and German.* Gerhardt Christian Amstutz (Amsterdam: Elsevier Publishing Co., 1971), 196p.

Aviation and Aerospace

74. *Astronautical Multilingual Dictionary of the International Academy of Astronautics.* English, Russian, German, French, Italian, Spanish, Czechoslovakian (New York: Elsevier, 1970), 936p.
75. *Aviation and Space Dictionary.* 5th edition. Ernest J. Gentle and Lawrence W. Reithmaier (Los Angeles, Calif.: Aero, 1974).
76. *A Dictionary of Aviation.* David W. Wragg (New York: Frederick Fell, 1973), 286p.
77. *Dictionary of Guided Missiles and Space Flight.* Principles of Guided Missile Design Series, v. 5 (Princeton, N.J.: D. Van Nostrand, 1959), 688p.
78. *Dictionary of Military and Naval Quotations.* Robert D. Heinl, Jr. (Annapolis, Md.: U.S. Naval Institute, 1966), 367p.
79. *Dictionary of Technical Terms for Aerospace Use.* William H. Allen (Washington, D.C.: U.S. National Aeronautics and Space Administration, 1965), 314p. NASA SP-7.
80. *Elsevier's Dictionary of Aeronautics in Six Languages: English/American, French, German, Italian, Portuguese, Spanish.* A. F. Dorian and James Osenton (Amsterdam: Elsevier, 1964), 842p.
81. *German-English English-German Astronautics Dictionary.* Charles J. Hayman (New York: Consultants Bureau, 1968), 237p. Approximately 12,000 terms are defined.
82. *The New Dictionary and Handbook of Aerospace with Special Sections on the Moon and Lunar Flight.* Robert W. Marks (New York: Frederick A. Praeger, 1969), 531p.
83. *Russian-English Aerospace Dictionary.* Harry L. Darcy (Berlin: Walter de Gruyter, 1965), 407p.
84. *Russian-English Space Technology Dictionary.* Michael M. Konarski (Oxford: Pergamon Press, 1970), 416p.

Civil Engineering

85. *A Dictionary of Architectural Science.* H. J. Cowan (New York: Halsted Press, 1973), 354p.
86. *A Dictionary of Water and Water Engineering.* Archibald Nelson and K. D. Nelson (Cleveland, Ohio: CRC Press, 1973), 271p.

87. *Pest Control in Buildings: A Guide to the Meaning of Terms.* P. B. Cornwell (London: Hutchinson & Co., 1973), 192p.
88. *Wood Preservation: A Guide to the Meaning of Terms.* Norman E. Hickin (London: Hutchinson & Co., 1971), 109p.

Mechanical Engineering

89. *Dictionary of Engineering and Technology,* with extensive treatment of the most modern techniques and processes. 3rd edition. Richard Ernst (New York: Oxford University Press, 1974), 2 v.
    v. 1: Deutsch-English, 1061p.
    v. 2: English-Deutsch, 1146p.
    A pair of bilingual dictionaries with over 142,000 entries covering the broad field of industrial technology and including entries in electronics, data processing, computer science, plastics and semiconductors.
90. *Dictionary of Fuel Technology.* Alan Gilpin (London: Butterworths, 1969), 275p.
91. *Dictionary of Mechanical Engineering.* Joseph L. Nayler and G. H. F. Nayler (New York: Hart Publishing Co., 1967), 406p.

Electrical Engineering and Electronics

92. *Dictionary of Electrical Engineering.* K. G. Jackson (London: Butterworths, 1973), 380p.
93. *A Dictionary of Electronic Terms. Concise Definitions of Words used in Radio, Television and Electronics.* 8th edition. Robert E. Beam (Chicago: Allied Radio Corporation, 1968), 112p.
94. *Dictionary of Electronics and Nucleonics* (New York: Barnes and Noble, 1970), 443p.
95. *Funk and Wagnalls Dictionary of Electronics* (New York: Funk and Wagnalls, 1969), 230p.
96. *IEEE Standard Dictionary of Electrical and Electronics Terms.* 2nd edition. IEEE Std 100-1977 (New York: Institute of Electrical and Electronics Engineers, 1977).
    Contains 20,254 definitions with identification of defining documents. Extensive list of abbreviations and acronyms on pp. 797-882.

Biology

97. *Common Plants: Botanical and Colloquial Nomenclature.* John J. Cunningham and Rosalie J. Cote (New York: Garland STPM Press, 1976), 120p.
    Part I traces the chronological development of the Linnaean binomial system and contemporary systematics from ancient Greek and Latin roots. A glossary of species epithets and an index to the botanical names of over 2000 common plants is included. In Part II, the origins of 150

Selected List of Dictionaries

colloquial names are detailed to illustrate the influence of folklore, commerce, medicine, and religion.

98. *A Dictionary of Biology.* 6th edition. Michael Abercrombie, C. J. Hickman, and M. L. Johnson (Hammondsworth, Middlesex, England: Penguin Books, 1977), 311p.
Definitions of about 3000 common biological terms.

99. *A Dictionary of Botany, including Terms Used in Biochemistry, Soil Science, and Statistics.* George Usher (Princeton, N.J.: D. Van Nostrand, 1966), 404p.

100. *A Dictionary of English Plant Names.* James Britten and Robert Holland (Millwood, N.Y.: Kraus Reprint Corporation, 1965), 618p.

101. *A Dictionary of Flowering Plants and Ferns.* 8th edition. John C. Willis. Revised by H. K. Airy Shaw (Cambridge: Cambridge University Press, 1973), 1245p.

102. *A Dictionary of Genetics.* 2nd revised edition. Robert C. King (Oxford: Oxford University Press, 1974), 375p.

103. *Dictionary of Microbiology.* P. Singleton and D. Sainsbury (New York: Wiley-Interscience, 1978), 481p.
"A survey of the subject indexes in two volumes of the *Journal of Microbiology* revealed that in each case about 75% of the index keywords were listed in this dictionary."

104. *A Dictionary of Useful and Everyday Plants and Their Common Names.* Frank N. Howes (Cambridge: Cambridge University Press, 1974), 290p.

105. *A Dictionary of Zoology.* A. W. Leftwich (New York: Crane, Russak & Co., 1973), 478p.

106. *A Guide to Vocabulary of Biological Literature* (Philadelphia, Penn.: BioSciences Information Service of Biological Abstracts, 1973).

107. *The Reston Encyclopedia of Biomedical Engineering Terms.* Rudolf F. Graf and George J. Whalen (Reston, Va.: Reston Publishing Co., 1977), 415p.

108. *A Science Dictionary of the Plant World: An Illustrated Demonstration of Terms Used in Plant Biology, with over 400 Color Pictures.* Michael Chinery (New York: F. Watts, 1969), 288p.
Title of an earlier (1967) edition: *A Pictorial Dictionary of the Plant World.*

Environment

109. *Common Environmental Terms: A Glossary.* Gloria J. Studdard (Washington, D.C.: U.S. Environmental Protection Agency, 1973), 23p.

110. *Dictionary of Development Terminology.* J. Robert Dumouchel (New York: McGraw-Hill, 1975), 278p.
A dictionary of terms used by builders, lenders, architects, planners, investors, real estate brokers and attorneys, appraisers, land taxing and zoning authorities, government officials, community organizers, housing managers, and urban renewal specialists.

111. *A Dictionary of the Natural Environment.* F. J. Monkhouse and J. Small (New York: John Wiley, 1978), 320p.
112. *Ecology Field Glossary: A Naturalist's Vocabulary.* Walter H. Lewis (Westport, Conn.: Greenwood Press, 1977), 153p.
113. *Environmental Glossary.* Revised edition (Washington, D.C.: U.S. Government Printing Office, 1977), 99p.
114. *Environmental Impact Statement Glossary.* Mark Landy (New York: Plenum Publishing Corporation, 1979), 525p.
115. *Fire Sciences Dictionary.* B. W. Kuvshinoff, R. M. Fristrom, G. L. Ordway, and R. R. Tuve (New York: John Wiley, 1977), 439p.
116. *Glossary of Terms Frequently Used in Air Pollution.* Ralph E. Huschke (Boston, Mass.: American Meteorological Society, 1968), 34p. Prepared for a seminar for science writers on global air pollution, San Francisco, Calif.: January 30, 1968.
117. *The New York Times Encyclopedic Dictionary of the Environment.* Paul Sarnoff (New York: Quadrangle Books, 1971), 352p.

Energy

118. *Energy Microthesaurus. A Hierarchical List of Indexing Terms Used by NTIS* (Springfield, Va.: National Technical Information Service, 1976). PB 254 800.
119. *Integrated Energy Vocabulary* (Springfield, Va.: National Technical Information Service, 1976), 447p.

13

DIRECTORIES AND YEARBOOKS

13.1 Directories

Directories are basically lists of companies, academic or other institutions, government agencies, products and services, or of individuals, arranged in some systematic order for easy reference. The data presented in directories are usually obtained from the sources themselves (i.e., individuals, companies, or other agencies) through a questionnaire, or compiled from catalogs or other published sources. Since factual data on companies, products, and people are likely to change frequently, directories are updated periodically by issuing supplements or by publishing entirely new revised editions.

Based on the type of material presented, three kinds of directories may be identified: biographical directories, institutional or company directories, and product directories. Some directories are composite in nature and contain more than one type of information. Biographical directories are useful for locating specialists on a given topic and for obtaining biographical information on known personalities. The *Directory of Physics and Astronomy Staff Members* issued annually by the American Institute of Physics contains listings of staff members in North American colleges and universities, federally funded research and development centers, and government laboratories. The information is obtained from chairpersons of academic departments and directors of research centers. The 1976-77 edition of the directory listed approximately 22,000 staff members from about 2400 institutions. Other sources of biographical information on scientists and engineers have been discussed in an earlier chapter.

Institutional or company directories provide information on academic institutions, government agencies, or private sector companies. Such directories are useful in locating companies offering a certain product or service or organizations engaged in a certain type of activity. The detail of the information varies considerably, ranging from a minimum of name, address, and telephone number to detailed descriptions of activities, products, personnel, and financial information. Some directories simply describe one organization; others contain descriptive information on numerous organizations. *NASA Factbook* (2nd edition, 1975), and *NSF Factbook* (2nd edition, 1975), both published by Marquis Who's Who, Inc., Chicago, are examples of directories that describe a single organization and its activities and products.

Encyclopedia of Associations, 14th edition (Detroit, Mich.: Gale Research Company, 1980), is a very comprehensive directory in three parts. The first part, *National Organizations of the U.S.* (1566 pages) contains descriptions of some 14,000 organizations. A geographic index and an executive index make up the second part (816 pages). The third part, entitled *New Associations and Projects*, is a periodical interedition supplement. The *Encyclopedia of Associations*, first published in 1956, provides information on the following types of agencies:

American membership organizations of national significance

Selected nonmembership groups

Foreign groups of interest to Americans

International groups with large American memberships

Citizen action groups, projects, and programs

The *Encyclopedia* also lists some local and regional groups of wide impact as well as missing organizations which were listed in earlier editions but whose current status is undetermined. The earlier 13 editions of the *Encyclopedia* are available on microfiche.

Another similar but complementary publication of the Gale Research Company is the *Research Centers Directory*. The sixth edition, published in 1979, describes 6268 research centers including university research departments and other nonprofit research organizations in the United States and Canada. This directory is periodically updated by a paper-bound interedition supplement entitled *New Research Centers*.

Industrial Research Laboratories of the United States, 16th edition (New York: R. R. Bowker, 1979), has information on 11,500 publicly and privately owned research facilities. The National Academy of Sciences-National Research Council initiated and published the first 11 editions of this directory.

The Scientific Institutions of Latin America (Stanford, Calif.: California Institute of International Studies, 1970), contains detailed descriptions of the organization and information facilities of scientific institutions in 13 Latin American countries and some international, European, and U.S. organizations promoting science and science information in Latin America.

Directory of Selected Scientific Institutions in Mainland China (Hoover Institution Publications Series: 96, 1970), was published for the National Science Foundation by the Hoover Institution on War, Revolution, and Peace, Stanford University, Stanford, Calif. The directory describes 490 scientific institutions in English, but with names of institutions, locations, personnel, and publications also in Chinese characters.

Product directories are usually composite directories that contain not only descriptions of products, but also information on manufacturing and distributing agencies and their executive personnel. *World Aviation Directory*, published semiannually in summer and winter by Ziff-Davis Publishing Company, Washington, D.C., contains information on aviation/aerospace companies, their officials, and their products. *Worldwide Directory of Computer Companies,* 2nd edition (Chicago: Marquis Who's Who, Inc., 1973) lists more than 4000 computer-industry firms, including those producing hardware and software products and those providing time-sharing, leasing, consultation, and other services.

Many product directories, or "buyers' guides" are special issues or supplements of journals. *Chemical Engineering Equipment Buyers' Guide, Chemical Week Buyers' Guide Issue*, and *Electronics Buyers' Guide* are annual special issues of the journals *Chemical Engineering, Chemical Week*, and *Electronics*, respectively, all published by McGraw-Hill, Inc. A directory of scientific instruments appears as a special issue of *Science* in November each year. Information on special issues and supplements can be found in the *Guide to Special Issues* and *Indexes of Periodicals*, 2nd edition, by Charlotte M. Devers, Doris B. Katz, and Mary Margaret Regan (New York: Special Libraries Association, 1976).

13.2 Yearbooks

Yearbooks are reference books that describe the events pertaining to a particular year. *The Yearbook of Astronomy* (New York: Norton, 1962-) contains data on the phases of the moon, orbits of planets, and similar information especially useful to the amateur astronomer. It also includes a directory of astronomical societies. *McGraw-Hill Yearbook of Science and Technology* summarises the developments of the preceding year and is a supplement to the *McGraw-Hill Encyclopedia of Science and Technology. Aerospace Facts and Figures* and *Aerospace Yearbook*, both published by the Aerospace Industries Association of America, are examples of yearbooks that contain narrative and statistical accounts of the events of the preceding year. *Statistical Abstracts of the United Nations* is a familiar example of a yearbook containing purely statistical information. *Europlastics Yearbook* (London: IPC Industrial Press), and *British Aviation Yearbook* (London: Hanover Press) are more akin to a directory than a yearbook. The *Yearbook of Agriculture* (1894-) of the U.S. Department of Agriculture is not a yearbook; it is a monographic series, issued annually, and each issue deals with a particular topic such as "Climate and Man" (1941), "Insects" (1952), "Water" (1955), "Soils" (1957), "Land" (1958), "Handbook for the Home" (1973), "The Face of Rural America" (1976), and "Living on a Few Acres" (1978).

13.3 Bibliographies of Directories

The directories and yearbooks mentioned in the preceding sections are typical of a very large number of such publications covering diverse disciplines, industries, and geographical regions. The following bibliographies of directories are useful in identifying an appropriate directory for a given purpose. *Guide to American Scientific and Technical Directories*, 2nd edition (Rye, N.Y.: Todd Publications, 1975), is a classified guide to over 2500 publications covering social and physical sciences and all industrial and technical areas. Information on each directory includes: title, number of entries, method of arrangement, indexes, frequency of publication, publisher, and price. *Guide to American Directories*, 9th edition (Detroit, Mich.: Gale Research Co., 1975) describes over 5200 directories arranged under 300 subject headings.

International Bibliography of Directories (volume 5 of the *Handbook of Documentation and Information*) edited by Helga Lengenfelder (New York: K. G. Saur Publishing, Inc., 1978) lists some 6500 directories published in 100 countries and dealing with such areas as aeronautics, telecommunications, medicine, and patents. A geographical index and a list of publishers' addresses are also provided.

Directory Information Service (Detroit, Mich.: Gale Research Co., 1977-) is a new reference periodical published three times a year. It describes business and industrial directories, professional and scientific rosters, and other lists. The 1977-78 issues contain descriptions of nearly 4000 new directories and current editions of established directories. Entries are grouped under broad subject categories and indexed by subject and title.

Yearbooks and directories are also listed in *Irregular Serials and Annuals: An International Directory* (5th edition, 1978-1979) issued biennially by R. R. Bowker Co. *Bowker Serials Bibliography* is an interedition supplement that updates both *Ulrich's International Periodicals Directory* and *Irregular Serials and Annuals*.

Selected List of Directories

Bibliographies of Directories

1. *Current British Directories, 1966-67.* 5th edition. G. P. Henderson and I. G. Anderson (Croydon, Surrey, England: C. B. D. Research Ltd., 1966). 214p.
2. *Current European Directories.* G. P. Henderson (Beckenham, Kent, England: C. B. D. Research Ltd., 1969). 222p.
3. *Directory Information Service* (Detroit, Mich.: Gale Research Co., 1977-). Published three times a year.

Selected List of Directories

"A reference periodical covering business and industrial directories, professional and scientific rosters, and other lists and guides of all kinds" (subtitle).

4. *Guide to American Directories.* 9th edition. Bernard Klein (Detroit, Mich.: Gale Research Co., 1975). 496p.
Formerly, *Guide to American Directories for Compiling Mailing Lists* (1954-1958).
5. *Guide to American Scientific and Technical Directories.* 2nd edition (Rye, N.Y.: Todd Publications, 1975), 271p.
6. *International Bibliography of Directories* (Vol. 5: *Handbook of Documentation and Information*). Helga Lengenfelder (New York: K. G. Saur Publishing, Inc., 1978).
7. *Serials for Libraries.* Joan K. Marshall (Santa Barbara, Calif.: ABC-Clio Press and Neal-Schuman Publishers, 1978).
An annotated guide to annuals, directories, yearbooks, and other nonperiodical serials. About 3000 entries.

General Directories

8. *Awards, Honors, and Prizes.* 4th edition (Detroit, Mich.: Gale Research Co., 1975). 2v.
    V. 1: United States and Canada
    V. 2: International and Foreign
An alphabetical directory of organizations sponsoring awards, honors, and prizes in almost all fields.
9. *Consultants and Consulting Organizations Directory.* 4th edition (Detroit, Mich.: Gale Research Co., 1979). 1136p.
Describes 6150 consulting firms and individuals in 135 specific fields including acoustics, advertising, aptitude testing, construction management, product design, etc. Typical entry gives name and address, telephone number, types of services performed, and other details. Indexes. Updated periodically by an interedition supplement entitled *New Consultants.*
10. *Directory of Federal Technology Transfer.* U.S. Federal Council for Science and Technology. Committee on Domestic Technology Transfer (Washington, D.C.: U.S. Government Printing Office, 1975). 202p.
A tool for state and local government officials and private industry to share the results of federal programs aimed at the development of knowledge and technologies. Descriptions of agencies contain: programs, research base, technology transfer policy, objectives, areas of responsibility, methods of implementation, accomplishments, user organizations, and contact points. Index.
11. *Directory of Federally Supported Information Analysis Centers* (Washington, D.C.: Library of Congress, Science and Technology Division, National Referral Center, 1970).

12. *A Directory of Information Resources in the United States: Federal Government, with a Supplement of Government-Sponsored Information Analysis Centers.* Revised edition (Washington, D.C.: Library of Congress, Science and Technology Division, National Referral Center, 1974). 416p.
13. *A Directory of Information Resources in the United States: Physical Sciences, Engineering* (Washington, D.C.: Library of Congress, Science and Technology Division, National Referral Center, 1971). 803p.
14. *Directory of Special Libraries and Information Centers.* 5th edition (Detroit, Mich.: Gale Research Co., 1979). 3 v.

    V. 1: *Special Libraries and Information Centers in the U.S. and Canada.* 1279p. Key to the holdings, services, and personnel of 14,000 special libraries and information centers; included are nearly 600 networks and consortia.

    V. 2: *Geographic and Personnel Indexes.* 715p.

    V. 3: *New Special Libraries.* A periodic interedition supplement.

15. *Subject Directory of Special Libraries and Information Centers.* 5th edition (Detroit, Mich.: Gale Research Co., 1977). 5 v. 1437p.

    V. 1: Business and law libraries, including military and transportation libraries.

    V. 2: Education and information science libraries.

    V. 3: Health sciences libraries.

    V. 4: Social sciences and humanities libraries.

    V. 5: Science and technology libraries.

16. *Encyclopedia of Associations*, 14th edition (Detroit, Mich.: Gale Research Co., 1980). 3 v.

    V. 1: *National Organizations of the United States.*

    V. 2: *Geographic and Executive Index.*

    V. 3: *New Associations and Projects.* Periodical interedition supplement.

    First and second editions (1956, 1959) published under the title *Encyclopedia of American Associations.*

17. *Encyclopedia of Governmental Advisory Organizations.* 2nd edition. Linda E. Sullivan and Anthony T. Kruzas (Detroit, Mich.: Gale Research Co., 1975). 668p.

    Periodically updated by *New Governmental Advisory Organizations.*

18. *Industrial Research Laboratories of the United States.* 16th edition (New York: R. R. Bowker, 1979).

    Description of 11,500 research facilities in nongovernment and nonacademic organizations. Some university, foundation, and cooperative laboratories are included where these have separate autonomy. Geographic index, personnel index, and subject index.

Selected List of Directories

19. *National Faculty Directory 1980* (Detroit, Mich.: Gale Research Co., 1980). 2 v. ca. 2700p.
   An alphabetical listing of about 480,000 names and addresses of teaching faculty members at over 2750 U.S. and selected Canadian junior colleges, colleges, and universities. Includes a list of schools covered.
20. *National Trade and Professional Associations of the United States and Canada, and Labor Unions, 1978.* 13th annual edition. Craig Colgate, Jr. (Washington, D.C.: Columbia Books, 1978). 388p.
   Over 5700 organizations are described. Information given for each organization includes address and telephone number, chief executive, date of establishment, membership, size of staff, annual budget, publications, meetings, and historical data. Four indexes: keyword index, geographic index, chief executives' names, and budget index. Introduction relates history, importance, and scope of U.S. trade associations and professional societies. Bibliography.
21. *NSF Factbook.* A guide to National Science Foundation Programs and Techniques. 2nd edition (Chicago: Marquis Who's Who, Inc., 1975). 561p.
22. *Research Centers Dictionary.* 6th edition (Detroit, Mich.: Gale Research Co., 1979).
   Over 6200 entries covering the United States and Canada. Subject index, institutional index, and index of research centers. Updated periodically by *New Research Centers.* First edition issued in 1960 under the title *Directory of University Research Bureaus and Institutes.*
23. *Scientific, Technical, and Related Societies of the United States.* 9th edition (Washington, D.C.: National Academy of Sciences, 1971). 213p. First edition published in 1927.
24. *Subject Collections.* 5th edition. Lee Ash (New York: R. R. Bowker, 1978). ca. 1250p.
   Description of collections housed in over 15,000 academic, public, and special libraries and 1000 museums in the United States and Canada. Entries arranged under LC subject headings. Under each subject heading, holding libraries are arranged alphabetically by state.

Directories of Foreign and International Agencies

25. *Directory of British Associations.* 5th edition. G. P. Henderson and S. P. A. Henderson. A C.B.D. Research Publication (Detroit, Mich.: Gale Research Co., 1977). 457p.
26. *Directory of European Associations.* 2nd edition. I. G. Anderson (Detroit, Mich.: Gale Research Co., 1975-76). 2 parts.
   Part 1: *National Industrial, Trade, and Professional Associations.* 1976. 557p. Contains information on over 9000 organizations in European countries excepting Great Britain and Ireland.

Part 2: *National Learned, Scientific and Technical Societies.* 1975. 315p.
27. *Directory of Hungarian Research Institutions* (Budapest, Hungary: Hungarian Central Technical Library and Documentation Center, 1975). Contains descriptions of 1047 research institutions.
28. *Directory of Selected Scientific Institutions in Mainland China.* R. J. Watkins. Hoover Institution Publications Series, 96. (Stanford, Calif.: Hoover Institution Press, for the National Science Foundation, 1971). 469p.
29. *East European Research Index* (Guernsey, British Isles: Francis Hodgson, 1976).
Directory of government, industrial, university, and other official bodies interested in scientific and technological research in Eastern Europe including the USSR. Index.
30. *European Research Index* (Guernsey, British Isles: Francis Hodgson, 1976). 2 v.
Includes descriptions of government, independent, industrial, and university establishments, and an international section. Limited to Western Europe. Index.
31. *Guide to European Sources of Technical Information.* 3rd edition. Colin H. Williams (Guernsey, Channel Islands: Francis Hodgson, 1970). 309p.
32. *Guide to International Science and Technology* (Guernsey, British Isles: Francis Hodgson, 1976).
33. *Guide to Science and Technology in the Asian/Pacific Area* (Guernsey, British Isles: Francis Hodgson, 1977).
34. *Guide to Science and Technology in the United Kingdom.* 2nd edition (Guernsey, British Isles: Francis Hodgson, 1976).
35. *Guide to Science and Technology in the USSR.* 2nd edition (Guernsey, British Isles: Francis Hodgson, 1976).
36. *Guide to World Science.* 2nd edition (Guernsey, British Isles: Francis Hodgson, 1976). 25 v.
37. *Industrial Research in Britain.* 8th edition (Guernsey, British Isles: Francis Hodgson, 1976).
38. *International Foundation Directory.* H. V. Hodson, consultant editor (Detroit, Mich.: Gale Research Co.; London: Europa Publications, 1974). 396p.
39. *International Organizations: A Guide to Information Sources.* Alexine L. Atherton (Detroit, Mich.: Gale Research Co., 1976). 2 parts.
Part 1: *Sources of Information.* Contains about 1500 annotated entries of bibliographies, guides, indexes, catalogs, periodicals, yearbooks, and directories concerned with international organizations.
Part 2: *Bibliography.* Lists primarily books under 9 broad subject headings and numerous subheadings.

40. *Trade Associations and Professional Bodies of the United Kingdom.* 6th edition. Patricia Millard (Oxford: Pergamon Press, 1979). 533p.
41. *World Guide to Scientific Associations and Learned Societies.* 2nd edition (New York: R. R. Bowker; Munich: Verlag Dokumentation, 1978). 510p.
  Describes approximately 10,200 associations and societies from all fields of science and technology in 134 countries. Entries are arranged alphabetically within countries grouped by continent. Each entry includes name of society or association, year of foundation, address, names of chief executives, and number of members. Cessations are listed separately. The preface, table of contents, running page headings and subject headings are in German and English.
42. *Yearbook of International Organizations, 1978-79.* 17 edition. A publication of the Union of International Associations (Detroit, Mich.: Gale Research Co., 1978). 1123p.
  Approximately 8000 entries describing government, nongovernment organizations, associations, and committees. 13 indexes.

Computerized Information Systems and Databases

43. *Computer-Readable Bibliographic Databases: A Directory and Data Source Book.* Martha E. Williams (Washington, D.C.: American Society for Information Science, 1979). Semiannual supplements.
  The 1979 edition describes over 500 databases from all over the world.
44. *Databases in Europe: A Directory of Machine-Readable Databases and Data Banks in Europe* (London: ASLIB, 1976). 66p.
45. *Directory of Computerized Data Files, Software, and Related Technical Reports* (Springfield, Va.: National Technical Information Service, 1974-). Annual.
46. *The Directory of Online Databases.* Ruth N. Landau and others (Santa Monica, Calif.: Cuadra Associates, 1979-). Quarterly.
  Describes both bibliographic and nonbibliographic databases.
47. *Directory of Online Information Resources.* 3rd edition (Rockville, Md.: CSG Press, 1979). 68p.
  Describes about 150 bibliographic and nonbibliographic databases. Subject and source indexes.
48. *Encyclopedia of Information Systems and Services.* Anthony T. Kruzas. 3rd edition (Detroit, Mich.: Gale Research Co., 1978). 1029p.
  Updated by intermittent interedition supplement entitled *New Information Systems and Services.*
49. *A Guide to Selected Computer-Based Information Services.* Ruth Finer (London: ASLIB, 1972). 115p.
50. *Selected Federal Computer-Based Information Systems.* Saul Herner and Matthew J. Vellucci (Washington, D.C.: Information Resources Press, 1972). 215p.

51. *The International Directory of Computer and Information System Services.* 3rd edition (London: Europa Publications, 1974). 636p. Revised biennially.

Mathematics

52. *Combined Membership List of the American Mathematical Society, Mathematical Association of America, and the Society for Industrial and Applied Mathematics* (Providence, R.I.: American Mathematical Society). Annual. Alphabetical and geographic listings of individual, institutional and corporate members. A total of 40,455 memberships have been listed in the 1976/77 edition.
53. *Mathematical Sciences: Administrative Directory* (Providence, R.I.: American Mathematical Society). Annual.

Computer Science

54. *Administrative Directory of College and University Computer Science Departments and Computer Centers.* 2nd edition (New York: Association for Computing Machinery, 1977). 138p.
55. *Directory of Top Computer Executives* (Phoenix, Ariz.: Applied Computer Research). Semiannual.
   Names and addresses of top EDP executives in industries and government agencies with an annual EDP budget of $250,000 or more. Entries are arranged geographically.
56. *Worldwide Directory of Computer Companies.* 2nd edition (Chicago: Marquis Who's Who, Inc., 1973). 633p.

Physics and Astronomy

57. *A Directory of Computer Software Applications. Physics. 1970-May 1978.* PB 281 642 (Springfield, Va.: National Technical Information Service, 1978).
58. *Directory of Physics and Astronomy Staff Members: North American Colleges and Universities, Federally Funded R&D Centers, Government Laboratories, 1976-77* (New York: American Institute of Physics, 1976). 295p.
59. *International Physics and Astronomy Directory, 1969-70* (New York: W. A. Benjamin, 1970). 802p.
   Directory of university professors and researchers, academic departments, scholarships and grants, and publishers.
60. *Observatories of the World.* Thornton Page (Cambridge, Mass.: Smithsonian Astrophysical Observatory, 1967). 41p.
61. *Yearbook of Astronomy* (New York: W. W. Norton, 1962-). Annual. Annual report of astronomical events, data, discussion articles. Directory of astronomical societies.

Selected List of Directories 203

Chemistry, Chemical Engineering, and Technology

62. *American Chemical Society. Lab Guide* (Washington, D.C.: American Chemical Society, 1966-). Annual.
    A directory of laboratory supply houses, manufacturing companies, instruments, equipment, chemicals, supplies, analytical and research services, trade names and new books in analytical chemistry.
63. *Chemical Engineering Catalog. Equipment and Materials for the Process Industries* (Stamford, Conn.: Reinhold Publishing Co., 1916-). Annual.
    Equipment and materials index; product directory; trade name index; sales offices/telephone directory.
64. *Chemical Engineering. Equipment Buyers' Guide Issue* (New York: McGraw-Hill). Annual.
    Annual special issue of *Chemical Engineering*.
65. *Chemical Week. Buyers' Guide Issue* (New York: McGraw-Hill, 1937-). Annual.
    Annual special issue of *Chemical Week*.
66. *Director of Chemical Producers, U.S.A.* (Menlo Park, Calif.: SRI International). Cumulative supplements and annual revisions.
    Free inquiry service to subscribers.
67. *Directory of Chemical Producers, Western Europe* (Menlo Park, Calif.: SRI International). Periodic supplements and annual revisions.
    Free inquiry service to subscribers.
68. *Directory of Graduate Research* (Washington, D.C.: American Chemical Society, Committee on Professional Training, 1953-). Annual.
    The directory lists faculties, publications, and doctoral theses in departments or divisions of chemistry, chemical engineering, biochemistry, and pharmaceutical and/or medical chemistry at universities in the United States and Canada. Index of instructional staff.

Mining and Metallurgy

69. *E/MJ International Directory of Mining and Mineral Processing Operations* (New York: McGraw-Hill, 1968-). Annual.
70. *Metal Finishing Guidebook-Directory* (Hackensack, N.J.: Metals and Plastics Publications, Inc., 1932-). Annual.
    Title varies; suspended during 1933-34.
71. *Metal Statistics: The Purchasing Guide of the Metal Industries* (New York: Fairchild Publications, 1904-). Annual.
    Metal prices are based on quotations published in *American Metal Market*.
72. *World Mines Register* (San Francisco, Calif.: Miller Freeman, 1975/76-). Irregular.

Aviation and Aerospace

73. *Aerospace Materials Buyers' Guide* (Los Angeles, Calif.: Technology Publishing Corporation, 1969-). Annual.

74. *Aerospace Yearbook.* Aerospace Industries Association of America (Washington, D.C.: American Aviation Publications, 1919–). Annual.
75. *Aircraft Yearbook* (New York: Aeronautical Chamber of Commerce of America, Inc., 1919–). Annual.
76. *Aviation Week and Space Technology Marketing Guide* (New York: Aviation Week and Space Technology; McGraw-Hill, 1955–). Annual.
77. *Defense Documentation Center Referral Data Bank Directory.* U.S. Department of Defense. Defense Documentation Center (Springfield, Va.: National Technical Information Service, 1972). DDC-TR-72-3. AD 750 400.
78. *NASA Factbook.* 2nd edition (Chicago: Marquis Who's Who, Inc., 1975). 613p.
79. *World Aviation Directory* (Washington, D.C.: Ziff-Davis Publishing Co., 1940–). Semiannual.
80. *World Space Directory* (Washington, D.C.: American Aviation Publications, 1962–). Semiannual.
    Lists U.S. and foreign missile/space companies, officials, and government agencies concerned with aviation and aerospace.

Civil Engineering and Architecture

81. *The Concrete Yearbook, 1977.* 53rd edition (Slough, England: Cement and Concrete Association, 1977). 640p.
82. *Directory of Computer Software Applications. Civil and Structural Engineering, 1970–January 1978* (Springfield, Va.: National Technical Information Service, 1978). PB 278 125.
83. *PROFILE: The Official AIA Directory of Architectural Firms.* Edited by Henry W. Schirmer with the cooperation of the American Institute of Architects and published by Archimedia, Inc. (Detroit, Mich.: Gale Research Co., 1978).
    Directory of nearly 6000 architectural firms. Also included are several articles: "How to Find, Evaluate, Select, and Negotiate with an Architect," "You and Your Architect," "The AIA: From the Client's Viewpoint."

Mechanical Engineering

84. *Best's Safety Directory. Industrial Safety, Hygiene, Security.* 18th annual edition, 1978 (Oldwick, N.J.: A. M. Best Co.). Annual.
85. *Machinery's Annual Buyers' Guide* (London: Machinery Publishing Co., 1926–). Annual.

Electrical Engineering and Electronics

86. *Electrical and Electronic Trader Yearbook* (London: Electrical-Electronic Press, 1965–). Annual.

Selected List of Directories

87. *Electrical and Electronics Trade Directory* (Stevenage, Herts., England: Peter Peregrinus, 1883-). Annual.
88. *Electronics Buyers' Guide* (New York: McGraw-Hill, 1945-). Annual. Annual special issue of *Electronics*.
89. *IEEE Membership Directory* (New York: Institute of Electrical and Electronics Engineers, 1966-). Annual.

Biology, Agriculture, and Fisheries

90. *A Directory of Information Resources in the United States: Biological Sciences*. U.S. Library of Congress. National Referral Center (Washington, D.C.: Library of Congress, 1972). 577p.
91. *Agricultural Research Index*. 6th edition (Guernsey, British Isles: Francis Hodgson, 1977).
A directory of research establishments in agriculture, forestry, and food and fisheries throughout the world.
92. *Guide to Graduate Study in Botany for the United States and Canada* (New York: Botanical Society of America, 1976).
Data on 108 departments in the United States and 21 in Canada which offer Ph.D. programs in botany.
93. *International Directory of Agricultural Engineering Institutions* (Rome, Italy: Food and Agriculture Organization of the United Nations, 1974).
94. *World Directory of Collections of Cultures of Microorganisms*. Compiled by S. M. Martin and C. Quadling of Canada, and M. L. Jones and V. B. Skerman of Australia, for the World Federation for Culture Collections of the International Association of Microbiological Societies (New York: Wiley-Interscience, 1972). 560p.
This directory provides a means for locating samples of a particular species of microorganism or a cell line. The introduction and certain sections of text are in English, French, German, Russian, Spanish, and Japanese.
In addition to a main list of collections, the directory contains a geographical index, an index of collections by type of microorganisms (algae, fungi, etc.), an index of the main interests of the collections (e.g., medical microbiology), and a personnel index. Also included are lists of microorganisms, keyed to the collections, with cross-references to indicate preferred nomenclature. Separate lists cover algae, bacteria, fungi, lichens, protozoa, tissue cultures, viruses, and yeasts.
95. *Yearbook of Fishery Statistics* (Rome, Italy: Food and Agriculture Organization of the United Nations, 1947-). Annual; two volumes per year since 1964.

Environment

96. *American Water Works Association Yearbook* (New York: American Water Works Association). Annual.
Up to 1968, included in *AWWA Journal*, and called *AWWA Directory*.

97. *Annual Directory of Environmental Information Sources* (Boston, Mass.: National Foundation for Environmental Control, 1971–).
98. *Chemical Engineering Desk Book: Environmental Engineering, Energy/ Environment* (New York: McGraw-Hill, 1974).
Special issue of *Chemical Engineering*, v. 8, No. 22 (October 21, 1974).
99. *Conservation Directory* (Washington, D.C.: National Wildlife Federation, 1956–). Annual.
Supersedes *Directory of Organizations Concerned with the Protection of Wildlife and Other Natural Resources.*
100. *Databases Suitable for Users of Environmental Information.* Gerda Yska and J. Martyn (London: ASLIB, 1976).
101. *Directory of Government Air Pollution Agencies* (Pittsburgh, Penn.: Air Pollution Control Association, 1972–). Annual.
102. *Directory of Pollution Control Equipment Companies in Western Europe.* R. M. Whiteside (London: Graham and Trottman Ltd., 1977). 614p.
103. *Environmental Protection Directory: A Comprehensive Guide to Environmental Organizations in the U.S. and Canada.* Thaddeus C. Trzyna and Sally R. Osberg. 2nd edition (Chicago: Marquis Academic Media, 1975). 526p.
Describes 2664 agencies, including government and private agencies concerned with consumer protection and environmental protection. Agencies are arranged under states. For each agency, address, telephone number, key personnel, functions and publications are listed.
104. *Guide to Ecology Information and Organizations.* John Gordon Burke and Jill Swanson Reddig (New York: H. W. Wilson, 1976). 292p.
Describes citizen action guides, indexes, reference books, histories, monographs, government publications, nonprint media, and periodicals. Highly technical material has been excluded. List of publishers and distributors appended.
105. *Pollution Control Directory* (Easton, Penn.: American Chemical Society). Annual.
106. *U.S. Directory of Environmental Sources.* 3rd edition (Washington, D.C.: U.S. Environmental Protection Agency, 1979). 856p.
107. *World Directory of Environmental Research Centers* (Scottsdale, Ariz.: Oryx Press; Distributor: R. R. Bowker, 1970–). Triennial.
Formerly, *Directory of Organizations Concerned with Environmental Research.*

Energy

108. *Energy Directory* (New York: Environment Information Center, 1974–). Annual.
109. *Nuclear News Buyers' Guide* (Hinsdale, Ill.: American Nuclear Society, 1969–).
Annual special issue of *Nuclear News.*

Selected List of Directories

110. *Nuclear Research Index.* 5th edition (New York: International Publications Service, 1975).
 Directory of over 2000 nuclear research agencies throughout the world.
111. *Information on International Research and Development Activities in the Field of Energy.* Smithsonian Institution. Science Information Exchange (Springfield, Va.: National Technical Information Service, 1976). 370p. PB 254 315/5PSA.
112. *Directory of Federal Energy Data Sources.* U.S. Federal Energy Administration. Office of Policy Analysis (Springfield, Va.: National Technical Information Service, 1976). PB 254 163/9PSA.
113. *Energy Information Activities at the FEA.* Gil Rodgers. U.S. Federal Energy Administration. Office of Data and Analysis (Springfield, Va.: National Technical Information Service, 1976). 51p. PB 253 962.

14
HANDBOOKS AND TABLES

14.1 Handbooks

Handbooks are reference books in which quantifiable primary data collected from a great many diverse sources are assembled, categorized, and presented for ready use. Handbooks are useful for answering "ready reference" questions (Voigt's "everyday approach") calling for a specific piece of information or numerical datum. From 50 to 80% of the factual questions asked in scientific and technical libraries can be answered from handbooks. Because of the vast amount of diverse data contained in a handbook, usually packed into a single volume, handbooks have been referred to as "one-volume libraries." Most handbooks contain quantitative data, usually presented in the form of tables, charts, and lists, with little expository text. Presentation of data in this format makes the handbook a very convenient reference tool. Elaborate indexes in handbooks further insure rapid location and retrieval of the desired data element from the vast bulk of data contained in handbooks.

Popularly known as the "Chemist's Bible," or the "Rubber Bible," the *CRC Handbook of Chemistry and Physics* (Cleveland, Ohio: Chemical Rubber Company Press, 1931-) is perhaps the most widely used handbook. It is revised annually. The 60th edition for 1979-1980, edited by Robert C. Weast, packs within one volume an incredible amount of physical and chemical data on organic and inorganic chemical substances, 128 pages of mathematical tables, and a very exhaustive (23) pages) table of conversion factors. Bibliographic references to original sources of data are given wherever applicable.

Another handbook that is extensively used is the *Chemical Engineer's Handbook*, 5th edition (McGraw-Hill, 1973). This is often referred to as "Perry's Handbook" after the late John H. Perry who edited the first editions of this handbook.

A multivolume compilation is the German series *Landolt-Börnstein Zahlenwerte und Funktionen aus Naturwissenschaften und Technik Neue Serie* (Numerical Data and Functional Relationships in Science and Technology),

14.1 Handbooks

published by Springer-Verlag. The new series began appearing in 1961 under the editorship of K. H. Hellwege. The work is still being published in numerous volumes as and when the appropriate data are accumulated. Each volume is issued in one of the following basic groups:

Nuclear and particle physics

Atomic and molecular physics

Crystal and solid state physics

Macroscopic and technical properties of matter

Geophysics and space research

Astronomy, astrophysics, and space research

Introduction, contents lists, and section headings are in both German and English.

Two other encyclopedic compilations, which have become classics in the literature of chemistry, are: *Handbuch der Anorganischen Chemie* (begun in 1819 by Leopold Gmelin), and *Handbuch der Organischen Chemie* (begun by Friedrich Konrad Beilstein in 1882), commonly referred to as *Gmelin* and *Beilstein*, respectively. Detailed descriptions of these may be found in many articles [1-5] and in guides to chemical literature such as R. T. Bottle's *The Use of Chemical Literature*, 3rd edition (London: Butterworths, 1980).

Because of the predominantly quantitative nature of the material contained in handbooks, and the tabular format in which the material is presented, most handbooks are not suitable for continuous reading. In this sense handbooks are different from monographs and treatises. However, some handbooks do consist of long expository articles, often supported with quantitative material and bibliographies. *CRC Handbook of Food Additives*, edited by Thomas E. Furia (Cleveland, Ohio: Chemical Rubber Company, 1968), consists of 16 chapters, including an introductory chapter, supported with numerous tables, charts, graphs, and extensive bibliographies. The chapters are written by different authors. The chapter on acidulants in food processing is 41 pages long and lists 395 bibliographic references. A 63-page chapter on nonnutritive sweeteners lists 534 bibliographic references and 90 patents. The Chemical Rubber Company, which produces numerous handbooks, has published a *Composite Index for CRC Handbooks*, 2nd edition (1971); this is a composite index to the contents of some 50 CRC handbooks.

Quantitative data are sometimes published in loose-leaf format or journal format. The *Aerospace Structural Metals Handbook* produced by the Mechanical Properties Data Center of the U.S. Department of Defense is a five-volume handbook in loose-leaf format. Additional sources of critically evaluated data are discussed in a later section entitled National Standard Reference Data System.

14.2 Tables

Data of a purely numerical character are often published in tabular format. *International Critical Tables of Numerical Data, Physics, Chemistry, and Technology* (seven volumes plus an index volume, 1926-1933), published by the McGraw-Hill Book Company for the National Research Council, represents the product of a major effort to compile critical data. A decision to prepare and publish a compilation of critical data was taken at a meeting of the International Union of Pure and Applied Chemistry (IUPAC) held in London in June 1919. The tables were prepared by the National Research Council of the United States under the auspices of the International Research Council and the National Academy of Sciences. The work is divided into 300 sections and contains "critical" data compiled by a large number of subject specialists from primary literature published up to 1924. "The word 'critical' in this connection means that the cooperating expert was requested to give in each instance the 'best' value which he could derive from all the information available, together, where possible, with an indication of its probable reliability" (from the Introduction). References to original sources of data are also given. The *International Critical Tables*, published half a century ago, are still being widely used.

Mathematical tables are the most common examples of tabular compilations. These may be found in general handbooks such as the *Handbook of Chemistry and Physics* and the American Institute of Physics *Handbook of Physics*, as well as in many textbooks on mathematics and statistics. *Table of Natural Logarithms* (National Bureau of Standards, 1941) is a four-volume compilation of tables of natural logarithms to 16 decimal places. *CRC Handbook of Tables for Mathematics* (since 4th edition, 1975) contains new sections entitled "Astrodynamics: Basic Orbital Equations," and "Astrodynamical Terminology, Notation and Usage." *A Million Random Digits with 100,000 Normal Deviates* (Glencoe, Ill.: The Free Press, 1955) prepared by the Rand Corporation is one of the most frequently used tables of random digits. Smaller tables of random numbers may be found in handbooks and textbooks of statistics. *Tables of Thomson Functions and Their First Derivatives* by L. N. Nosova (Pergamon Press, 1961), originally issued by the Computer Center of the Academy of Sciences of the USSR, and *Tables of Bessel Transforms* by Fritz Oberhettinger (Springer-Verlag, 1972) are examples of tables of special functions.

Thousands of mathematical tables have been computed and published during the last 300 years. There are several indexes and guides to mathematical tables; the following are three notable examples:

1. *An Index to Mathematical Tables*, 2nd edition, by A. Fletcher, J. C. P. Miller, L. Rosenhead, and L. J. Comrie (Addison-Wesley, 1962). The first volume is an index to mathematical tables according to functions; the second volume lists the bibliographic sources of the tables indexed, and also contains a section listing the errors noted in the mathematical tables.

14.2 Tables

2. *Guide to Tables in Mathematical Statistics*, by J. Arthur Greenwood and H. O. Hartley (Princeton University Press, 1962).
3. *A Guide to Mathematical Tables*, by A. V. Lebedev and R. M. Fedorova (English edition published by Pergamon Press, Oxford, 1960), contains a much fuller account of Russian mathematical tables than those found in the *Index to Mathematical Tables* by Fletcher and others. A supplement to this *Guide* was issued in 1960.

Tables of astronomical and meteorological data and steam tables are other types of data compilations frequently published in tabular format. The following are typical examples of these types of tables:

1. *Weather Almanac.* 2nd edition (Detroit, Mich.: Gale Research Co., 1977), contains climatic data, climatological narratives for cities, list of earthquakes in the United States, data on air quality trends, and a glossary of weather terms.
2. British Electrical and Allied Industries Research Association. *1967 Steam Tables* (New York: St. Martin's Press, 1967).
3. *ASME Steam Tables. Thermodynamic and Transport Properties of Steam, Comprising Tables and Charts for Steam and Water* (New York: American Society of Mechanical Engineers, 1967).

Smaller steam tables are usually appended to textbooks of physics and thermodynamics.

The American Ephemeris and Nautical Almanac (Washington, D.C.: U.S. Government Printing Office), is an annual publication, and contains tables showing the projected positions of the sun, moon, and other planets and satellites of the solar system for the coming year. Since 1960, this has been a unified Anglo-American publication, incorporating the *American Ephemeris and Nautical Almanac* issued by the Nautical Almanac Office, U.S. Naval Observatory, and the *Astronomical Ephemeris* issued by H. M. Nautical Almanac Office, Royal Greenwich Observatory. With the exception of the introductory pages, the publication is printed separately in the two countries from reproducible material prepared partly in the United States and partly in the United Kingdom. The *Astronomical Ephemeris* was a successor to the *Nautical Almanac and Astronomical Ephemeris* introduced by Nevil Maskelyne in 1767.

U.S. Standard Atmosphere, 1976, published by the National Oceanic and Atmospheric Administration, the National Aeronautics and Space Administration, and the U.S. Air Force, contains tables of atmospheric properties (temperature, pressure, density, acceleration due to gravity, sound speed, viscosity, composition, etc.) up to an altitude of 1000 kilometers in the international system of metric units. *Climatological Data* published by the Environmental Data Service, National Oceanic and Atmospheric Administration, is a monthly serial containing tables of daily and monthly precipitation, wind movements, temperatures, and other climatological data.

14.3 National Standard Reference Data System (NSRDS)

The importance to the scientific community of critically evaluated physical and chemical data can hardly be overestimated. Several countries including the United States, the United Kingdom, and the USSR have set up national-level organizations to collect, evaluate, organize, and publish the vast amount of numerical data widely scattered in published and unpublished sources. In the United Kingdom, the Office for Scientific and Technical Information (OSTI), established in 1965 as a part of the Department of Education and Science, is responsible for coordinating and stimulating data compilation activities. The OSTI publishes periodically a list of ongoing numerical data projects in the United Kingdom.

In the USSR, the State Service for Standard and Reference Data (GSSSD) is responsible for determining the data requirements of science and industry, developing standards for testing and evaluating data, and providing standard and reference data to the Soviet scientific and technical community. A number of products of the Russian standard reference data system have been translated by the Israel Translation Service, Jerusalem, and the American Publishing Company Private Limited, New Delhi, with support of the National Science Foundation at the request of the National Bureau of Standards. Most of these translations are available from the National Technical Information Service, 5285 Port Royal Road, Springfield, Va. 22161.

In the United States, the National Standard Reference Data System (NSRDS) was established in May 1963 by the Federal Council for Science and Technology. The aims of the NSRDS are twofold:

1. To provide critically evaluated numerical data, in a convenient and accessible form, to the scientific and technical community; and
2. To provide feedback into experimental work to help raise the general standards of measurement

Further impetus to this program was given by the Standard Reference Data Act which became a law on July 11, 1968 (P.L. 90-396). The Act states: "It is the policy of the Congress to make critically evaluated data readily available to scientists, engineers, and the general public. . . . The Secretary of Commerce is authorized and directed to provide or arrange for the collection, compilation, critical evaluation, publication, and dissemination of standard reference data."

NSRDS is administered by the Office of Standard Reference Data at the NBS, and consists of a network of over 25 data centers located throughout the country in government agencies, academic institutions, and laboratories in the private sector. The Office of Standard Reference Data does not operate or directly supervise the activities of the data centers; it coordinates their technical

14.3 National Standard Reference Data System (NSRDS)

programs pertaining to the generation, collection, evaluation, and dissemination of standard reference data in the following seven program areas:

1. Nuclear properties
2. Atomic and molecular properties
3. Solid state properties
4. Thermodynamic and transport properties
5. Chemical kinetics
6. Colloid and surface properties
7. Mechanical properties

The principal output of NSRDS consists of compilations of evaluated data, and is disseminated through the following channels:

Journal of Physical and Chemical Reference Data, a quarterly journal containing data compilations and critical data reveiws, published jointly by NBS, AIP, and the ACS

NSRDS-NBS series of publications distributed by the Superintendent of Documents, U.S. Government Printing Office, Washington, D.C.

Publications of professional societies and commercial publishers

Responses by NSRDS data centers to inquiries for specific data

NSRDS also makes available selected technical data on magnetic tape as well as computer programs on magnetic tape for handling the technical data.

A survey of 2700 special librarians was conducted in 1971 by Herman Weisman, manager of information services at the Office of Standard Reference Data, to ascertain the extent of use made of NSRDS publications, and the problems encountered in using the publications [6]. The survey revealed that:

Almost 88% of the respondents had heard of NSRDS and its publications.

About 95% indicated that their libraries carried NSRDS publications.

Less than 44% had actually used an NSRDS publication.

About 94% of those who had used NSRDS publications had found them "very useful" or "useful to some extent."

Information on NSRDS publications may be obtained from the Office of Standard Reference Data, National Bureau of Standards, Washington, D.C. 20234. Partial lists of NSRDS publications may be seen in the following publications:

CRC Handbook of Chemistry and Physics (Cleveland, Ohio: Chemical Rubber Company).

R. Aluri and P. A. Yannarella, NBS: A compilation of the NBS data and related publications since 1969, *Special Libraries*, 65:77-82 (February 1974).

U.S. Department of Commerce. National Bureau of Standards. Critical Evaluation of Data in the Physical Sciences: A Status Report on the NSRDS. NBS Technical Note 947 (January 1977).

National Standard Reference Data System. *Publication List 1964-1977.* LP 81 (Washington, D.C.: U.S. Department of Commerce, National Bureau of Standards, 1978).

Current activities relating to reference data and new products of NSRDS are announced in the bimonthly newsletter *NSRDS Reference Data Report* (formerly, *NSRDS News*). NSRDS publications can be purchased from the Superintendent of Documents, U.S. Government Printing Office, Washington, D.C. 20402.

At the international level, the Committee on Data for Science and Technology of the International Council of Scientific Unions (CODATA), established in 1966, is concerned with promoting the production and distribution of compendia and other forms of collections of critically selected numerical data on substances of interest and importance to science and technology. CODATA has produced the *International Compendium of Numerical Data Projects: A Survey and Analysis* (New York: Springer-Verlag, 1969). This is a directory of over 160 numerical data projects and centers in 26 countries that systematically extract, evaluate, and publish scientific and technical data.

Selected List of Handbooks and Tables

General Science and Engineering

1. *Composite Index for CRC Handbooks.* 2nd edition (Cleveland, Ohio: Chemical Rubber Company Press, 1977). 1111p.
2. *Continuing Numerical Data Projects: A Survey and Analysis.* Prepared under the auspices of the Office of Critical Tables, National Academy of Sciences and National Research Council. Available from the National Academy of Sciences, 2101 Constitution Avenue, Washington, D.C. 20418. Excerpts from this compendium can be seen in the *CRC Handbook of Chemistry and Physics.*
3. *CRC Handbook of Tables for Applied Engineering Science.* 2nd edition. Ray E. Bolz and George L. Tuve (Cleveland, Ohio: Chemical Rubber Company Press, 1973). 1166p.
4. *A Dictionary for Unit Conversion.* Yi Shu Chiu (Washington, D.C.: George Washington University, School of Engineering and Applied Science, 1975). 451p.
5. *Engineering Mathematics Handbook: Definitions, Theorems, Formulas, Tables.* Jan J. Tuma (New York: McGraw-Hill, 1970). 334p.
6. *Engineer's Guide to High-Temperature Materials.* Francis J. Clauss (Reading, Mass.: Addison-Wesley, 1969), 401p.

Selected List of Handbooks and Tables 215

7. *Handbook of the Engineering Sciences.* James H. Potter (Princeton, N.J.: Van Nostrand, 1967). 2 v.
8. *International Critical Tables of Numerical Data. Physics, Chemistry, and Technology.* Prepared under the auspices of the International Research Council and the National Academy of Sciences (New York: McGraw-Hill, 1926-1933). 8 v.
9. *Kempe's Engineers' Yearbook* (London: Morgan Bros., 1894-). Annual.
10. Rand Corporation. *A Million Random Digits with 100,000 Normal Deviates* (Glencoe, Ill.: The Free Press, 1955), 200p.

Mathematics

Indexes to Mathematical Tables

11. *A Guide to Mathematical Tables.* A. V. Lebedev and R. M. Fedorova. English edition prepared from the Russian by D. G. Fry (Oxford: Pergamon Press, 1960). 586p.
12. *A Guide to Mathematical Tables.* N. M. Burunova. English edition prepared from the Russian by D. G. Fry (Oxford: Pergamon Press, 1960). 190p.
    Supplement to *A Guide to Mathematical Tables* by Lebedev and Fedorova.
13. *Guide to Tables in Mathematics and Statistics.* Joseph A. Greenwood and H. O. Hartley (Princeton, N.J.: Princeton University Press, 1962). 1014p.
14. *An Index to Mathematical Tables.* A. Fletcher and others. 2nd edition (Reading, Mass.: Addison-Wesley, 1962). 2 v. 994p.

Mathematical and Statistical Tables

15. *CRC Handbook of Tables for Mathematics.* 4th edition (Cleveland, Ohio: Chemical Rubber Company Press, 1965). 1152p.
    New sections on astrodynamics, basic orbital equations, and astrodynamical terminology, notation, and usage have been added in the 4th edition.
16. *CRC Handbook of Tables for Probability and Statistics.* 2nd edition. William H. Beyer (Cleveland, Ohio: Chemical Rubber Company Press, 1968). 642p.
17. *Eight-Place Tables of Trigonometric Functions for Every Second of Arc, with an Appendix on the Computation to Twenty Places.* Jean Peters (Bronx, N.Y.: Chelsea Publishing Co., 1965). 954p.
18. *Handbook of Mathematical Functions with Formulas, Graphs, and Mathematical Tables.* Milton Abramowitz and Irene A. Stegun (Washington, D.C.: National Bureau of Standards, 1964). 1046p. (NBS Applied Mathematics Series, 55).
19. *Handbook of Mathematical Tables and Formulas.* Richard S. Burington. 5th edition (New York: McGraw-Hill, 1973). 500p.
20. *Handbook of Probability and Statistics with Tables.* 2nd edition. Richard S. Burington and Donald C. May (New York: McGraw-Hill, 1970). 462p.

21. *Handbook of Statistical Tables.* Donald B. Owen (Reading, Mass.: Addison-Wesley, 1962). 58p.
22. *Mathematical Handbook for Scientists and Engineers: Definitions, Theorems, and Formulas for Reference and Review.* 2nd edition. Granino A. Korn and Theresa M. Korn (New York: McGraw-Hill, 1968). 1130p.
23. *Smithsonian Elliptic Functions Tables.* G. W. Spenceley and R. M. Spenceley (Washington, D.C.: Smithsonian Institution, 1947). 366p. (Smithsonian Miscellaneous Collections, 109).
24. *Smithsonian Logarithmic Tables to Base e and Base 10.* George W. Spenceley, Rheba M. Spenceley, and Eugene R. Epperson (Washington, D.C.: Smithsonian Institution, 1952), 402p. (Smithsonian Miscellaneous Collections, 118).
25. *Smithsonian Mathematical Formulae and Tables of Elliptic Functions.* Edwin P. Adams and R. L. Hippisley (Washington, D.C.: Smithsonian Institution, 1957). 314p. (Smithsonian Miscellaneous Collections, 74/1).
26. *Statistical Functions.* A Source of Practical Derivations Based on Elementary Mathematics. Buddy L. Myers and Norbert L. Enrick (Kent, Ohio: Kent State University Press, 1970). 174p.
27. *A Table of Indices and Power Residues for All Primes and Prime Powers Below 2000.* Computed by the University of Oklahoma Mathematical Tables Project (New York: W. W. Norton & Co., 1962).
28. *Tables of Factorials 0! to 9999!* Joseph B. Reid and G. Montpetit (Washington, D.C.: National Academy of Sciences-National Research Council, 1962). Publication 1039.

Computer Science

29. *Handbook of Industrial Control Computers.* Thomas J. Harrison (New York: Wiley-Interscience, 1972). 1056p.
30. *Minicomputer Handbook.* Charles J. Sippl (New York: Petrocelli/Charter, 1976). 454p.

Physics and Astronomy

Handbooks and Tables

31. *The Amateur Astronomer's Handbook.* Revised edition. James Muirden (New York: Crowell, 1974). 404p.
32. *American Institute of Physics Handbook.* 3rd edition. Dwight E. Gray (New York: McGraw-Hill, 1972).
33. *Basic Tables in Physics.* John Robson (New York: McGraw-Hill, 1967). 365p.
34. *Bibliography of Infrared Spectroscopy Through 1960* (Washington, D.C.: U.S. Department of Commerce, National Bureau of Standards, 1975). (NBS Special Publication 428).
    Annotated index to spectral data on about 25,000 compounds. For each compound the empirical formula, range of wavelength reported,

Selected List of Handbooks and Tables

state of material, type of data presented in the paper, and literature citation are given.

35. *CRC Handbook of Chemistry and Physics* (Cleveland, Ohio: Chemical Rubber Company, 1913-). Revised almost every year.
36. *Crystal Data. Determinative Tables.* 3rd edition. National Standard Reference Data System. Obtainable from: National Standard Reference Data System, Joint Committee on Powder Diffraction Standards, 1601 Park Land, Swarthmore, PA 19081.

    v. 1:   *Organic Compounds,* 1977.

    v. 2:   *Inorganic Compounds,* 1977.
37. *Electronic Properties Research Literature Retrieval Guide: 1972-1976.* J. F. Chaney and T. M. Putnam (New York: Plenum Publishing Corporation, 1979). 4 v.

    Lists 19,104 references on 15 electronic, electrical, magnetic, and optical properties and seven property groups for some 9634 materials. Synonyms and trade names are also included.
38. *Handbook of Fluorescence Spectra of Aromatic Molecules.* 2nd edition. Isadore B. Berlman (New York: Academic Press, 1971). 473p.
39. *Handbook of Physical Calculations.* Jan J. Tuma (New York: McGraw-Hill, 1976). 370p.

    "Definitions, formulas, technical applications, physical tables, conversion tables, graphs, dictionary of physical terms, etc." (Subtitle).
40. *Handbook of Heat Transfer.* Warren H. Rohsenow and J. P. Hartnett (New York: McGraw-Hill, 1973).
41. *Handbook of Physics.* 2nd edition. E. U. Condon and Hugh Odishaw (New York: McGraw-Hill, 1967).
42. *Handbook of Thermodynamic Tables and Charts.* Kuzman Raznjevic. Translated by Marijan Boskovic and Richard Podhorsky (Washington, D.C.: Hemisphere Publishing Corporation, 1976). 392p.
43. *Handbook of X-Rays for Diffraction, Emission, Absorption and Microscopy.* Emmett F. Kaelble (New York: McGraw-Hill, 1967).
44. *Infrared Radiation: A Handbook for Applications, with a Collection of Reference Tables.* Mikael A. Bramson. Translated by William L. Wolfe (New York: Plenum Press, 1968). 623p.
45. *Landolt-Börnstein Zahenwerte Und Funktionen Aus Physik, Chemie, Astronomie, Geophysik, Und Technik* (Numerical Data and Functional Relationships in Science and Technology) (Berlin: Springer-Verlag, 1950-, New Series, 1974-).
46. *Mössbauer Effect Data Indexes* (New York: Plenum Publishing Corporation, 1966-1976). 9 v.
47. *Cumulative Index to the Mössbauer Effect Data Indexes.* John G. Stevens, Virginia Stevens, and William Gettys, Mössbauer Effect Data Center, University of North Carolina, Ashville, N.C. (New York: Plenum Publishing Corporation, 1980). 358p.

    Isotope, subject, and author indexes to nine volumes of *Mössbauer Effect Data Indexes.*

48. *Smithsonian Physical Tables.* 9th edition. William Elmer Forsythe. (Washington, D.C.: Smithsonian Institution, 1956). 827p. (Smithsonian Miscellaneous Collection, 120).
49. *Scpectroscopic Data.* S. N. Suchard (New York: IFI/Plenum, 1975-76).
50. *Tables of Physical and Chemical Constants and Some Mathematical Functions.* 14th revised edition. George W. C. Kaye and T. H. Laby (New York: John Wiley, 1973). 386p.
    All tables and values are in SI units.
51. *Thermophysical Properties of Matter.* Y. S. Touloukian and others (New York: IFI/Plenum, 1970-1977). 13 v.
    Data compiled at the Thermophysical Properties Research Center, Purdue University, West Lafayette, Ind.
52. *Master Index to Materials and Properties.* Y. S. Touloukian and C. Y. Ho (New York: IFI/Plenum, 1979). 180p.
    Index to 6362 individual materials and properties reported in the 13 volumes of *Thermophysical Properties of Matter.*
53. *Thermophysical Properties Research Literature Retrieval Guide.* 2nd revised edition. Y. S. Touloukian. Thermophysical Properties Research Center, Purdue University, West Lafayette, Ind. (New York: Plenum Press, 1967-). 3 v. Supplements.

Astronomical Atlases, Almanacs, and Catalogs

54. *Astronomical Catalogues, 1951-1975.* Mike Collins (London: Institution of Electrical Engineers, INSPEC, 1978). 325p.
    List of nearly 2500 catalogs and listings of stars, nebulae, galaxies, sources of electromagnetic radiation, black holes, and other astronomical phenomena and objects. Some of these catalogs are available on machine-readable magnetic tape.
55. *The Atlas of the Universe.* Patrick Moore (Chicago: Rand McNally, 1970). 272p.
    Over 1500 photographs, maps, and other illustrations (largely in color) of earth, the moon, the solar system, and other cosmic bodies. Appendices contain a glossary, a catalog of stellar objects, and a "Beginner's Guide to the Heavens." Index.
56. *Concise Atlas of the Universe.* Patrick Moore (New York: Rand McNally, 1974). 172p.
57. *A New Photographic Atlas of the Moon.* Zdenek Kopal (New York: Taplinger, 1971). 310p.
    About 200 pictures of the moon, starting from the earliest telescopic pictures of Galileo and including recent pictures obtained in the Apollo Mission.
58. *The American Ephemeris and Nautical Almanac.* U.S. Nautical Almanac Office (Washington, D.C.: U.S. Government Printing Office, 1855-).
    Since 1960, the *American Ephemeris and Nautical Almanac* issued by the Nautical Almanac Office, U.S. Naval Observatory, has been unified with the

Selected List of Handbooks and Tables

Astronomical Ephemeris issued by H. M. Nautical Almanac Office, Royal Greenwich Observatory.
59. *The Nautical Almanac* (Washington, D.C.: U.S. Government Printing Office, 1909-).
   Produced jointly by H. M. Nautical Almanac Office, Royal Greenwich Observatory and by the Nautical Almanac Office of the U.S. Naval Observatory.
60. *Star Catalog* (Washington, D.C.: Smithsonian Institution, Astrophysical Observatory, 1966). 4 v.
   Catalog of 258,997 stars with astronomical data and positions for the epoch 1950.0.

Spectra

61. *The Aldrich Library of Infrared Spectra.* 2nd edition. Charles J. Pouchert (Milwaukee, Wisc.: Aldrich Chemical Co., 1975). 1575p.
62. *Atlas of Electron Spin Resonance Spectra.* Benon H. J. Bielski and Janusz M. Gebicki (New York: Academic Press, 1967). 665p.
63. *CRC Atlas of Spectral Data and Physical Constants for Organic Compounds.* Jeanette G. Grasselli (Cleveland, Ohio: Chemical Rubber Company Press, 1973).
64. *The Sadtler Guide to NMR Spectra.* W. W. Simons and M. Zanger (Philadelphia, Penn.: Sadtler Research Laboratories, 1972). 542p.
65. *The Sadtler Standard Spectra* (Philadelphia, Penn.: Sadtler Research Laboratories, 1961-). IR, UV, and NMR series.

Chemistry, Chemical Engineering, and Technology

66. *The Chemical Formulary.* Harry Bennett (New York: The Chemical Publishing Co., 1933-). 19th volume published in 1976.
   A condensed collection of formulae for making thousands of products in all fields of industry.
67. *Chemical Tables for Laboratory and Industry.* W. Helbing and A. Burkart. Translation of *Chemie-Tabellen fur Labor and Betrieb* (New York: John Wiley, Halsted Press, 1980). 270p.
68. *Chemical Technicians' Ready Reference Handbook.* Gershon J. Shugar, Ronald A. Shugar, and Lawrence Bauman (New York: McGraw-Hill, 1973). 463p.
69. *Computer Compilation of Molecular Weights and Percentage Compositions for Organic Compounds.* Michael J. S. Dewar and Richard Jones (Oxford: Pergamon Press, 1969). 476p.
70. *CRC Handbook of Biochemistry and Molecular Biology.* 3rd edition. Gerald D. Fasman (Cleveland, Ohio: Chemical Rubber Company Press, 1975). 4 Sections, 9 v.
   I. Proteins: amino acids, peptides, polypeptides, and proteins. 3 v.
   II. Nucleic Acids: purines, pyrimidines, nucleotides, oligonucleotides, tRNA, DNA, RNA. 2 v.

III. Lipids, Carbohydrates, Steroids. 1 v.
IV. Physical and Chemical Data, Miscellaneous: ion exchange, chromatography, buffers, miscelleneous, e.g., vitamins. 2 v.
Cumulative series index. 1 v.

71. *CRC Handbook of Chemistry and Physics* (Cleveland, Ohio: Chemical Rubber Company Press, 1913-). Revised annually.
72. *CRC Handbook of Chromatography.* Gunter Zweig and Joseph Sherma (Cleveland, Ohio: Chemical Rubber Company Press, 1972). 2v.

   v. 1: Chromatographic data, 784p

   v. 2: Principles and techniques, practical applications, 343p

73. *CRC Handbook of Food Additives.* 2nd edition. Thomas E. Furia (Cleveland, Ohio: Chemical Rubber Company Press, 1972). 998p.
74. *Dangerous Properties of Industrial Materials.* 4th edition. Irving Sax (New York: D. Van Nostrand Reinhold, 1975), 1258p.
75. *CRC Handbook of Laboratory Safety.* 2nd edition. Norman V. Steere (Cleveland, Ohio: Chemical Rubber Company Press, 1971). 854p.
76. *CRC Handbook of Radioactive Nuclides.* Yen Wang (Cleveland, Ohio: Chemical Rubber Company Press, 1969). 960p.
77. *Food Products Formulatory.* Stephan L. Komarick and others (Westport, Conn.: AVI Publishing Co., 1974-76). 3 v.

   v. 1: *Meats, Poultry, Fish, Shellfish.* 1974. 348p.

   v. 2: *Cereals, Baked Goods, Dairy and Egg Products.* 1975. 431p.

   v. 3: *Fruit, Vegetable and Nut Products.* 1976. 278p.

78. *Handbook of Industrial Chemistry.* 7th edition. James A. Kent (New York: Van Nostrand Reinhold, 1974). 902p.
79. *Handbook of Metal Ligand Heats and Related Thermodynamic Quantities.* 2nd edition (New York: Marcel Dekker, 1975). 512p.
80. *Handbook of Moisture Determination and Control: Principles, Techniques, Applications.* A. Pade (New York: Marcel Dekker, 1974-75). 4 v.
81. *Handbook of the Physiochemical Properties of the Elements.* G. V. Samsonov. Akademia Nauk USSR, Kiev. Instytut Problem Materialoznavstva. Translated from Russian (New York: IFI/Plenum, 1968). 941p.
82. *Handbook of Plastics and Elastomers.* Charles A. Harper (New York: McGraw-Hill, 1975).
83. *Handbook of Pulp and Paper Technology.* 2nd edition. Kenneth W. Britt (New York: Van Nostrand Reinhold, 1970). 723p.
84. *Handbook of Reactive Chemical Hazards.* L. Bretherick (Cleveland, Ohio: Chemical Rubber Company Press, 1975). 976p.
85. *Handbook of Silicone Rubber Fabrication.* Wilfred Lynch (New York: Van Nostrand Reinhold, 1978). 257p.
86. *Handbook of Spectroscopy.* J. W. Robinson (Cleveland, Ohio: Chemical Rubber Company Press, 1974). 2 v.

Selected List of Handbooks and Tables 221

87. *Industrial Solvents Handbook.* 2nd edition. Ibert Mellan (Park Ridge, N.J.: Noyes Data Corporation, 1977). 567p.
88. *Lange's Handbook of Chemistry.* 11th edition. Norbert A. Lange. John A. Dean (New York: McGraw-Hill, 1973).
89. *Melting Point Tables of Organic Compounds.* Walther Utermark and Walter Schicke. 2nd edition (New York: Interscience, 1963). 715p.
90. *Merck Index of Chemicals and Drugs.* An encyclopedia for chemists, pharmacists, physicians, and members of allied professions. 9th edition. Martha Windholz (Rahway, N.J.: Merck & Co., 1976).
91. *Mixing and Excess Thermodynamic Properties: A Literature Source Book.* Jamie Wisniak and Abraham Tamir (Amsterdam: Elsevier Scientific Publishing Co.; New York: Elsevier-North Holland, 1978). 935p.
92. *NMR Data Tables for Organic Compounds.* Frank A. Bovey (New York: Interscience, 1967-).
93. *Petroleum Processing Handbook.* William F. Bland and Robert L. Davidson (New York: McGraw-Hill, 1967).
94. *Polymer Handbook.* 2nd edition. J. Brandup and E. H. Immergut (New York: John Wiley, 1975).
95. *Process Instruments and Controls Handbook.* 2nd edition. Douglas M. Considine (New York: McGraw-Hill, 1974).
96. *Solubilities: Inorganic and Metalorganic Compounds.* A. Seidell and W. F. Linke. 4th edition (Washington, D.C.: American Chemical Society, 1958-1965). 2 v.
97. *Solubilities of Inorganic and Organic Compounds.* H. Stephan and T. Stephan (New York: MacMillan, 1963-64). 2 v., 4 parts.
98. *The Stereo Rubbers.* William M. Saltman (New York: John Wiley, 1977). 897p.
99. *Tables of Standard Electrode Potentials.* Guilio Milazzo, S. Caroli, and V. K. Sharma (New York: John Wiley, 1977). 440p.
100. *The Vapor Pressure of Pure Substances.* Thomas Boublik, Vojtech Fried and Eduard Hala (Amsterdam: Elsevier Scientific Publishing Co., 1973). 626p.
"Selected values of the temperature dependence of the vapor pressures of some pure substances in the normal and low pressure region" (Subtitle).

Earth Sciences, Materials Sciences, and Metallurgy

101. *CRC Handbook of Marine Science.* Frederick G. Walton Smith (Cleveland, Ohio: Chemical Rubber Company Press, 1974).
102. *CRC Handbook of Materials Science.* Charles T. Lynch (Cleveland, Ohio: Chemical Rubber Company Press, 1974).
103. *Handbook of Stainless Steels.* Donald Peckner and I. M. Bernstein (New York: McGraw-Hill, 1977).

Contains 48 chapters written by experts on various aspects of stainless steel including constitution, effects of alloying elements, metallurgical reactions, properties, corrosion, heat treatment, and applications.

104. *Guide to Uncommon Metals.* Eric N. Simons (London: F. Muller, 1967). 244p.
105. *Metal Bulletin Handbook* (London: Metal Bulletin Limited, 1968–). Annual.
106. *Metals Reference Book.* Colin James Smithells. 5th edition (Wolburn, Mass.: Butterworths, 1976), 1566p.
107. *Mineral Facts and Problems.* 5th edition (Washington, D.C.: U.S. Bureau of Mines, 1976). U.S. Bureau of Mines Bulletin 667, 1975 edition.
Contains data on reserves and resources, uses, production and consumption, demand and supply, and economic data on mineral commodities. Each commodity chapter includes a bibliography and sources of information.
108. *The Underwater Handbook: A Guide to Physiology and Performance for the Engineer.* Charles W. Shilling, Margaret F. Werts, and Nancy Schandelmier (New York: Plenum Publishing Corporation, 1976). 912p.

Aviation, Aerospace, and Meteorology

109. *Aerospace Facts and Figures* (Washington, D.C.: Aerospace Industries Association of America, 1945–). Annual.
110. *Aerospace Structural Metals Handbook* (Traverse City, Mich.: Properties Data Center of Belfour Stulers, Inc., 1969–). Irregular.
Looseleaf format, periodically revised and expanded.
111. *AGARD Flight Test Manual.* NATO Advisory Group on Aerospace Research and Development (New York: Pergamon Press, 1959–). 4 v.
112. *AGARD Manual on Aeroelasticity* (NATO Advisory Group on Aerospace Research and Development, 1968). 6 v.
113. *Air Facts and Feats. A Record of Aerospace Achievement.* John Taylor and others (New York: Two Continents Publishing Group, 1974).
114. *Aircraft Engines of the World.* Paul H. Wilkinson (Washington, D.C.: Paul H. Wilkinson, 1941–). Biennial.
115. *Astronautics and Aeronautics: An American Chronology of Science and Technology in the Exploration of Space, 1915-1960* (Washington, D.C.: U.S. National Aeronautics and Space Administration, 1961). 240p. Annual Supplements.
116. *Handbook of Soviet Space Science Research.* George E. Wukelic. Bettelle Memorial Institute, Columbus, Ohio (New York: Gordon and Breach Science Publishers, 1968). 505p.
117. *Climate and Man* (Detroit, Mich.: Gale Research Co., 1977). 1248p.
Republication of data compiled by the U.S. Department of Agriculture, 1941 (House Document No. 27, 77th Congress, 1st Session). Illustrations, including charts, graphs, and maps; bibliographies and index.

Selected List of Handbooks and Tables

118. *The Climates of the States.* James A. Ruffner and Frank E. Bair (Detroit, Mich.: Gale Research Co., 1978), 1150p.
     Part 1: Climates of the States. Tables, graphs, maps and bibliographies, based on data originally published by the National Oceanic and Atmospheric Administration of the U.S. Department of Commerce.
     Part 2: A guide to Federal and State public services in climate and weather.
119. *Handbook of Astronautical Engineering.* Heinz H. Koelle (New York: McGraw-Hill, 1961).
120. *Handbook of Rockets and Guided Missiles.* 2nd edition. Norman John Bowman (Newton Square, Penn.: Perastadion Press, 1963). 1008p.
121. *Origins of NASA Names.* Helen T. Wells, Susan H. Whiteley, and Carrie E. Karegeannes (Washington, D.C.: U.S. National Aeronautics and Space Administration, 1976). 227p. NASA SP-4402.
122. *Space Materials Handbook.* 3rd edition. John B. Rittenhouse and John B. Singletary. Lockheed Aircraft Corporation. Lockheed Missiles and Space Company (Washington, D.C.: U.S. National Aeronautics and Space Administration, 1969). 734p. NASA SP-3051.
123. *Smithsonian Meteorological Tables.* 6th edition. Robert J. List (Washington, D.C.: The Smithsonian Institution, 1966). 527p. (Smithsonian Miscellaneous Collections, v. 114).
124. *Spacecraft and Missiles of the World, 1966.* James Barr and William E. Howard (New York: Harcourt, 1966). 104p.
125. *Sunrise and Sunset Tables* (Detroit, Mich.: Gale Research Co., 1977). 369p.
     A collection of the U.S. Naval Observatory's comprehensive tables providing the hour and minute of surnrise and sunset for every day of the year for the 369 key locations in the United States; valid for the 20th century.
126. *U.S. Standard Atmosphere, 1962.* U.S. Committee on Extension to the Standard Atmosphere. Prepared under the sponsorship of NASA, USAG, and the U.S. Weather Bureau (Washington, D.C.: U.S. Government Printing Office, 1962). 278p. Supplement, 1966, 188p. Supplement, 1976, 227p.
127. *FAA Statistical Handbook of Aviation.* U.S. Federal Aviation Agency (Washington, D.C.: U.S. Government Printing Office, 1944-). Annual.
128. *Weather Almanac.* 2nd edition. James A. Ruffner and Frank Bair (Detroit, Mich.: Gale Research Co., 1977). 728p.
     A reference guide to weather, climate and air quality of the United States and its key cities, comprising statistics, principles, and terminology. Includes maps, tables, charts, and a glossary.
129. *Weather Atlas of the United States* (Detroit, Mich.: Gale Research Co., 1975). 262p.

Republication of *U.S. Environmental Data Service* (Washington, D.C.: U.S. Government Printing Office, 1968). Originally published under the title *Climatic Atlas of the United States*. 271 maps and 15 statistical tables.

Civil Engineering

130. *Concrete Construction Handbook.* 2nd edition. Joseph J. Waddell (New York: McGraw-Hill, 1974).
131. *Handbook of Concrete Engineering.* Mark Fintel (New York: Van Nostrand Reinhold, 1974). 801p.
132. *Handbook of Dam Engineering.* Alfred R. Golze (New York: Van Nostrand Reinhold, 1977). 793p.
133. *Handbook of Highway Engineering.* Robert F. Baker (New York: Van Nostrand Reinhold, 1975). 894p.
134. *Handbook of Municipal Administration and Engineering.* William S. Foster (New York: McGraw-Hill, 1978).
135. *Standard Details for Fire-Resistive Building Construction.* Louis Przetak (New York: McGraw-Hill, 1977). 365p.
136. *Standard Handbook for Civil Engineers.* 2nd edition. Frederick S. Merritt (New York: McGraw-Hill, 1976).
137. *Transportation and Traffic Engineering Handbook.* 3rd edition. John E. Baerwald (Englewood Cliffs, N.J.: Prentice-Hall, for the Institute of Traffic Engineers, 1976). 1080p.

Mechanical Engineering

138. *ASHRAE Handbook and Product Directory* (New York: American Society of Heating, Refrigerating and Air-Conditioning Engineers, 1973–). Annual.
139. *ASME Steam Tables.* 3rd edition (New York: American Society of Mechanical Engineers, 1977). 329p.
    Tables and charts showing thermodynamic and transport properties of steam and water.
140. *Handbook of Industrial Metrology.* American Society of Tool and Manufacturing Engineers (Englewood Cliffs, N.J.: Prentice-Hall, 1967). 492p.
141. *Handbook of Industrial Engineering Management.* William G. Ireson and Eugene L. Grant. 2nd edition (Englewood Cliffs, N.J.: Prentice-Hall, 1971). 907p.
142. *Handbook of Industrial Noise Control.* L. L. Faulkner (New York: McGraw-Hill, 1976). 584p.
143. *Handbook of Industrial Pipework Engineering.* Ernest Holmes (New York: John Wiley, 1974). 570p.
144. *Handbook of Noise and Vibration Control.* R. H. Warring (Morden, Surrey, England: Trade and Technical Press, 1970). 617p.

Selected List of Handbooks and Tables 225

145. *Handbook of Precision Engineering.* A. Davidson (London: Macmillan; New York: McGraw-Hill, 1971/72-). In progress; to be completed in 12 volumes.
146. *Handbook of Thermodynamic Tables and Charts.* Kuzman Raznjevic (Washington, D.C.: Hemisphere Publishing Corporation, 1976). 329p.
147. *Handbook of Valves.* Philip Schweitzer (New York: Industrial Press, 1972). 180p.
148. *ISA Handbook of Control Valves.* 2nd edition. J. W. Hutchison (Pittsburgh, Penn.: Instrument Society of America, 1976). 533p.
149. *Industrial Engineering Handbook.* 3rd edition. Harold B. Maynard (New York: McGraw-Hill, 1971).
150. *Industrial Noise Control Handbook.* Paul N. Cheremisinoff and Peter P. Cheremisinoff (Ann Arbor, Mich.: Ann Arbor Science Publishers, 1977). 361p.
151. *Industrial Safety Handbook.* 2nd edition. William Handley (New York: McGraw-Hill, 1977). 400p.
152. *Lyons Encyclopedia of Valves.* Jerry L. Lyons and Carl L. Askland (New York: Van Nostrand Reinhold, 1975). 290p.
153. *Machinery's Handbook.* Eric Oberg and others (New York: Industrial Press, 1914-). (20th edition, 1975. 2482p.)
154. *Maintenance Engineering Handbook.* 3rd edition. Lindley R. Higgins and L. C. Morrow (New York: McGraw-Hill, 1977).
155. *Marks' Standard Handbook for Mechanical Engineers.* 8th edition. Theodore Baumister, editor-in-chief; Eugene A. Avallone, associate editor; and Theodore Baumister, III, associate editor (New York: McGraw-Hill, 1978).
156. *Piping Handbook.* 5th edition. Reno C. King (New York: McGraw-Hill, 1967).
157. *Production Handbook.* Gordon B. Carson, Harold A. Bolz, and Hewitt H. Young, editorial consultants (New York: Ronald Press, 1972).
158. *Production and Inventory Control Handbook.* J. H. Greene (New York: McGraw-Hill, 1970).
159. *SAE Handbook* (New York: Society of Automotive Engineers, 1926-). Revised almost every year.
160. *Shock and Vibration Handbook.* Cyril M. Harris and Charles E. Crede (New York: McGraw-Hill, 1961). 3 v.
    v. 1: *Basic Theory and Measurements*
    v. 2: *Data Analysis, Testing and Methods of Control*
    v. 3: *Engineering Design and Environmental Conditions*
161. *Tool and Manufacturing Engineers Handbook.* 3rd edition. Society of Manufacturing Engineers. Daniel B. Dallas (New York: McGraw-Hill, 1976).

Electrical Engineering and Electronics

162. *Buchsbaum's Complete Handbook of Practical Electronic Reference Data.* 2nd edition. Walter H. Buchsbaum (Englewood Cliffs, N.J.: Prentice-Hall, 1978). 645p.
163. *Directory of Electronic Circuits, with a Glossary of Terms.* Matthew Mandl (Englewood Cliffs, N.J.: Prentice-Hall, 1978). 321p.
164. *Electronic Components Handbook.* Thomas H. Jones (Reston, Va.: Reston Publishing Co., 1977). 391p.
165. *Electronic Engineer's Reference Book.* 4th edition. L. W. Turner (London: Butterworths, 1976).
166. *Electronics: A Handbook for Engineers and Scientists.* 2nd edition. G. H. Olsen (London: Butterworths, 1974). 482p.
167. *Guidebook of Electronic Circuits.* John Markus (New York: McGraw-Hill, 1974). 1067p.
168. *Handbook of Components for Electronics.* Charles A. Harper (New York: McGraw-Hill, 1977).
169. *Handbook of Electronic Charts, Graphs, and Tables.* John D. Lenk (Englewood Cliffs, N.J.: Prentice-Hall, 1970). 224p.
170. *Handbook of Electronic Circuit Design.* Robert G. Middleton (Reston, Va.: Reston Publishing Company, 1978). 276p.
171. *Handbook of Electronic Circuit Design Analysis.* Harry E. Thomas (Reston, Va.: Reston Publishing Company, 1972). 502p.
172. *Handbook of Electrical Systems Design Practices.* John E. Traister (Reston, Va.: Reston Publishing Company, 1978). 212p.
173. *Handbook of Electronic Systems Design.* Frank Weller (Reston, Va.: Reston Publishing Co., 1978). 288p.
174. *Handbook of Electronic Test Equipment.* John D. Lenk (Englewood Cliffs, N.J.: Prentice-Hall, 1971). 460p.
175. *Handbook of Plastics in Electronics.* Dean Grzegorczyk and George Feineman (Reston, Va.: Reston Publishing Co., 1974). 444p.
176. *Handbook of Semiconductor Electronics.* 3rd edition. Lloyd P. Hunter (New York: McGraw-Hill, 1970).
177. *Illustrated Handbook of Electronic Tables, Symbols, Measurements, and Values.* Raymond H. Ludwig (Englewood Cliffs, N.J.: Prentice-Hall, 1978). 352p.
178. *Newnes Radio and Electronics Engineers Pocket Book.* 15th edition. H. W. Moorshead (London: Butterworths, 1978). 176p.

Biology and Agriculture

179. *CRC Handbook of Microbiology.* Allen I. Lastein and Hubert A. Lechevalier (Cleveland, Ohio: Chemical Rubber Company Press, 1973-74). 4 v.

    v. 1: *Organismic Microbiology*, 1973.
    v. 2: *Microbial Composition*, 1973.

Selected List of Handbooks and Tables 227

      v. 3: *Microbial Products*, 1974.

      v. 4: *Microbial Metabolism, Genetics and Immunology*, 1974.

180. *Handbook of Engineering in Medicine and Biology.* David G. Fleming and Barry N. Feinberg (Cleveland, Ohio: Chemical Rubber Company Press, 1976). 421p.
A collection of 13 expository articles on various aspects of biomedical engineering. Index.

181. *Mycotoxic Fungi, Mycotoxins, Mycotoxicoses: An Encyclopedic Handbook.* T. D. Wyllie and L. G. Morehouse (New York: Marcel Dekker, 1977-78). 3 v.

      v. 1: *Mycotoxic Fungi and Chemistry of Mycotoxins.* 568p. 1977.

      v. 2: *Mycotoxicoses of Domestic and Laboratory Animals, Poultry, and Aquatic Invertebrates.* 600p. 1978.

      v. 3: *Mycotoxicoses of Man and Plants: Mycotoxin Control and Regulatory Practices.* 232p. 1978.

182. *Handbook of Agricultural Charts.* U.S. Department of Agriculture (Washington, D.C.: U.S. Government Printing Office). U.S. Department of Agriculture Handbook No. 359.

Environment

183. *Air Pollution Control: Guide Book to U.S. Regulations.* Martin S. Hertzendorf (Westport, Conn.: Technomic Publishing Co., 1973).
Describes Federal regulations and requirements concerning air pollution control measures.

184. *Architectural Handbook: Environmental Analysis, Architectural Programming, Design and Technology, and Construction.* A. M. Kemper (New York: Wiley-Interscience, 1979). 591p.

185. *Environmental Impact Handbook.* Robert W. Burchell and David Listokin (New Brunswick, N.J.: Rutgers University, Center for Urban Policy Research, 1973).
Describes federal regulations and requirements concerning air pollution control measures.

186. *CRC Handbook of Environmental Control.* Richard G. Bond and Conrad P. Straub (Cleveland, Ohio: Chemical Rubber Company Press, 1972).
Tables and charts, illustrations, and bibliographic references in four volumes:

      v. 1: *General*

      v. 2: *Solid Waste*

      v. 3: *Water Supply and Treatment*

      v. 4: *Waste Water Treatment and Disposal*

Each volume has an index

187. *Environmental Engineers' Handbook.* Bela G. Liptak (Radnor, Penn.: Chilton Book Co.). 3 v.
    v. 1:   *Water Pollution,* 1974
    v. 2:   *Air Pollution,* 1974
    v. 3:   *Land Pollution,* 1975
188. *Air Pollution Control: Guide Book to U.S. Regulations.* Martin S. Hertzendorf (Westport, Conn.: Technomic Publishing Co., 1973). Describes federal regulations and requirements concerning air pollution control measures.
189. *Environmental Law Handbook.* 3rd edition. J. Gordon Arbuckle and others (Bethesda, Md.: Government Institutes, 1975). 308p.
190. *Environmental Phosphorous Handbook.* Edward J. Griffith and others (New York: John Wiley, 1973). 718p.
191. *Environmental Science Handbook for Architects and Builders.* S. V. Szokolay (New York: John Wiley, 1980). 550p.
192. *Environmental Sources and Emissions Handbook.* Marshall Sittig (Park Ridge, N.J.: Noyes Data Corporation, 1975). 523p. (Environmental Technology Handbook, No. 2).
193. *Handbook of Air Pollution.* James P. Sheehy, William C. Achinger, and Regina A. Simon (Durham, N.C.: U.S. Department of Health, Education, and Welfare, Public Health Service. Bureau of Disease Prevention and Environmental Control, National Center for Air Pollution Control, 1968).
194. *Handbook of Air Pollution Control* (Westport, Conn.: Technomic Publishing Co., 1973).
195. *Handbook of Environmental Civil Engineering.* Robert G. Zilly (New York: Van Nostrand Reinhold, 1975). 1029p.
196. *Handbook of Environmental Data on Organic Chemicals.* Karel Verschueren (New York: Van Nostrand Reinhold, 1977). 659p.
197. *Handbook of Industrial Toxicology.* Edmond R. Plunkett (New York: Chemical Poblishing Co., 1976). 552p.
198. *Handbook for Monitoring Industrial Waste Water* (Washington, D.C.: Associated Water and Air Resources Engineers, Inc., 1973).
199. *Handbook of Pollution Control Management.* Herbert F. Lund (Englewood Cliffs, N.J.: Prentice-Hall, 1978). 448p.
200. *Industrial Pollution Control Handbook.* Herbert F. Lund (New York: McGraw-Hill, 1971).
201. *Industrial Waste Water Management Handbook.* H. S. Azad (New York: McGraw-Hill, 1976). 555p.
202. *Toxic Substances Sourcebook* (New York: Environment Information Center, 1979). 550p.
    Over 2000 digests of reports and articles on toxic substances; listing of NIOSH and EPA documents on toxic substances; articles on new Toxic Substances Control Act; and directory of databanks, periodicals, books, and films.

203. *Treatise on Environmental Law.* Frank P. Grad (New York: M. Bender, 1973-). Loose-leaf format.
204. *Water and Water Pollution Handbook.* Leonard L. Ciaccio (New York: Marcel Dekker, 1971-73). 4 v.
    - v. 1: *Environmental Systems*, 1971.
    - v. 2: *Chemical, Physical, Bacterial, Viral, Instrumental, and Bioassay Techniques*, 1971.
    - v. 3: *Measurement of BOD; Analysis of Water for Organic and Inorganic Pollutants*, 1972.
    - v. 4: *Instrumental and Automated Methods of Water Analysis*, 1973.
205. *Water Quality Criteria Data Book* (Washington, D.C.: U.S. Environmental Protection Agency, Water Quality Office, 1970-). A continuing series of volumes entitled Water Pollution Control Research Series.
206. *Water Treatment Handbook.* 5th edition (New York: John Wiley, Halsted Press, 1979). 1186p.

Energy

207. *Basic Petroleum Data Book, Petroleum Industry Statistics* (Washington, D.C.: American Petroleum Institute, 1977). Loose-leaf service.
208. *Energy Technology Handbook.* Douglas M. Considine (New York: McGraw-Hill, 1977). 1884p.
209. *Industrial Energy Conservation: A Handbook for Engineers and Managers.* David A. Reay (Oxford: Pergamon Press, 1977). 358p.
210. *Plant Engineers and Managers Guide to Energy Conservation.* Albert Thurman (New York: Van Nostrand, 1977). 365p.
211. *Reactor Handbook.* 2nd edition (New York: Wiley Interscience, 1960-). 4 v.
212. *The Solar Energy Handbook.* A practical engineering approach to the application of solar energy to the needs of humans and the environment, including a section on terrestrial cooling. 5th edition. Henry C. Landa. 1977. 173p.
213. *Solar Energy Technology Handbook.* W. C. Dickinson and P. N. Cheremisinoff (New York: Marcel Dekker, 1980).
    - Part A: Engineering Fundamentals. 912p. 1980.
    - Part B: Applications, Systems Design and Economics. 848p. 1980.

References

1. Walter Lippert, Gmelin's Handbook for Inorganic Chemistry, *Journal of Chemical Documentation*, 10(3):174-180 (August 1970).

2. Reiner Luckenbach, Der Beilstein, *Chemtech*, 9(10):612–621 (October 1979).
3. M. G. Mellon, Beilstein's Handbuch, *Encyclopedia of Library and Information Science* (New York: Marcel Dekker, 1969), v. 2, pp. 283–291.
4. D. L. Roth, Guide to the use of Beilstein, Gmelin and Landolt-Bornstein, *Herald of Library Science*, 11:325–333 (October 1972).
5. Oskar Weissbach, *The Beilstein Guide: A Manual for the Use of Beilstein's Handbuch der Organischen Chemie* (Berlin: Springer-Verlag, 1976).
6. Herman M. Weisman, Technical Librararians and the National Standard Reference Data System, *Special Libraries*, 63(2):69–76 (February 1972).

15
ENCYCLOPEDIAS

15.1 Specialized Encyclopedias

The aim of a general encyclopedia is to present in a concise and easily accessible form the whole corpus of knowledge. Encyclopedia articles usually contain an adequate amount of data for obtaining a general orientation on a wide variety of topics. Articles in general encyclopedias are sufficiently exhaustive to provide background information on specific topics not only to the general reader but also to the subject specialist seeking background information in an area peripheral to his specialization. A specialized encyclopedia differs from a general encyclopedia mainly in two respects: (1) The scope of a specialized encyclopedia is limited to a clearly defined branch of knowledge, such as physics, biochemistry, or chemical technology; and (2) the level of treatment is likely to be highly technical or scholarly. Thus, a specialized encyclopedia is designed primarily for the use of subject specialists. However, there are several general science encyclopedias addressed to nonspecialist readers. Some examples of this type will be discussed in the following section on single-volume encyclopedias.

Specialized encyclopedias contain articles written (and usually signed) by experts. The length of an article may range from a few lines to several pages. The articles on ketones is less than one page in *Van Nostrand's Scientific Encyclopedia*, 5th edition. On the same topic, the *Kirk-Othmer Encyclopedia of Chemical Technology*, 2nd edition, has a 69-page article with a bibliography of 289 items. The articles are usually illustrated, sometimes with color photographs, and may contain substantial amounts of data in the form of tables and charts. Each article is generally followed by a short bibliography. Besides topical articles, specialized encyclopedias may also contain biographical articles and articles describing organizations and historical events.

15.2 Single-Volume Encyclopedias

The *Harper Encyclopedia of Science*, revised edition (Harper & Row, 1967), is primarily aimed at the general reader. "We have insisted that articles be written with a minimum of jargon, with maximum clarity consistent with accuracy. The needs of the common reader—the student, the teacher, the non-specialist—have been our measuring rod" (from the editor's introduction). There are nearly 4000 articles written by about 450 scientists and engineers. The articles are illustrated, but do not have bibliographies. However, there is a long bibliography, arranged under subject headings, at the end. There is also an alphabetical index.

A somewhat more technical one-volume encyclopedia is *Van Nostrand's Scientific Encyclopedia* (5th edition, 1976). The first edition was published in 1938. The second, third, and fourth editions were brought out in 1947, 1958, and 1968, respectively. When work on the first edition began in 1935, the planners decided that "all topics were to be presented in depth sufficient for the needs—outside their fields of specialization—of scientists, engineers, medical doctors, mathematicians, students of those subjects, and anyone who wanted comprehensive information" (from preface to the fourth edition). This same pattern has been maintained in the subsequent editions. The fifth edition has over 7200 editorial entries, and 2450 diagrams, graphs, and photographs in 2370 pages. The articles are illustrated, but there is no index or bibliography.

Modern Science and Technology (Van Nostrand, 1965) is a collection of 81 articles taken from the journal *International Science and Technology*. The articles have been written "for professional scientists and engineers to read when they want to inform themselves about technical progress outside their own particular field of specialization" (from the preface). The writers have assumed that the reader would have a vigorous professional interest in technical matters and a familiarity with the fundamental principles of science and mathematics. The articles are illustrated, and each article has an abstract. There is an alphabetical index at the end.

The above three encyclopedias attempt to cover all branches of science and technology. The *Encyclopedia of Chemistry*, 3rd edition (New York: Van Nostrand Reinhold, 1973) and *Kingzett's Chemical Encyclopedia: A Digest of Chemistry and Industrial Applications*, 9th edition (Princeton, N.J.: Van Nostrand, 1966), are examples of single-volume encyclopedias limited in scope to one branch of science. The *Encyclopedia of Chemistry* has over 800 articles, with bibliographic references, prepared by 600 specialists to meet the needs of students, informed laymen, and experts in fields other than chemistry.

There are many one-volume encyclopedias that are highly specialized and technical in both content and presentation. The *Encyclopedia of X-Rays and*

Gamma Rays (New York: Reinhold, 1963), and *Chemical Process Industries*, 3rd edition (New York: McGraw-Hill, 1967) are examples of this type. The *Encyclopedia of Chemical Elements* (New York: Reinhold, 1968) contains concise articles by 104 contributors on the occurrence, properties, behavior, and uses of 103 chemical elements. In addition, there are 20 general articles on such topics as the periodic law and the periodic table, transuranium elements, isotopes, etc. The articles are 1400-1700 words long, and are accompanied by references to related literature for further reading. This encyclopedia has an alphabetical index.

15.3 Multivolume Encyclopedias

Multivolume encyclopedias are more likely to be meant for the subject specialist than the general reader. However, there is a very important multivolume encyclopedia meant for the nonspecialist; the *McGraw-Hill Encyclopedia of Science and Technology* (4th edition, 1977). In this encyclopedia, "each article is designed and written to be understandable to the nonspecialist.... Most articles, and at least the introductory parts of all of them, are within the comprehension of the interested high school student" (from the preface). The fourth edition has 14 volumes of text and an index volume; it contains about 7800 articles written by 2700 contributors. Most of the articles have illustrations (some have color plates), and are followed by a bibliography. The first three editions were brought out in 1960, 1966, and 1971, respectively. A *Readers' Guide* and a *Study Guide* have also been published to facilitate use of the *Encyclopedia*. The *McGraw-Hill Basic Bibliography of Science and Technology* is a companion publication designed to supplement the *Encyclopedia*. The entries in the *Bibliography* are arranged under the same subject headings as are the articles in the *Encyclopedia*. The following are some examples of multivolume encyclopedias in chemistry and chemical technology that are very technical in both content and style.

Chemical Technology: An Encyclopedic Treatment (New York: Barnes & Noble, 1968-), in eight volumes, describes the sources, manufacture, processing and use of materials, both natural and synthetic. An unusual feature of this encyclopedia is that its arrangement is systematic, not alphabetical, so as to facilitate collocation of related material. Each volume has an index. The text is illustrated, and each chapter has a short bibliography for further reading.

Encyclopedia of Chemical Reactions (New York: Reinhold, 1946-1959), eight volumes: Information on chemical reactions is arranged under the principal chemical elements and reactants. For each reaction, the conditions and reagents of the reaction, balanced equations, and the bibliographical details of the source document are given. Each volume has an index to reagents and an index to substances obtained.

Encyclopedia of Industrial Chemical Analysis (New York: John Wiley, 1966-1974), in 19 volumes "plans to give a comprehensive coverage of the methods and techniques used in industrial laboratories throughout the world for the analysis and evaluation of chemical products. Raw materials, intermediates, and finished products are included, and the treatment covers not only chemical analysis in the widest sense, but also evaluation of finished products for their intended functions" (from the preface). The first three volumes cover general techniques of industrial chemical analysis; the remaining volumes contain articles on the analysis of specific materials. The articles are illustrated and are followed by bibliographies.

Elsevier's Encyclopedia of Organic Chemistry (Amsterdam: Elsevier, 1940-1969), 20 volumes and several supplements: The articles are arranged according to a systematic (rather than an alphabetical) sequence. The articles contain literature references. Each volume has a subject index and a formula index.

Kirk-Othmer Encyclopedia of Chemical Technology, 2nd edition (New York: Interscience, 1963-1972), 24 volumes: The 22nd volume is a supplement, and the last volume is an index to the encyclopedia. The articles are exhaustive, and are aupported by illustrations, tables, charts, and extensive bibliographies. The article "Aluminum and aluminum alloys," for example, is 62 pages long. There are two exhaustive articles on patents: One on patents (practice and management), 31 pages, and another on patents (literature), 53 pages. The latter has a bibliography of 294 references. A third edition of this encyclopedia is in progress, and so far eleven volumes have been released. A combined index to the first four volumes of the third edition has also been published.

15.4 Encyclopedic Dictionaries

Like other forms of publications, encyclopedias also suffer from the problem of confusing titles. Some encyclopedias do not have the word "encyclopedia" in their titles, and are therefore difficult to recognize as such. For example, *Chemical Process Industries*, 3rd edition (New York: McGraw-Hill, 1967), and *Handbuch der Physik* (Berlin: Springer-Verlag, 1956-), 54 volumes, are encyclopedias though they are not so named. *Thorpe's Dictionary of Applied Chemistry*, 4th edition (Longmans, Green, 1937-1956), is really a 12-volume encyclopedia, the word "dictionary" in its title notwithstanding.

Some "encyclopedias" are really dictionaries or other types of reference works, despite the word "encyclopedia" in their titles. *The Concise Encyclopedia of Astronautics* (Chicago: Follett, 1968) and the *Encyclopedia of Biomedical Engineering Terms* (Reston, Va.: Reston Publishing Co., 1977), are, in fact, dictionaries. The *International Encyclopedia of Physical Chemistry and*

Chemical Physics (Oxford: Pergamon Press, 1960-), is the collective title of a series of monographs on different subjects.

The difference between a dictionary and an encyclopedia is fairly clear. A dictionary is basically a word list. It gives definitions and meanings of words in the same language as the words or in one or more different languages, with or without illustrations. The entries are usually short and unsigned. An encyclopedia contains longer and more informative articles usually signed by the author, and supported by illustrations and bibliographies. There are some publications which combine the features of a dictionary and an encyclopedia. The *Encyclopedic Dictionary of Physics* (Oxford: Pergamon Press, 1961-1971) is an example of this genre. This nine-volume publication contains short, signed articles; many articles have bibliographies. The last volume is a multilingual glossary in English, French, German, Japanese, Russian, and Spanish. Four supplementary volumes have so far been issued. The *Atomic Energy Desk Book* (New York: Reinhold, 1963), in three volumes, is another example of a publication that combines the features of both a dictionary and an encyclopedia.

15.5 Updating Encyclopedias

Information presented in encyclopedias rapidly becomes obsolete, especially in the fast-developing technical subjects and technologies. Updating an encyclopedia is an expensive and time-consuming process, and requires continuous effort. There are primarily two ways in which encyclopedias are updated: (1) Publication of a completely revised edition, at intervals of five or ten years, and (2) publication of supplements or "yearbooks" to update multivolume encyclopedias. The first method is usually followed for updating single-volume encyclopedias. The fifth edition of *Van Nostrand's Scientific Encyclopedia* (1976) has 300 more pages than the fourth edition, and is "essentially a new book, with less than approximately 20% of prior text and illustration taken from the earlier edition." The first edition was published in 1938, and since then, a new edition has been published approximately every ten years.

Some large (multivolume) encyclopedias issue supplements or "yearbooks" between successive editions. The *McGraw-Hill Encyclopedia of Science and Technology* was first published in 1960, and subsequent editions have been published at intervals of five or six years. Yearbooks have been issued to update the encyclopedia during the periods between two successive editions. The *Encyclopedia of Associations* (which is really a directory) is updated quarterly between editions. Part 3 of this work, called *New Associations and Projects*, is a quarterly interedition supplement to the first two parts entitled, respectively,

National Organizations of the United States and *Geographic and Executive Index.*

Selected List of Encyclopedias

General Science and Technology

1. *The Harper Encyclopedia of Science* (New York: Harper & Row, 1967). 1379p. Bibliography, pp. 1282-1297.
2. *McGraw-Hill Encyclopedia of Science and Technology.* 4th edition (New York: McGraw-Hill, 1977). 15 v. 11,650p.
    7800 articles by 2700 contributors. Illustrations, subject index, and topical index.
    Reader's Guide to the McGraw-Hill Encyclopedia of Science and Technology (New York: McGraw-Hill, 1977). 33p.
    Study Guide to the McGraw-Hill Encyclopedia of Science and Technology (New York: McGraw-Hill, 1977).
    The six study guides—biology, chemistry, earth and space sciences, health, physics, and science and technology in society—are especially useful for school and college students and science teachers.
3. *Modern Science and Technology.* Robert Colborn (Princeton, N.J.: Van Nostrand, 1965). 746p.
    A collection of articles originally published in the monthly journal *International Science and Technology.*
4. *The Realm of Science.* Stanley B. Brown and Barbara Brown (Louisville, Ky.: Touchstone Publishing Co., 1972). 21 v. 4140p.
    A collection of 324 readings from both monographic and periodical works, ranging in length from one-page excerpts to complete articles of several pages.

Part I:	*The Nature of Science*	(v. 1-4)
Part II:	*The Nature of Matter and Energy*	(v. 5-10)
Part III:	*The Nature of Space*	(v. 11-14)
Part IV:	*The Nature of the Earth: Environment*	(v. 15-16)
Part V:	*The Nature of Life*	(v. 17-20)
v. 21:	*Scienthesis: Science Concept Index and Cumulative Index*	

5. *Van Nostrand's Scientific Encyclopedia.* 5th edition (Princeton, N.J.: Van Nostrand, 1976). 2370p.
    Definitions and explanations of topics in earth and space sciences, energy technology, life sciences, materials sciences, mathematics and information sciences, physics, and chemistry. About 7200 editorial entries contributed by 190 authors; 2450 illustrations; 550 tables; 8000 cross references.

Mathematics and Statistics

6. *Encyclopedia of Mathematics and its Applications* (Reading, Mass.: Addison-Wesley, 1976-). In progress.

Selected List of Encyclopedias 237

   v. 1: *Integral Geometry and Geometric Probability.* Luis A. Santalo. 404 p.
   v. 2: *The Theory of Partitions.* George E. Andrews. 255p.
7. Mathematical Society of Japan. *Encyclopedic Dictionary of Mathematics.* Shokichi Iyanaga and Yukiosi Kawada (Cambridge, Mass.: The MIT Press, 1977). 2 v. 1750p.
8. *International Encyclopedia of Statistics.* William H. Kruskal and Judith M. Tanur (New York: Macmillan, 1978). 2 v.
9. *The Universal Encyclopedia of Mathematics* (New York: Simon and Schuster, 1964). 715p.

Computer Science

10. *Encyclopedia of Computer Science.* Anthony Ralston and C. L. Meek (New York: Petrocelli/Charter, 1976). 1522p.
11. *Encyclopedia of Computer Science and Technology.* Jack Belzer, Albert G. Holzman, and Allen Kent (New York: Marcel Dekker, 1975-). In progress.

Physics and Astronomy

12. *A Concise Encyclopedia of Astronomy.* A. Weigert and H. Zimmermann (New York: American Elsevier, 1968). 368p.
    Translated from German *ABC der Astronomie*. Approximately 1500 entries including biographical entries.
13. *A Concise Encyclopedia of Heat Transfer.* S. S. Kutateladze and V. M. Borishanskii. Translated by Henry Cohen (Oxford: Pergamon Press, 1966). 489p.
14. *A Dictionary of Applied Physics.* Sir Richard Glazebrook (New York: Peter Smith, 1950). 5 v.

   v. 1: *Mechanics, Engineering, Heat*
   v. 2: *Electricity*
   v. 3: *Meteorology, Metrology, Measuring Apparatus*
   v. 4: *Light, Sound, Radiology*
   v. 5: *Aeronautics, Metallurgy* General index.

15. *The Encyclopedia of Physics.* 2nd edition. Robert M. Besancon (New York: Van Nostrand Reinhold, 1974). 1067p.
    344 articles written at three levels: Articles on the main divisions of physics are meant for the general reader; those on subdivisions of physics are intended for readers with some background in science; and those on more specific topics are meant for readers with a background in physics and mathematics.
16. *The Encyclopedia of X-Rays and Gamma Rays.* George L. Clark (New York: Reinhood, 1963), 1149p.
17. *Encyclopedic Dictionary of Physics.* James Thewlis (Oxford: Pergamon Press, 1961-1967). 9v. plus supplements.

18. *Handbuch der Physik.* 2nd edition. S. Flugge (Berlin: Springer-Verlag, 1956-1972). 54 v.
    Articles are in English, French, or German. All volumes have title pages in English. Subject index in each volume.
19. *The New Space Encyclopedia: A guide to Astronomy and Space Exploration* (New York: Dutton, 1974). 326p.

Chemistry, Chemical Engineering, and Technology

20. *Chemical and Process Encyclopedia.* Douglas M. Considine (New York: McGraw-Hill, 1974). 1261p.
21. *Chemical Process Industries.* 3rd edition. R. Norris Shrive (New York: McGraw-Hill, 1967). 905p.
22. *Chemical Engineer's Handbook.* 5th edition. Robert H. Perry and C. H. Cyilton (New York: McGraw-Hill, 1973).
23. *Chemical Technology: An Encyclopedic Treatment* (New York: Barnes & Noble, 1968-). 8 v.

    v. 1: Air, water, inorganic chemicals, and nucleonics. 1968. 703p.

    v. 2: Nonmetallic ores, silicate industries, and solid minerals and fuels. 1971. 828p.

    v. 3: Metals and ores. 1970. 918p.

    v. 4: Petroleum and organic chemicals. 1972. 792p.

    v. 5: Natural organic materials and related synthetic products. 1972. 898p.

    v. 6: Wood, paper, textiles, plastics, and photographic materials. 1973. 686p.

    v. 7: Vegetable food products and luxuries.

    v. 8: Edible oils and fats and animal food products.
24. *The Encyclopedia of Biochemistry.* Roger J. Williams and Edwin M. Landsford (New York: Robert E. Krieger, 1977). 876p. (Reprint of 1967 Reinhold edition).
25. *Encyclopedia of the Chemical Elements.* Clifford A. Hampel (New York: Reinhold, 1968). 849p.
26. *Encyclopedia of Chemical Reactions.* C. A. Jacobson (New York: Reinhold, 1946-49). 8 v.
27. *Kirk-Othmer Encyclopedia of Chemical Technology.* 3rd edition (New York: Wiley-Interscience, 1978-). In progress. (24 volumes in 2nd edition).
28. *The Encyclopedia of Chemistry.* 3rd edition. Clifford A. Hampel and Gessner G. Hawley (New York: Van Nostrand Reinhold, 1973). 1198p. Over 800 articles, with bibliographic references, prepared by 600 specialists. Mainly for students, intelligent laymen, and experts in other fields of specialization.

Selected List of Encyclopedias

29. *Encyclopedia of Common Natural Ingredients Used in Food, Drugs and Cosmetics.* A. Y. Leung (New York: Wiley-Interscience, 1980). 368p.
30. *Encyclopedia of Food Technology.* Arnold H. Johnson and Martin S. Peterson (Westport, Conn.: AVI Publishing Co., 1974). 993p. (Encyclopedia of Food Technology and Food Science Series, v. 2).
31. *Encyclopedia of Industrial Chemical Analysis.* Foster D. Snell and C. L. Hilton (v. 1-7); Foster D. Snell and Leslie S. Ettre (v. 8-20). (New York: John Wiley, 1966-74). 20 v.
32. *Encyclopedia of Polymer Science and Technology.* H. F. Mark and others (New York: Wiley-Interscience, 1964-72). 16 v.
33. *Kingzett's Chemical Encyclopedia: A Digest of Chemistry and Its Industrial Applications.* 9th edition (Princeton, N.J.: Van Nostrand, 1966). 1092p.

Earth Sciences, Geology, and Metallurgy

34. *1001 Questions Answered About Earth Science.* Richard Maxwell Pearl (New York: Dodd, Mead, 1969). 327p.
35. *A Concise Encyclopedia of Metallurgy.* Arthur Douglas Merriman (New York: American Elsevier, 1965). 1178p.
36. *The Encyclopedia of Earth Sciences.* Rhodes W. Fairbridge (New York: Halsted Press, 1966-). Continuing series in progress.
    v. 1: *Encyclopedia of Oceanography.* 1966.
    v. 2: *The Encyclopedia of Atmospheric Sciences and Astrogeology.* 1967.
    v. 3: *The Encyclopedia of Geomorphology.* 1975.
    v. 8: *The Encyclopedia of World Regional Geology.* 1975.
37. *Encyclopedia of Minerals.* Willard Lincoln Roberts, George R. Rapp, Jr., and Julius Weber (New York: Van Nostrand Reinhold, 1974). 693p.
38. *Jane's Ocean Technology 1978.* Robert L. Trillo (New York: Franklin Watts, 1978). 820p.
39. *Watkin's Encyclopedia of the Steel Industry.* 13th edition (Pittsburgh, Penn.: Steel Publications, Inc., 1971).

Aviation and Aerospace

40. *Audel's Encyclopedia of Space Science* (New York: Theodore Audel & Co., 1964). 4 v.
41. *Concise Encyclopedia of Astronautics.* Thomas De Galiana (Chicago: Follett, 1968). 294p.
42. *Jane's All the World's Aircraft* (London: B. P. C. Publishing Co., 1909-). Annual.
43. *Jane's Fighting Ships* (London: B. P. C. Publishing Co., 1897-). Annual.
44. *Jane's Weapon Systems* (London: B. P. C. Publishing Co., 1969-). Annual.

45. *The McGraw-Hill Encyclopedia of Space* (New York: McGraw-Hill, 1968). 831p.
46. *Rocket Encyclopedia Illustrated.* John W. Herrick and Eric Burgess (Los Angeles, Calif.: Aero Publishers, 1959). 607p.

Automobile Engineering

47. *The Complete Encyclopedia of Motor Cars, 1885 to the Present.* 2nd edition. G. N. Georgano (New York: Dutton, 1973). 751p.

Biology

48. *Butterflies of the World.* H. L. Lewis (Chicago: Follett, 1973). 312p. A pictorial encyclopedia with full-color illustrations and descriptions of 5000 butterflies. 208 pages of color plates.
49. *Encyclopedia of Bioethics.* Warren T. Reich. Center for Bioethics, Kennedy Institute of Ethics, Georgetown University (New York: Macmillan, 1978). 4 v.
50. *The Encyclopedia of the Biological Sciences.* 2nd edition. Peter Gray (New York: Van Nostrand Reinhold, 1970). 1027p.
51. *The Encyclopedia of Plant Portraits.* Arthur George Lee Hellyer (London: W. H. and L. Collingridge; New York: Transatlantic Arts, 1953). 322p. Illustrations of cultivated plants, showing in alphabetical sequence portraits of hardy and half-hardy plants, trees and shrubs, orchids, ferns, hothouse and greenhouse plants, etc., with a short description of each (subtitle).
52. *The Illustrated Encyclopedia of the Animal Kingdom* Danbury, Conn.: Danbury Press, a division of Grolier Enterprises, Inc., 1972). 20 v. 2372p. Intended for use by elementary or secondary school students and lay persons not trained in science.
53. *Insects of the World.* Walter Linsenmaier (New York: McGraw-Hill, 1972). 392p. Translated from German by Leigh E. Chadwick.
54. *Marine Life: An Illustrated Encyclopedia of Invertebrates in the Sea.* J. D. George and J. George (New York: John Wiley, 1979). 288p. Describes the general morphology and ecology of each phylum with marine representatives, followed by the most recent classification of each phylum into classes, orders, and families. 1300 color photographs.
55. *Yearbook of Agriculture.* U.S. Department of Agriculture (Washington, D.C.: U.S. Government Printing Office, 1894-). Annual. Title varies. 1894-1919: *Yearbook of the United States Department of Agriculture.* 1920-1922: *Yearbook.* 1923-25: *Agriculture Yearbook.* Since 1936 each volume has a special title. Beginning with 1930, the yearbook is designated by the year in which it is published. Consequently, there is no issue bearing the date 1929.

Selected List of Encyclopedias

Environment

56. *The Encyclopedia of Geochemistry and Environmental Sciences.* Rhodes Whitmore Fairbridge (New York: Halsted Press, 1972). 1321p. (Encyclopedia of Earth Sciences Series, v. 4A).
57. *Encyclopedia of Urban Planning.* Arnold Whittick (New York: McGraw-Hill, 1974), 1218p.
58. *McGraw-Hill Encyclopedia of Environmental Science*, 2nd edition (New York: McGraw-Hill, 1980). 858p.
59. *Topics and Terms in Environmental Problems.* John R. Houlm (New York: John Wiley, 1977). 729p.
    239 main entries consisting of basic definitions and introductory essays on topics and terms emphasizing energy resources, chemicals, physical forces, air pollution, water pollution, wastes, and pesticides.
60. *The Water Encyclopedia: A Compendium of Useful Information on Water Resources.* David Keith Todd (Port Washington, N.Y.: Water Information Center, 1970). 559p.

Energy

61. *International Petroleum Encyclopedia.* John McCaslin (Tulsa, Okla.: Petroleum Publishing Co., 1976). 456p. Indexes.
62. *The Woodburners Encyclopedia: An Information Source of Theory, Practice and Equipment Related to Wood as Energy.* John W. Shelton and others (Waitsfield, Vt.: Vermont Crossroads Press, 1976). 155p.

16

REVIEW LITERATURE

16.1 Introduction

The rapid growth in the volume of scientific and technical literature in recent decades has been noted in an earlier chapter. This growth is accompanied by two interesting phenomena in scientific research: first, the phenomenon of excessive specialization, and second, the emergence of interdisciplinary team effort. In an attempt to cope with the ever-broadening horizons of science, scientists concentrate on narrower and narrower branches of science, so that they can absorb the information generated in their specialities and keep abreast of current developments. This narrow specialization gives rise to very specialized scientific papers and reports which fall within the field of interest and competence of only a few scientists. In view of the limited area of interest and competence of any one scientist, scientific research is becoming more and more a collaborative effort, in which the knowledge and skills of a team of specialists are brought to bear on the research process. Because of this collaborative nature of modern scientific research, the individual scientist specializing in a narrow area finds that his specialization is contingent upon and overlaps with the specializations of his fellow scientists. This interrelationship among specialities or subdisciplines makes it imperative that a specialist in any area keep abreast of advances in fields adjacent to his own. But the quantity of published scientific literature is so huge, and its quality so uneven, that no one individual can hope to be able to screen the deluge of literature and then read and digest the items pertinent to his own and related specialities. The review article is a corrective to this situation. In a recent study of review literature in the sciences, Julie A. Virgo highlighted the role of the review article thus [1]:

> Not all that is published is of high quality, nor does every contribution necessarily add something new to the state of knowledge in a particular field. Much that is reported is repetitious to varying extents and of low enough quality that it could be ignored without a loss of significant

16.1 Introduction

information. One possible solution, then, to the problem of digesting large masses of published material, could lie in the provision of reviews or summaries of papers dealing with the same topic, synthesizing the pertinent and useful facts from each and sifting out material which contributes nothing new to the subject. Hence one review article could replace a number of primary articles and the reader would need to read only one article in place of many, or he would be directed from the review article to those primary articles that appeared to be especially worthwhile, when viewed in perspective with other articles on the subject.

The concept of the review article is not new; it existed even in the age of the encyclopedic scholar, when the magnitude of published scientific literature was such that a dedicated scholar could read all or most of the papers published in a discipline. Many of the learned periodicals of the 17th and 18th centuries, such as the *Acta Eruditorum* and the *Journal des Scavans*, contained book reviews. These were typical of many of the early review journals, which were preponderantly book-reviewing media. With the appearance of the *Critical Review or Annals of Literature* in 1726, the review journal became an established entity. According to Kronick, who has traced the history of the review journal from 1665 onwards, the first scientific review journal of any extent or duration was the *Commentarii de Rebus in Scientia Naturali et Medicina Gestis*, which was published in Leipzig between 1752 and 1798 [2]. This contained reviews of scientific books, dissertations and journals.

Although review journals have existed from the beginning of scientific journalism, their importance to scientists has been accentuated in recent decades by the three current trends noted earlier: the proliferation of all forms of published and unpublished scientific and technical literature, the fragmentation of science into narrow subdisciplines, and the interdisciplinary nature of scientific research. All of these have made it difficult for the scientist and the scholar to remain aware of the developments in the wavefront of knowledge. In the Royal Society Scientific Information Conference in 1948, there was a great deal of discussion concerning the role and importance of review literature. The Conference made the following final recommendations [3]:

> The Conference has concluded that both critical and constructive reviews, written in particular fields, and reviews by specialists for other workers in science are of great importance, and therefore invites the Royal Society to bring the following recommendations to the notice of those concerned:
>
> Critical, general and specialist reviews should be made informative to nonspecialists by a general introduction and conclusion.
>
> Senior scientists should regard the provision of reviews as an important ancillary to the pursuit of knowledge.

The Washington Conference (1958) also emphasized the importance of review literature for promoting cross-fertilization and browsing in related fields, and for indicating directions for further research [4]. The Weinberg Report (1963) underscored the important role of scholarly reviews in easing the information crisis: "They serve the special needs of both the established workers in a field and the graduate student entering the field, as well as the general needs of the nonspecialist" [5]. The report also urged the government to promote technical reviews of report literature: "Because there are so many technical reports, and most of them are unrefereed and of uneven quality, reviews of technical reports can be particularly useful as discriminating guides to literature" [6].

The SATCOM Report also concluded that the functions performed by critical reviews and compilations, namely, digestion, consolidation, simplification, and repackaging for speciific categories of users, are essential for the effective utilization of scientific and technical information [7]. As a result of these influences, there is at the present time a substantial investment of effort and resources in the production of review literature. According to Woodward, well over 20,000 reviews are published annually at a cost estimated at between $100 and $200 million [8].

16.2 Review Authors and Review Preparation

Preparation of reviews is an intellectually challenging task. It involves assembling, digesting, and evaluating scattered primary documents and synthesizing their contents into coherent and concise packages. Authors of reviews should be scholars or experts in their own disciplines, and should also be willing to invest a great deal of creative thought and effort to preparing reviews. The qualifications of an ideal author of a review were enumerated by Cuadra in his introduction to the first volume of the *Annual Review of Information Science and Technology* [9]:

1. He must have a strong grasp of the basic issues in his field and he must be able to understand and express them in their historical perspective.
2. He must have an established habit of keeping informed by reading reports and published literature and by making effective use of his contacts in the "invisible college."
3. He must be able to write lucid, incisive prose and must be willing and able to make objective value judgment—in public—about the merit and implications of given lines of reported work, research, and experience.
4. He must have, in addition to his technical and literary talent, sufficient prestige in the field to invite the reader's respectful attention to his contribution.
5. He must be willing to do an immense amount of sifting, reading and evaluation on an extremely tight schedule.

16.2 Review Authors and Review Preparation

Preparation of scholarly reviews is a labor-intensive process. Once an author is committed to writing a review, he is faced with the immediate task of identifying and physically acquiring all the documents to be considered for the review. Indexing and abstracting services are only partially helpful because of the time-lag of six months to two years in their coverage of primary documents. The review author has to use a variety of additional sources including his own personal files, current issues of journals and their indexes, book review media, and nonformal channels.

Rowena Swanson has presented experiential data on the work activities, performance times, and costs involved in the preparation of an analytical review for the *Annual Review of Information Science and Technology (ARIST)* [10]. Preparation of a chapter "Design and Evaluation of Information Systems" for volume 10 of the *Annual Review of Information Science and Technology* (Washington, D.C.: American Society for Information Science, 1975, v. 10, pp. 43-101) consumed a total of 712 hours (including 114 hours of secretarial effort). The total personnel cost for this effort was estimated to be $7527.36 assuming an hourly rate of $12.00 for the reviewer's time. The author of a review chapter for the *Annual Review of Information Science and Technology* receives an honorarium of $300. Swanson estimates that the total cost for the preparation of an *ARIST*-type review would be between $17,000 and $20,000. She also contends that a reference budget of about $2000 to $2500 would be needed towards the cost of documents (or photocopies) to be acquired and read for the preparation of the review. Audrey Grosch who wrote a review entitled "Library Automation" for volume 11 of *ARIST* (1976, pp. 225-226), feels that Swanson's estimates of time needed for review preparation may be on the modest side [11].

Research scientists who have the ability to synthetsize volumes of primary literature and prepare reviews appear to be reluctant to undertake the writing of reviews. Both physicists and chemists tend to regard the writing of reviews as less rewarding than original research in terms of prestige [12,13]. Scientists who do engage in writing reviews view this activity with mixed feelings. The following two responses made by reviewers for the *Annual Review of Psychology* typify the attitude of scientists who were invited to write review articles [14]:

1. Digesting the material so that it could be presented on some conceptual basis was plain torture. I spent over 200 hours on that job all told. I wonder if 200 people spent even one hour reading it.
2. The *Annual Review* invitation is a miserable job to do. The rewards are 50 reprints, and some intangibles. The selection, however, is a suggestion by the Board of Editors (representing the scientific public) that one should know and be interested in a certain arena of our science.

Various incentives such as fellowships, sabbatical leave, and prizes for outstanding reviews have been suggested to make review writing more attractive to capable scientists. In an editorial in *Science*, Lewis Branscomb suggested that grant support be made available to scientists for writing scholarly reviews and data compilations [15].

Scientists have been exhorted from time to time to regard the writing of reviews as a worthwhile activity essential for the effective utilization of recorded knowledge. The Weinberg Panel of the President's Science Advisory Committee made explicit recommendations, placing the responsibility for preparing reviews squarely on scientists and engineers [16]:

> We shall cope with the information explosion, in the long run, only if some scientists and engineers are prepared to commit themselves deeply to the job of sifting, reviewing, and synthesizing information; i.e., to handling information with sophistication and meaning, not merely mechanically. Such scientists must create new science, not just shuffle documents: their activities of reviewing, writing books, criticizing, and synthesizing are as much a part of science as traditional research. We urge the technical community to accord such individual the esteem that matches the importance of their jobs and to reward them well for their efforts.... Review writing is a task worthy of the deepest minds, able to recast, critically analyze, synthesize, and illuminate large bodies of results. The relation of the reviewer to the existing but widely scattered bits of knowledge resembles the relation of the theorist to available pieces of experimental information. In order to emphasize the growing importance of the reviewer and also the growing difficulties that he faces, scientific and technical societies should reward his work with good pay and with the regard that has been reserved heretofore for the discoverer of experimental information. Those asked to write reviews or to give invited papers reviewing a subject should be selected by the scientific societies with the same care as are recipients of honors or of appointments to the staff of a university.

As noted earlier, the preparation of reviews is a time-consuming and tedious process. In terms of intellectual effort, it is no less challenging than original research. "After all, science consists in the creation of simplicity out of the complexity of nature, and it is scarcely less of a feat to create new simplicity out of the complexity of the literature" [17]. It is therefore necessary to prevent wasteful duplication of reviewing effort, and direct the energies of reviewers to areas where gaps exist. The Royal Society Scientific Information Conference suggested that the organization and production of reviews could best be linked with abstracting and indexing services and learned societies [18]. Secondary services could easily detect areas where reviews were needed, and the learned societies could recommend specialists with appropriate qualifications to prepare the needed reviews.

16.3 Functions of Reviews

The SATCOM Report also recommended that the responsibility for the stimulation of critical reviews should be borne by professional societies [19]. The Weinberg Panel contended that the specialized information centers, staffed by subject specialists, could provide a better organizational framework for identifying the need for reviews and coordinating their preparation [20]. Specialized information centers could also assist review authors by compiling bibliographies and making available copies of needed documents and translations. The Brain Information Service, a specialized information center of the National Institute of Neurological Diseases and Blindness, has proposed a procedure for comprehensive reviews and analysis of the literature of the neurological sciences [21].

16.3 Functions of Reviews

The universe of primary scientific literature consists of an ever-increasing mass of unrelated journal articles, reports, dissertations, conference papers, patents, and other forms of primary literature scattered across temporal, geographic, and linguistic dimensions. Each one of these documents is an isolated package constituting the basic building block of scientific literature. Much of it is of an ephemeral or tentative nature and is likely to be of temporary significance. For the effective utilization of scientific information, it is necessary that this unrelated mass of primary scientific literature be sifted, organized, and synthesized into a coherent and validated corpus of recorded scientific knowledge. Abstracting and indexing services, catalogs, and bibliographies provide the necessary organizational structure to the universe of primary documents, and so are essential for the identification and location of individual documents by cutting across linguistic and geographic barriers and variations of format and physical medium of the documents. The critical evaluation, sifting, synthesis, and integration of the organized primary literature are accomplished in reviews. Reviews are therefore secondary literature based on primary documents. Further compaction and repackaging of primary scientific literature for specific categories of uses takes place in textbooks, monographs, treatises, and encyclopedias and other derived publications. The individual research paper and other types of primary document then become archival records of facts, figures, and minutiae. Monographs, textbooks, encyclopedias, and similar derived publications constitute records of evaluated and integrated knowledge for transmission to students, scholars, and scientists for assimilation and utilization. Review publications constitute an essential intermediate link in this gradual evolutionary process of scientific and technical information. The three major functions of review articles are: (1) current awareness, (2) tutorial, and (3) bibliographical.

16.3.1 Current Awareness Function

Reviews enable scientists and scholars to maintain current awareness of present activity in their own and related fields of interest. This is necessary in order to avoid unintended duplication of research effort and to identify areas for further research. Harris and Katter found that the *Annual Review of Information Science and Technology* had suggested new research topics to 45% of its readers [22]. Besides researchers and specialists engaged in research and education, two other groups need reviews: (1) managers and administrators of scientific research who need to know what is being done, and where, in the fields with which they are concerned and in related fields, and (2) engineers and industrialists who are concerned with the practical application of current advances in science.

16.3.2 Tutorial Function

Reviews are useful to students in understanding new subjects, and to specialists and researchers in continuing education or in obtaining an overview of a subject outside their field of specialization before commencing a new project [23,24]. In an attempt to promote cross-fertilization of ideas among the subdisciplines of biochemistry, the editors of the *Annual Review of Biochemistry* have instituted a new section entitled "Perspectives and Summary" beginning with volume 45 (1976). In this section, which appears in each review chapter, the review authors "are invited to locate their particular district on the biochemical map for those customarily residing elsewhere, and to summarize in more general terms than those provided for the specialist those developments they feel are most noteworthy during the period under review." (From the preface to the *Annual Review of Biochemistry*, volume 45, 1976.)

16.3.3 Bibliographic Function

Reviews are almost invariably accompanied by comprehensive bibliographies of the primary documents reviewed, and are therefore useful as bibliographic tools in making retrospective literature searches. Review articles with 600 or 700 literature references are not uncommon. For example, the article "Analytical chemistry of the sulphur acids" in *Talanta*, 21:1-44 (1974), has 883 bibliographic references. In a paper presented at the Washington Conference (1958) Brunning [25] reported on the bibliographic function of reviews:

> A good review contains a good bibliography so that a bibliography of reviews is in large measure a bibliogtaphy of bibliographies and as such represents a key to a large volume of literature and is often a quicker means of obtaining a number of references on a specific subject than are the abstract journals.

16.3 Functions of Reviews

Earlier studies have indicated that reviews are used fairly extensively for current awareness and tutorial purposes and also as sources of bibliographic references. In general, scientists use reviews in a wider range of subjects than the primary literature they read. Eugene Garfield has reported that the average review article published in the *Accounts of Chemical Research* was cited five times as often as the average primary article in the *Journal of the American Chemical Society* [26]. However, the extent of use and the type of review used depend on the category of user. In a survey on the use of scientific literature, Bernal found that "an overwhelming proportion of scientists (76%) read and appreciate reviews and that in fact they must form a very important and increasing part of general background reading" [27]. The scientists surveyed by Bernal were employed in academic, industrial, and government research laboratories. Brunning's survey indicated that a majority of chemists read reviews in their own and related fields, but that they read reviews of a general scientific nature to a much lesser extent. It was also indicated that chemists engaged entirely in research require comprehensive reviews on specialized topics with extensive bibliographies, whereas those engaged in teaching ask for general reviews with key bibliographic references [28].

A survey of 3201 physicists and chemists carried out by the Advisory Council on Scientific Policy in England demonstrated the widespread use of reviews. Over 90% of the sample had read or consulted a review within the previous month; well over one-half rated reviews as the most useful source for current awareness—more useful than abstracts or conferences; and between 46% and 55% said they would like more reviews [29]. Another survey of 2702 mechanical engineers (who are traditionally far less 'papyrocentric' than pure scientists), showed that reviews are used consistently by engineers engaged in all types of activity (e.g., management, research, design, testing, sales, and production). Reviews are used far more frequently than abstracts and indexes by all except those engaged in research. Over one-half of the respondents expressed the need for more review articles [30].

An inquiry carried out by the secretary-general of Editerra (European Association of Earth Science Editors) revealed that students and scientists below 40 years of age use reviews primarily for a first orientation, to become better informed and to keep up to date about their own professional fields [31]. The study also showed that on an average research workers in commercial companies use reviews more frequently to keep abreast of advances made in fields adjacent to their own than do academic colleagues. As pointed out in the introduction to the 1966 volume of *Macromolecular Reviews*, ". . . the review article is becoming the primary (i.e., the principal) source of information to a large majority of scientists."

16.4 Characteristics of Reviews

Reviews are essentially a secondary form of scientific literature; they are not based on original research, but on other publications that contain primary information. The results of research reported in primary documents such as journal articles, conference papers, technical reports, and patents are sifted, evaluated, and synthesized in reviews. Woodward has identified six characteristics of review literature [32]. Listed in order of decreasing importance, these are:

1. No original research is reported in reviews.
2. Reviews appear in publications devoted to publishing reviews or in review sections of periodicals.
3. Reviews have a title stating that the item is a review.
4. The title or abstract of a review contains one or more of the following words: review, progress, survey, overview, advances.
5. Reviews contain a large number of references.
6. Reviews are indexed or flagged as reviews by secondary services.

Many of these characteristics are "external" features that enable one to determine whether or not a particular document is a review. Woodward used the number of references cited in documents as a means of identifying reviews. It was hypothesized that documents citing a large number of references were more likely to be review documents than those that cited fewer references. A study based on the citation data obtained from the Institute for Scientific Information led to the conclusion that to rely solely on the number of references to identify reviews would miss 10-17% of reviews at the 30-references level, and 15-25% of reviews at the 40-reference level [23].

In Brunning's survey of industrial chemists, the respondents quoted the following criteria as most desirable in a good review: (1) expert writer, (2) critical approach, (3) comprehensiveness, (4) clarity and balance, (5) good bibliography, and (6) synopsis, with tables where suitable [25].

There are three inherent characteristics that are generally applicable to most reviews. These are: (1) integration, (2) evaluation, and (3) compaction of primary literature. Unlike an annotated bibliography or an abstracting journal, a review article places each of the articles reviewed in a context, and it integrates them with the existing corpus of knowledge in a field. Some degree of evaluation is also an essential characteristic of a review. While some reviews (e.g., the critical review) are primarily and explicitly evaluative in nature, others (e.g., a descriptive review) only indirectly involve some degree of evaluation; in these there is an implicit evaluation hidden in the very process of selecting or rejecting items for the review.

16.5 Types and Sources of Reviews

Compaction of primary literature is another important feature of a review. The degree of compaction can be estimated by determining (1) the average number of references cited per page of text in the review, or (2) the ratio of the number of pages of primary documents reviewed to the number of pages in the review document. The degree of compaction varies from field to field, and within a given field between review articles and long treatises. Herring found that review articles in biology average five or more references per page; those in chemistry, four per page; and those in physics, 2 or 2.5 per page [33].

In another study, Cottrell estimated the degree of compaction achieved in the state-of-the-art reports of the Nuclear Safety Information Center by dividing the total number of pages in the primary literature reviewed by the number of pages in the review report. The "compression ratio" in the case of five reports was found to vary from 50 to 216 [34]. Density of references and compression ratios such as those suggested by Herring and Cottrell are only approximate indicators of the degree of compaction achieved in reviews. Cottrell's note of caution is appropriate [34]:

> It must be noted that compression of pages of information may or may not be synonymous with compression of information, and furthermore the original information pages being compressed will, in many cases, contain information relevant to other subjects, and thus are not involved in the actual compression of information on the subject in question.

"Terse literatures" and "ultraterse literatures" suggested by Charles Bernier are a relatively new type of secondary literature designed to achieve a very high degree of compaction [35-39]. However, undue compaction makes the review incomprehensible except to the specialist.

16.5 Types and Sources of Reviews

Review literature may be classified according to its length, function, expected readership, and other characteristics. A review may be a one-time or occasional publication, or a periodical review published at regular or irregular intervals. Based on intended readership, reviews may be written for subject specialists, students, or general readers. Reviews range from a short review article published in a journal to a review monograph or a multivolume macroreview in which the chapters or volumes are written by different authors. In 1970, the secretary-general of Editerra (European Association of Earth Science Editors) surveyed 253 natural scientists on their usage habits and preferences regarding reviews. For 36% of the respondents, the length of a specialized review did not matter very much, if the quality was good; 18% had a preference for articles between 6000 and 9000 words; 15% for articles of 3000 to 6000 words; 13% for articles of 9000 to 12,000 words, and 6% for articles of 12,000 to 18,000 words. In the life and earth sciences, specialized reviews may be somewhat longer than those in the chemical and physical sciences. Those who had an opinion on general reviews answered that these could run up to twice the length of specialized reviews [31].

Woodward has provided the following classification of reviews [23]:

1. Critical, evaluative, interpretive, speculative
2. State-of-the-art
3. Historical, biographical
4. Tutorial
5. Technical, application
6. Article and book reviews, comments

Ignoring finer distinctions, reviews may be broadly categorized into two classes: (1) critical, and (2) descriptive. Preparation of a critical review involves a careful and impartial examination of all significant publications in a specific area, with critical evaluation as needed in view of advances in the area [40]. The critical review is aimed at the specialist to satisfy his current awareness need. Descriptive reviews are predominantly expository rather than evaluative in approach, and they are ideally suited for the tutorial function.

Annual reviews summarize recent developments in a field. The literature published during a calendar year is assembled and reviewed, often with critical comments by the review author, and the review is published in the following year. These are typically entitled *Annual Review of_____*, *Progress in _____*, *Advances in _____*, etc. State-of-the-art reviews summarize the current state of practice or thinking in a given field, and are usually nonevaluative. "Literature surveys" found in dissertations are similar to state-of-the-art reviews; these are interpretive rather than critical in approach.

Reviews of books and articles are mainly intended to serve as book selection aids; they are also useful as current awareness aids.

Reviews may be found in a variety of sources: review serials, primary journals, conference proceedings, monographs, technical reports, dissertations, and secondary services.

16.5.1 Review Serials

These include serial publications exclusively devoted to the publication of review articles, ranging from general review journals such as *Scientific American* to annual reviews published in book format. *Annual Reports on the Progress of Chemistry, Russian Chemical Reviews, Progress in Industrial Microbiology, Essays in Toxicology, Physics Reports,* and *Comments on Atomic and Molecular Physics* are examples of review serials. The reviews in these serials are usually written by experts and are aimed at specialists.

16.5 Types and Sources of Reviews

The Chemical Society, London, conducted a survey of the subscribers to its *Annual Reports on the Progress of Chemistry* in 1966. The average subscriber claimed to read only 30% of the volume, and admitted to ignoring 30%. In order to improve the usefulness of the *Annual Reports*, and also to cope with the increasing volume of primary literature to be reviewed, the Chemical Society started publishing the *Annual Reports* in two volumes, each separately priced, starting from volume 64 (1967): Section A, General, Physical and Inorganic Chemistry; and Section B, Organic Chemistry [41].

Faced with the expanding scope of biochemistry and a mounting volume of literature in molecular biology and biophysics, the publishers of the *Annual Review of Biochemistry* (1932-) were forced to take a similar step in 1966: "This year our *Annual Review* has shown behavior akin to a primary biological phenomenon: It has undergone binary fission" (from the Preface to volume 35, 1966). In order to preserve the integrity of the review, each volume is provided with author and subject indexes covering both volumes. These two examples illustrate the current trend in the growth of review serials.

16.5.2 Primary Journals

A large number of reviews are published in primary journals along with papers reporting original research. Woodward estimated the output of all reviews in primary journals to be between 10,200 and 11,300 items during 1972; this represents between 1.02 and 1.13% of all articles published in primary journals [23]. Reviews published in primary journals may be unsolicited articles submitted by authors, or invited papers prepared by specialists at the request of the journal editorial board. *Talanta* frequently published invited review articles. "Talanta Reviews" are also sold as separates. The April issue of *Analytical Chemistry* is a special issue dedicated to review papers. Publication of reviews along with original research papers in primary journals is viewed by scientists with mixed feelings. An overwhelming majority of chemists surveyed by Brunning were in favor of confining reviews to review journals and against having review articles interspersed among reports of original work. This objection did not apply to general scientific journals such as *Nature* and *Science* which report preliminary communications and current research [25].

16.5.3 Conference Papers and Proceedings

Although the majority of papers presented at conferences report the results of research, very often good review papers are also presented at conferences. Conference organizers sometimes solicit review papers to be presented as an introduction to each session. Conference proceedings are occasionally published as monographs or as review serials of the *"Advances in _____"* type. The plenary lectures presented at the Fourth Polish Conference on

Analytical Chemistry, Warsaw, August 1974, were published in *Pure and Applied Chemistry* (volume 44, No. 3), and supplied to subscribers to this journal as part of their subscription. The same material was again published as a monograph entitled *Analytical Chemistry* (Oxford: Pergamon Press, 1976). The publication practices and bibliographic control of conference papers and proceedings have been discussed in an earlier chapter. A major criticism of conference reports is that they often appear a year or more after the meetings have been held, and that they are published without adequate indexing or editing [25].

16.5.4 Review Monographs

These range from single-volume, advanced-level textbooks to multivolume scholarly treatises such as the *Treatise on Analytical Chemistry* edited by I. M. Kolthoff and Philip J. Elving (New York: Interscience, 1959-); over 30 volumes have been published so far. The chapters are written by various authors. The Introduction to this treatise states:

> The aims and objectives of this treatise are to present a concise, critical, comprehensive, and systematic, but not exhaustive, treatment of all aspects of classical and modern analytical chemistry. The treatise is designed to be a valuable source of information to all analytical chemists, to stimulate fundamental research in pure and applied analytical chemistry, and to illustrate the close relationship between academic and industrial analytical chemistry. . . . The treatise as a whole is intended to be a unified, critical, and stimulating treatment of the theory of analytical chemistry, of our knowledge of analytically useful properties, of the theoretical and practical fundamentals of the techniques for their measurement, and of the ways in which they are applied to solving specific analytical problems.

Another typical multivolume review treatise is Rodd's *Chemistry of Carbon Compounds: A Modern Comprehensive Treatise*, 2nd edition (Amsterdam: Elsevier Publishing Co., 1964-); over 22 volumes have so far been published.

A large proportion of reviews in pure physics in 1966 were found to be treatises. Also, treatises were found to have been cited most often in a sample of papers in the *Physical Review* [13].

16.5.5 Technical Reports

A small number of critical and state-of-the-art reviews are issued in the form of technical reports. A sample survey of government technical reports released in 1972 indicated that approximately 450 were reviews [23]. Most technical reports contain a well-documented introduction summarizing the previous developments as a preamble to the investigation described in the report. In a multivolume technical report, the first volume is likely to be a literature survey with bibliography.

16.5.6 Secondary Services

Some secondary services such as *Applied Mechanics Reviews, Computing Reviews*, and *Nutrition Abstracts and Reviews* include commissioned review articles from time to time. These are usually state-of-the-art reviews, but their number is rather small. It should be mentioned that these are really abstracting journals, despite the word "reviews" in their title.

16.6 Bibliographic Control of Reviews

Woodward's survey of review literature indicated an estimated output of 22,000 reviews in 1972. Of these, 4500 were found in review serials, 10,500 in primary journals, and 4000 in conference proceedings. About 2500 were books and 500 were technical reports [23]. Another survey, in the fields of biology and medicine, showed that less than one-quarter appeared in review publications, and over two-thirds were found in primary journals [42]. The identification and location of reviews are difficult in view of their dispersion in diverse sources and their uneven coverage in secondary services. Some abstracting services just mention reviews but do not abstract them. A further handicap is that it is not always easy to recognize reviews; out of 8601 reviews in the field of chemistry, only two contained the word "review" in the title [42]. The following three directories are very useful in locating review serials:

1. *Irregular Serials and Annuals. An International Directory.* 5th edition (New York: R. R. Bowker, 1978).
 Published biennially since 1972, and updated between editions through *Ulrich's Quarterly.*
2. Emanuel B. Ocran. *Scientific and Technical Series: A Select Bibliography* (Metuchen, N.J.: Scarecrow Press, 1973).
3. A. M. Woodward. *Directory of Review Serials in Science and Technology, 1970-1973: A Guide to Regular or Quasi-Regular Publications Containing Critical, State-of-the-Art and Literature Reviews* (London, ASLIB, 1974).

Other useful guides to review publications are: *List of Annual Reviews of Progress in Science and Technology*, 2nd edition (Paris: UNESCO, 1969), and *KWIC Index to Some of the Review Publications in the English Language*, published in 1966 by the National Lending Library for Science and Technology (now, the British Library Lending Division), Boston Spa, England.

Individual review articles and review monographs are covered by secondary services such as *Biological Abstracts/RRM, Chemical Abstracts, Computing Reviews*, and *Physics Abstracts*. It is estimated that 6% of the entries in *Chemical Abstracts* are for review articles [42]. *Science Citation Index* covers reviews and indicates them by the letter R. Technical reports are abstracted and indexed in *Government Reports Announcements and Index* (National Technical Information Service) and *Scientific and Technical Aerospace Reports* (National Aeronautics and Space Administration).

Many review serials publish cumulative indexes separately in order to facilitate retrospective searching for reviews. The cumulative index to the first ten volumes of the *Annual Review of Information Science and Technology* and the cumulative index to *Scientific American* for the years 1948-1978 are examples of separately published cumulative indexes. As an alternative to separately published indexes, it is a more common practice for review serials to include cumulative indexes in the volumes themselves. *Reports on Progress in Physics* (London: The Physical Society, 1938-) published cumulative subject and author indexes to the first 15 volumes in Volume 15 (1952). Since then, cumulative subject and author indexes have been published every five years. Cumulative subject and author indexes to the first 60 volumes of *Chemical Reviews* (American Chemical Society, 1924-) were published in Volume 60 (1960); cumulative indexes to Volumes 61 to 70 were included in Volume 70 (1970). Each volume of *Advances in Enzymology* contains author and subject indexes to the volume as well as cumulated author and subject indexes to all the volumes beginning with the first volume. Each volume of the *Annual Review of Biochemistry* contains cumulative indexes of contributing authors and chapter titles for the current volume and the preceding four volumes; this is in addition to the regular author and subject indexes to each volume of the series.

The need for specialized indexes to review literature has been recognized recently. The following are important examples of indexes dedicated to review literature:

Norman Kharasch and others. *Index to Reviews, Symposia Volumes and Monographs in Organic Chemistry* (Oxford: Pergamon Press, 1962-). Covers review papers published in English, French, and German from 1940 onwards.

Monthly Bibliography of Medical Reviews (National Library of Medicine, 1968-).
This is a continuation of the *Bibliography of Medical Reviews* (1955-1967).

Index to Scientific Reviews (Philadelphia, Penn.: Institute for Scientific Information, 1974-).
This is a semiannual index to over 28,000 review articles and state-of-the-art reviews in various scientific disciplines; cumulated annually..

Selected List of Review Serials

Mathematics and Statistics

1. *Advances in Applied Probability* (Sheffield, England: University of Sheffield, 1969-). Annual.
2. *Advances in Mathematical Systems Theory* (University Park, Penn.: The Pennsylvania State University Press, 1969). 174p.
3. *Advances in Mathematics* (New York: Academic Press, 1965-). Monthly.
4. *Advances in Probability and Related Topics* (New York: Marcel Dekker, 1970-). Irregular.
5. *Lecture Notes in Mathematics* (Berlin: Springer-Verlag, 1964-). Irregular. This series contains preliminary drafts of original papers, and monographs, lectures, seminar work-outs, and reports of meetings devoted to a single topic of exceptional interest. Papers in English, German, or French are published.
6. *Progress in Mathematics* (Translation of *Itogi Nauki, Seriya Matematika*) (New York: Plenum Press, 1968-). Annual.
7. *Progress in Statistics.* European Meeting of Statisticians, Budapest, Hungary, 1972 (Amsterdam: North-Holland Publishing Co., 1974). 2 v., 912p.
8. *SIAM Review* (Philadelphia, Penn.: Society for Industrial and Applied Mathematics, 1959-). Quarterly.

Computer Science

9. *Advances in Computers* (New York: Academic Press, 1960-). Annual.
10. *Annual Review in Automatic Programming* (Oxford: Pergamon Press, 1960-). Annual.
11. *Computing Surveys* (New York: Association for Computing Machinery, 1969—). Quarterly.

Physics and Astronomy

12. *Advances in Astronomy and Astrophysics* (New York: Academic Press, 1962-). Annual.
13. *Advances in Atomic and Molecular Physics* (New York: Academic Press, 1964-). Annual.
14. *Advances in Magnetic Resonance* (New York: Academic Press, 1965-). Annual.
15. *Advances in Nuclear Physics* (New York: Plenum Publishing Corporation, 1968-). Annual.
16. *Advances in Physics* (London: Taylor and Francis, Ltd., 1952-). Bimonthly.
17. *Advances in Theoretical Physics* (New York: Academic Press, 1965-). Annual.
18. *Annual Review of Astronomy and Astrophysics* (Palo Alto, Calif.: Annual Reviews, Inc., 1963-). Annual.

19. *Annual Review of Nuclear and Particle Science* (Palo Alto, Calif.: Annual Reviews, Inc., 1951-).
20. *CRC Critical Reviews in Solid State and Material Sciences* (Cleveland, Ohio.: Chemical Rubber Company Press, 1970-). Quarterly.
    Formerly, *CRC Critical Reviews in Solid State Sciences*.
21. *Physical Review* (Lancaster, Penn.: American Institute of Physics, 1970-).
    A. General Physics, monthly.
    B. Solid State, semimonthly.
    C. Nuclear Physics, monthly.
    D. Particles and Fields, semimonthly.

Chemistry, Chemical Engineering, and Technology

22. *Advances in Activation Analysis* (New York: Academic Press, 1969-). Irregular.
23. *Advances in Analytical Chemistry and Instrumentation* (New York: John Wiley, 1960-). Irregular.
24. *Advances in Carbohydrate and Biochemistry* (New York: Academic Press, 1945-). Annual.
25. *Advances in Chemical Engineering* (New York: Academic Press, 1956-). Irregular.
26. *Advances in Chemistry Series* (Washington, D.C.: American Chemical Society, 1950-). A continuing series of books.
27. *Advances in Chromatography* (New York: Marcel Dekker, 1966-). Irregular.
28. *Advances in Electrochemistry and Electrochemical Engineering* (New York: Wiley Interscience, 1961-). Irregular.
29. *Advances in Free Radical Chemistry* (New York: Academic Press, 1966-). Irregular.
30. *Advances in Gas Chromatography* (Niles, Ill.: Preston Technical Abstracts, 1965-). Biennial.
31. *Advances in Heterocyclic Chemistry* (New York: Academic Press, 1963-). Irregular.
32. *Advances in High Temperature Chemistry* (New York: Academic Press, 1967-). Irregular.
33. *Advances in Inorganic Chemistry and Radiochemistry* (New York: Academic Press, 1959-). Irregular.
34. *Advances in Organic Chemistry* (New York: Wiley Interscience, 1960-). Irregular.
35. *Advances in Organometallic Chemistry* (New York: Academic Press, 1964-). Irregular.
36. *Advances in Polymer Science* (New York: Springer-Verlag, 1958-). Annual.

Selected List of Review Serials 259

37. *A.I.Ch.E. Symposium Series* (New York: American Institute of Chemical Engineering, 1951-). Irregular.
38. *Annual Reports on NMR Spectroscopy* (New York: Academic Press, 1968-). Irregular.
39. *Annual Report on the Progress of Chemistry* (London: The Chemical Society, 1904-).
40. *Annual Review of Biochemistry* (Palo Alto, Calif.: Annual Reviews, Inc., 1932-).
41. *Annual Review of Physical Chemistry* (Palo Alto, Calif.: Annual Reviews, Inc., 1950-).
42. *Catalysis Reviews: Science and Engineering* (New York: Marcel Dekker, 1967-). Semiannual.
43. *Chemistry and Physics of Carbon: A Series of Advances* (New York: Marcel Dekker, 1966-). Irregular (v. 15 in 1979).
44. *CRC Critical Reviews in Analytical Chemistry* (Cleveland, Ohio: Chemical Rubber Company Press, 1970-). Quarterly.
45. *CRC Critical Reviews in Macromolecular Sciences* (Cleveland, Ohio: Chemical Rubber Company Press, 1972-). Quarterly.
46. *Electroanalytical Chemistry: A Series of Advances* (New York: Marcel Dekker, 1966-). Irregular (v. 11 in 1979).
47. *Fluorine Chemistry Reviews* (New York: Marcel Dekker, 1967-). Irregular (v. 8 in 1977).
48. *Ion Exchange and Solvent Extraction: A Series of Advances* (New York: Marcel Dekker, 1966-). Irregular (v. 7 in 1977).
49. *Progress in Boron Chemistry* (Oxford: Pergamon Press, 1964-). Quarterly.
50. *Progress in Reaction Kinetics* (Oxford: Pergamon Press, 1961-). Quarterly.
51. *Russian Chemical Reviews* (London: Chemical Society, 1960-). Monthly. English translation of the Russian review journal *Uspekhi Khimii.*
52. *Transition Metal Chemistry: A Series of Advances* (New York: Marcel Dekker, 1966-). Irregular (v. 7 in 1972).

Earth Sciences, Geology, and Metallurgy

53. *Advances in Fisheries and Oceanography* (Tokyo: Japanese Society of Fisheries and Oceanography, 1966-). Irregular.
54. *Advances in Geophysics* (New York: Academic Press, 1952-). Irregular.
55. *Advances in Hydroscience* (New York: Academic Press, 1964-). Irregular.
56. *Annual Review of Earth and Planetary Sciences* (Palo Alto, Calif.: Annual Reviews, Inc., 1973-).
57. *Annual Review of Material Science* (Palo Alto, Calif.: Annual Reviews, Inc., 1971-).
58. *Annual Summary of Information on Natural Disasters* (Paris: UNESCO, 1966-). In English and French.

59. *Developments in Geotechnical Engineering* (Amsterdam: Elsevier Scientific Publishing Co., 1972-). Irregular.
60. *Developments in Geotectonics* (Amsterdam: Elsevier Scientific Publishing Co., 1965-). Irregular.
61. *Developments in Solid Earth Physics* (Amsterdam: Elsevier Scientific Publishing Co., 1964-). Irregular.
62. *Earth Science Reviews* (Amsterdam: Elsevier Scientific Publishing Co., 1966-). Quarterly.
"The international geological journal bridging the gap between research articles and textbooks." (Subtitle).
63. *International Hydrographic Review* (Monte Carlo, Monaco: International Hydrographic Organization, 1923-). Semiannual.
64. *International Geology Review* (Washington, D.C.: American Geological Institute, 1959-). Monthly.
65. *International Metals Reviews* (London: Metals Society; Metals Park, Ohio: American Society for Metals, 1956-). Quarterly.
Formerly: *International Metallurgical Reviews*.
66. *Methods in Geochemistry and Geophysics* (Amsterdam: Elsevier Scientific Publishing Co., 1964-). Irregular.
67. *Progress in Extractive Metallurgy* (New York: Gordon and Breach Science Publishers, 1973-). Irregular.
68. *Progress in Geophysics* (Hyderbad, India: National Geophysical Research Institute, 1966-). Annual.
69. *Progress in Oceanography* (Oxford: Pergamon Press, 1963-). Quarterly.
70. *Progress in Powder Metallurgy. Powder Metallurgy Technical Conference* (Princeton, N.J.: American Powder Metallurgy Institute, 1947-). Annual.
71. *Reviews of Geophysics and Space Physics* (Washington, D.C.: American Geophysical Union, 1963-). Quarterly.
Formerly: *Reviews of Geophysics*.
72. *Reviews on High Temperature Materials. A Quarterly Review of Progress in High Temperature Materials* (Tel Aviv, Israel: Freund Publishing House, 1971-). Quarterly.

Aviation and Aerospace

73. *Advances in Aeronautical Sciences. Proceedings of the International Council of the Astronautical Sciences* (Washington, D.C.: Spartan Books, 1958-). Irregular.
74. *Advances in the Astronautical Sciences* (San Diego, Calif.: American Astronautical Society, 1957-). Irregular.
75. *Advances in Satellite Meteorology* (New York: John Wiley, 1973-). Irregular.
76. *Advances in Space Science and Technology* (New York: Academic Press, 1959-).
77. *Aeronautical Engineering Review* (Easton, Penn.: Institute of the Aeronautical Sciences, 1942-). Monthly.

Selected List of Review Serials

Supersedes the Aeronautical Review Section of the *Journal of Aeronautical Sciences* (later called *Journal of the Aerospace Sciences*).
78. *Air University Review* (Alabama: Maxwell Air Force Base, 1947–). Bimonthly.
Formerly: *Air University Quarterly Review.*
79. *Progress in Aeronautical Sciences* (Oxford: Pergamon Press, 1961–). Irregular.
80. *Progress in Astronautics and Aeronautics Series* (New York: American Institute of Aeronautics and Astronautics, 1968–). Irregular.
v. 1–8: *Progress in Astronautics and Rocketry* (v. 1–23 published by Academic Press).

Mechanical Engineering

81. *Advances in Applied Mechanics* (New York: Academic Press, 1948–). Irregular.
82. *Advances in Automobile Engineering* (Oxford: Pergamon Press, 1963–). Irregular.
83. *Advances in Heat Transfer* (New York: Academic Press, 1964–). Irregular.
84. *Advances in Machine Tool Design and Research* (Oxford: Pergamon Press, 1964–). Irregular.
85. *Annual Review of Fluid Mechanics* (Palo Alto, Calif.: Annual Reviews, Inc., 1969–).
86. *Progress in Control Engineering* (New York: Academic Press, 1962–). Irregular.

Electrical Engineering and Electronics

87. *Advances in Electronics and Electron Physics* (New York: Academic Press, 1948–). Irregular.
88. *Advances in Microwaves* (New York: Academic Press, 1966–). Irregular.
89. *Advances in Quantum Electronics* (New York: Academic Press, 1970–). Annual.
90. *Advances in Radio Research* (New York: Academic Press, 1964–). Irregular.
91. *Electronics Express* (New York: International Physics Index, 1958–). Monthly.
"Comprehensive digest of current Russian literature dealing with electronics topics." (Subtitle).
92. *IEE Reviews* (London: Institution of Electrical Engineers, 1970–). Annual. Special issue of *IEE Proceedings.*
93. *Progress in Quantum Electronics* (Oxford: Pergamon Press). Quarterly.
94. *Progress in Semiconductor Science and Technology* (Oxford: Pergamon Press, 1975–). Quarterly.

Biology

95. *Advances in Applied Microbiology* (New York: Academic Press, 1959-). Annual.
96. *Advances in Cell Biology* (New York: Appleton-Century-Crofts, 1970-). Annual.
97. *Advances in Food Research* (New York: Academic Press, 1948-). Irregular.
98. *Annual Review of Biophysics and Bioengineering* (Palo Alto, Calif.: Annual Reviews, Inc., 1972-).
99. *Annual Review of Entomology* (Palo Alto, Calif.: Annual Reviews, Inc., 1956-). Published in cooperation with the Entomological Society of America.
100. *Annual Review of Genetics* (Palo Alto, Calif.: Annual Reviews, Inc., 1967-).
101. *Annual Review of Microbiology* (Palo Alto, Calif.: Annual Reviews, Inc., 1947-).
102. *Annual Review of Phytopathology* (Palo Alto, Calif.: Annual Reviews, Inc., 1963-).
103. *Annual Review of Plant Physiology* (Palo Alto, Calif.: Annual Reviews, Inc., 1950-).
104. *Bacteriological Reviews* (Washington, D.C.: American Society for Microbiology, 1937-). Quarterly.
105. *CRC Critical Reviews in Bioengineering* (Cleveland, Ohio: Chemical Rubber Company Press, 1971-). Quarterly.
106. *CRC Critical Reviews in Clinical Laboratory Sciences* (Cleveland, Ohio: Chemical Rubber Company Press, 1970-). Quarterly.
107. *CRC Critical Reviews in Microbiology* (Cleveland, Ohio: Chemical Rubber Company Press). Quarterly.
108. *CRC Critical Reviews in Toxicology* (Cleveland, Ohio: Chemical Rubber Company Press, 1971-). Quarterly.
109. *The Kew Record of Taxonomic Literature Relating to Vascular Plants* (London: HMSO, 1971-). Annual.
110. *Progress in Molecular and Subcellular Biology* (Berlin; New York: Springer-Verlag, 1969-). Irregular.
111. *The Quarterly Review of Biology* (Stony Brook, N.Y.: State University of New York at Stony Brook; Stony Brook Foundation, Inc., 1926-). Quarterly.

Environment

112. *Advances in Ecological Research* (New York: Academic Press, 1962-). Irregular.
113. *Advances in Environmental Science and Technology* (New York: Wiley Interscience). Irregular.

114. *Advances in Water Pollution Research* (Oxford: Pergamon Press). Proceedings of the biennial conference of the International Association on Water Pollution Research.
115. *Annual Review of Ecology and Systematics* (Palo Alto, Calif.: Annual Reviews, Inc., 1970-).
116. *CRC Critical Reviews in Environmental Control* (Cleveland, Ohio: Chemical Rubber Company Press, 1970-). Quarterly.
117. *Ecological Studies: Analysis and Synthesis* (New York: Springer-Verlag, 1970-). Annual.
118. *International Pollution Control* (Chicago, Ill.: Scranton Publishing Co., 1972-). Annual.

Energy

119. *Advances in Nuclear Science and Technology* (New York: Academic Press, 1962-). Irregular.
120. *Annual Review of Energy* (Palo Alto, Calif.: Annual Reviews, Inc., 1976-).
121. *Annual Review of Nuclear Science* (Palo Alto, Calif.: Annual Reviews, Inc., 1952-).
122. *Energy Developments: Atomic, Petroleum, and Others* (New York: International Review Series, 1957-). Monthly.
123. *Energy Info* (Dana Point, Calif.: Robert Morey Associates, 1965-). Monthly.
"A monthly executive summary of events occurring in the fields of power generation and research" (subtitle).
124. *Monthly Energy Review* (Washington, D.C.: U.S. Federal Energy Administration, National Energy Information Center, 1974-). Monthly.
125. *Progress in Nuclear Energy. Section 2: Reactors* (Oxford: Pergamon Press). Quarterly.
126. *Progress in Nuclear Energy. Section 4: Technology, Engineering and Safety* (Oxford: Pergamon Press, 1956-). Quarterly.
127. *Progress in Nuclear Energy. Section 8: Economics of Nuclear Power* (Oxford: Pergamon Press). Quarterly.
128. *Progress in Nuclear Physics* (Oxford: Pergamon Press, 1950-). Quarterly.
129. *Solar Energy Progress in Australia and New Zealand* (Highett, Victoria, Australia: International Solar Energy Society, Australia and New Zealand Section, 1962-). Irregular.
130. *Solar Energy Update* (Oak Ridge, Tenn.: U.S. Department of Energy, Office of Public Affairs, Technical Information Center, 1976-). Monthly.

References

1. Julie A. Virgo, The review article: Its characteristics and problems, *Library Quarterly*, 41(4):275-291 (October 1971).

2. David A. Kronick, *A History of Scientific and Technical Periodicals. The Origins and Development of the Scientific and Technical Press, 1665-1790.* 2nd edition (Metuchen, N.J.: Scarecrow Press, 1976). pp. 184-201.
3. *The Royal Society Scientific Information Conference 21 June-2 July 1948. Report and Papers Submitted* (London: The Royal Society, 1948). p. 201.
4. *Proceedings of the International Conference on Scientific Information, Washington, D.C., November 16-21, 1958* (Washington, D.C.: National Academy of Sciences-National Research Council, 1959). p. 649.
5. *Science, Government, and Information. The Responsibilities of the Technical Community and the Government in the Transfer of Information. A report of the President's Science Advisory Committee* (Washington, D.C.: The White House, 1963). p. 27.
6. *Science, Government, and Information* (Ref. 5), p. 42.
7. *Scientific and Technical Communication. A Pressing National Problem and Recommendations for Its Solution* (Washington, D.C.: National Academy of Sciences, 1969). p. 40.
8. A. M. Woodward, The role of reviews in information transfer, *Journal of the American Society for Information Science*, 28(3):175-180 (May 1977).
9. Carlos A. Cuadra, *Annual Review of Information Science and Technology* (New York: Interscience, 1966), v. 1, p. 8.
10. Rowena W. Swanson, A work study of the review production process, *Journal of the American Society for Information Science*, 27(1):70-72 (January-February 1976).
11. Audrey N. Grosch, A workstudy of the review production process, *Journal of the American Society for Information Science*, 29(1):48 (January 1978).
12. *International Conference on Scientific Information* (Ref. 4), p. 650.
13. Conyers Herring, Distill or drown: The need for reviews, *Physics Today*, 21(9):27-33 (September 1968).
14. American Psychological Association, An informal study of the preparation of chapters for the *Annual Review of Psychology, Project on Scientific Information Exchange in Psychology. Report No. 2* (Washington, D.C.: American Psychological Association, 1963). p. 34.
15. Lewis M. Branscomb, Support for reviews and data evaluations, *Science*, 187(4177):603 (February 21, 1975).
16. *Science, Government, and Information* (Ref. 5), p. 2.
17. Herring (Ref. 13), p. 30.
18. *Royal Society Scientific Information Conference* (Ref. 3), p. 655.
19. *Scientific and Technical Communication* (Ref. 7), pp. 41, 44.
20. *Science, Government, and Information* (Ref. 5), pp. 32-33.
21. Virgo (Ref. 1), pp. 286-287.
22. Linda Harris and Robert V. Katter, Impact of the *Annual Review* of Information Science and Technology, In *Proceedings of the ASIS Annual*

References

   Meeting, No. 5 (New York: Greenwood Publishing Co., 1968), pp. 331–333.
23. A. M. Woodward, Review literature: Characteristics, sources, and output in 1972, *ASLIB Proceedings*, 26(9):367–376 (September 1974).
24. Virgo (Ref. 1), p. 277.
25. Dennis A. Brunning, Review literature and the chemist, *Proceedings of the International Conference on Scientific Information* (Ref. 4), pp. 545–570.
26. Eugene Garfield, Is citation frequency a valid criterion for selecting journals? *Current Contents*, 5–6 (April 5, 1972).
27. J. D. Bernal, Preliminary analysis of pilot questionnaire on the use of scientific literature, *Royal Society Scientific Information Conference* (Ref. 3), p. 599.
28. Brunning (Ref. 25), p. 546.
29. B. H. Flowers, Survey of information needs of physicists and chemists, *Journal of Documentation*, 21:83–112 (1965).
30. D. N. Wood and D. R. L. Hamilton, *The Information Requirements of Mechanical Engineers* (London: The Library Association, 1967).
31. A. A. Manten, Scientific review literature, *Scholarly Publishing*, 5:75–89 (October 1973).
32. Woodward (Ref. 23), p. 369.
33. Conyers Herring, Critical reviews: The user's point of view, *Journal of Chemical Documentation*, 8(4):232–236 (November 1968).
34. William B. Cottrell, Evaluation and compression of scientific and technical information at the Nuclear Safety Information Center, *American Documentation*, 19(4):375–380 (October 1968).
35. Charles L. Bernier, Terse literatures: 1. Terse conclusions, *Journal of the American Society for Information Science*, 21(5):316–319 (September–October 1970).
36. Charles L. Bernier, Terse literatures: 2. Ultraterse literatures, *Journal of Chemical Information and Computer Sciences*, 15(3):189–192 (August 1975).
37. Charles L. Bernier, Terse literature viewpoint of wordage problems, *Journal of Chemical Documentation*, 12(2):81–84 (May 1972).
38. Charles L. Bernier, Condensed technical literatures, *Journal of Chemical Documentation*, 8(4):195–197 (November 1968).
39. Irving Gordon and Russell L. K. Carr, Utilization of terse conclusions in an industrial research environment, *Journal of Chemical Documentation*, 12(2):86–88 (May 1972).
40. Leroy B. Townsend, Critical reviews: The author's point of view, *Journal of Chemical Documentation*, 8(4):239–241 (November 1968).
41. Preface: A major development in Chemical Society Publications, *Annual Reports on the Progress of Chemistry* (London: The Chemical Society), v. 64, pp. vii–ix (1967).
42. Denis J. Grogan, *Science and Technology: An Introduction to the Literature*, 2nd edition (Hamden, Conn.: Shoe String Press, 1973), p. 164.

17

TRANSLATIONS

17.1 Introduction

Scientific information is a global entity; no country, however advanced technologically, can afford to ignore the scientific information produced in other countries. According to Locke, the belated discovery of a Russian paper on the application of Boolean matrix algebra to the study of relay contact networks cost the United States an estimated $200,000 and caused a five-year delay in certain switching circuit developments [1]. Approximately one-half of the scientific and technical literature of the world is produced in languages other than English. Table 17.1 shows the distribution by language of the literature covered in six major English-language indexing and abstracting services [2].

More recent data for *Chemical Abstracts* show a decrease in the proportion of French and German chemical literature [3]:

Language	Percentage of references cited in *Chemical Abstracts*
English	62.8
Russian	20.4
German	5.0
French	2.4
Japanese	4.7
43 Other languages	4.7
Total	100.0

Russian scientific literature accounts for almost one-quarter of chemical and chemical engineering literature reported in *Chemical Abstracts*, and for 17% of physics literature abstracted in *Physics Abstracts*. This dispersion of scientific literature in numerous languages has been of much concern to scientists and

17.1 Introduction

Table 17.1 Language Distribution of Literature Covered in Abstracting and Indexing Services

	Percentage of literature covered in					
Language	Chemical Abstracts	Biological Abstracts	Physics Abstracts	Eng. Index	Index Medicus	Math. Reviews
English	50.3	75.0	73.0	82.3	51.2	54.8
Russian	23.4	10.0	17.0	3.9	5.6	21.4
German	6.4	3.0	4.0	8.6	17.2	8.7
French	7.3	3.0	4.0	2.4	8.6	7.8
Japanese	3.6	1.0	0.5	0.1	0.9	0.7
Chinese	0.5	1.0	0.1	0.0	0.4	0.2
Other	8.5	7.0	1.4	2.7	16.1	6.4
Totals	100.0	100.0	100.0	100.0	100.0	100.0

science bibliographers alike. An eminent American scientist is reported to have said, after Sputnik, "Either we will have to learn physics, or we will have to learn Russian" [4].

17.2 Translated Journals

Journal literature constitutes the bulk of translations. Three categories of translations of journal literature may be identified: (1) cover-to-cover translations of journals, (2) journals containing translations of individual articles chosen from different sources, and (3) ad hoc translations of single articles. Over 300 scientific and technical journals are translated from cover to cover, mostly from Russian into English, by commercial publishing houses (e.g., Plenum Publishing Corporation, Allerton Press), and scientific societies (e.g., the American Institute of Physics, the Chemical Society, London). The American Institute of Physics alone publishes over 20 cover-to-cover translations of Russian physics journals. Most of the journals translated are primary journals, e.g., *Soviet Physics: Crystallography* (American Institute of Physics), and *Soviet Mathematics* (American Mathematical Society). A few reviews and abstracting services are also translated: *Russian Chemical Reviews* (The Chemical Society, London) and *Cybernetics Abstracts* (Scientific Information Consultants, London). The latter is an English translation of *Referativnyi Zhurnal—Kibernetika*.

A Guide to Scientific and Technical Journals in Translation, 2nd edition, by Carl J. Himmelsbach and Grace E. Brociner (New York: Special Libraries Association, 1972) lists cover-to-cover translations and journals containing selected translations. A list of 162 cover-to-cover translations may also be found in "A study of 162 cover-to-cover English translations of Russian scientific and technical journals" by V. K. Rangra, *Annals of Library Science and Documentation*, 15(1):7-23 (March 1968). A further list of 12 titles not covered in the above article appears in "Cover-to-cover translations" by B. K. Sen, *Annals of Library Science and Documentation*, 15(4):216-218 (December 1968). A selected list of translated journals is included at the end of this chapter.

Because of the huge costs involved in translating scientific and technical literature, translated journals are far more expensive than the original editions. Table 17.2 shows the annual subscription rates (for 1979) of a few selected translated journals.

The timelag between the publication of the original journal and the translated version varies from six months to two years or more. Some publishers (e.g., Consultants Bureau/Plenum Publishing Corporation) have concluded agreements with the Copyright Agency of the USSR (VAAP) to receive advance copies of the Russian journal as well as original glossy photographs and artwork. This agreement decreases the delay in the publication of the translated journal and also helps to improve its quality of production.

17.2 Translated Journals

Table 17.2 Subscription Rates of Translated Journals

Journal	Annual Subscription
Hydrotechnical Construction Translation of the Russian monthly journal *Gidrotekhnicheskoe Stroitel'stvo* published by Consultants Bureau (Plenum Publishing Corporation) for the American Society of Civil Engineers.	$260
Journal of Organic Chemistry of the USSR Translation of the Russian monthly journal *Zhurnal Organicheskoi Khimii* of the Academy of Sciences of the USSR, published by Consultants Bureau (Plenum Publishing Corporation).	$345
Soviet Progress in Chemistry Translation of the Russian monthly journal *Ukrainskii Khimicheskii Zhurnal* published by Allerton Press.	$160
The Soviet Chemical Industry Translation of the Russian monthly journal *Khimicheskaya Promyshlennost'* of the Soviet Ministry of Chemical Industry, published by Ralph McElroy Company, Austin, Texas.	$245
Soviet Journal of Nuclear Physics Translation of the Russian monthly journal *Yadernaya Fizika* published by the American Institute of Physics.	$285
Soviet Physics JETP Translation of the Russian monthly journal *Zhurnal Eksperimental'noi i Teoreticheskoi Fiziki* of the Academy of Sciences of the USSR, published by the American Institute of Physics.	$270
JETP Letters Translation of the Russian "letters" journal *Pis'ma v Zhurnal Eksperimental'noi i Teoreticheskoi Fiziki* of the Academy of Sciences of the USSR, published by the American Institute of Physics	$110
Chemistry of Heterocyclic Compounds Translation of the Russian monthly journal *Khimiya Geterotsiklicheskikh Soedinenii* of the Academy of Sciences of the Latvian SSR, published by Consultants Bureau (Plenum Publishing Corporation).	$295

17.3 Bibliographic Control of Translations

Translation of scientific and technical literature is an expensive and time-consuming process, and calls for the relatively rare combination of expertise in two languages and in-depth knowledge of the subject. It is therefore important for scientists and science bibliographers to be able to locate both existing translations and those that are in progress in order to prevent wasteful effort and expense of duplicate translation. Several national and international translation centers are striving to disseminate information on translated documents and also to supply copies of the translations. The activities and products of some of these translation centers will be briefly described in this section.

17.3.1 The National Translations Center

The National Translations Center (NTC), housed in the John Crerar Library, Chicago, is a depository and referral center for unpublished English translations of the world literature in the natural, physical, medical, and social sciences. The NTC is a cooperative nonprofit enterprise, and its services are designed to:

Eliminate costly duplication of translation effort, thus freeing funds for translating new material

Disseminate information on available translations

Provide copies of translations on file or refer inquiries to other known sources

The NTC was formally organized in 1953 after having existed for several years as a volunteer project of the Science and Technology Division of the Special Libraries Association. The Center receives translations from scientific and professional societies, government agencies, industrial and other special libraries, academic institutions, and commercial translation bureaus all over the world. The identity of the depositor is obliterated to assure donor anonymity. Index files are maintained by author, journal citation, report number, standard number, and patent number. New additions to the Center's collections are announced in the monthly journal *Translations Register Index*; this index journal was begun by the Special Libraries Association in 1967 and transferred to the NTC in 1971. The entries in the *Index* are arranged under 22 subject fields endorsed by the Committee on Scientific and Technical Information (COSATI) of the Federal Council for Science and Technology. Government-sponsored translations announced in *Government Reports Announcements and Index* of the National Technical Information Service are also indexed in the *Translations Register Index*. The NTC distributes copies of translations available in its collections as well as those available from the British Library Lending Division and the International Translations Center in Delft. Translations made prior to 1967 may be found through the following indexes:

17.3 Bibliographic Control of Translations

Author List of Translations (New York: Special Libraries Association, 1953). Supplement, 1954.

Bibliography of Translations of Russian Scientific and Technical Literature (Washington, D.C.: Library of Congress, 1953-1956).

Translations Monthly (New York: Special Libraries Association, 1955-1958).

Technical Translations (Springfield, Va.: Clearinghouse for Federal Scientific and Technical Information, 1959-1967).

Consolidated Index of Translations into English (New York: Special Libraries Association, 1969).

Translations required by government agencies and sponsored by the Joint Publications Research Service (JPRS) are listed in *Transdex: Bibliography and Index to the United States JPRS Translations* (New York: CMC Information Corporation, 1971-). *Transdex* is published monthly and cumulated semiannually. The NTIS collects and distributes all government-sponsored translations and announces them in the *Government Reports Announcements and Index*.

17.3.2 The International Translations Center

The International Translations Center (ITC), formerly known as the European Translations Center, is a cooperative organization established in 1960 jointly by about 20 Western European countries in cooperation with the Organization for Economic Cooperation and Development (OECD) and the United States. The participating countries have national centers which, together with the ITC, form an international translations network. The ITC acts as a referral center, maintains a central information file, distributes translations, and maintains lists of translators and translation agencies. The national centers perform similar functions within their national jurisdictions and cooperate with the ITC.

From 1967 to 1978 the ITC published a monthly announcement service entitled *World Index of Scientific Translations and List of Translations Notified to ETC*. The *World Index* served as a finding list of translations of periodical articles and patents primarily from non-Western languages: Russian and other Slavic languages, Finnish, Hungarian, Rumanian, Chinese, Japanese, and Arabic.

The *World Index* is now superseded by a new international translations index called the *World Transindex*. This is a monthly service jointly published by three agencies: (1) the International Translations Center in Delft, (2) the Directorate General for Scientific and Technical Information Management of the Commission of the European Communities in Luxembourg, and (3) the Scientific and Technical Documentation Center of the Centre National de la Recherche Scientifique (CNRS) in Paris. The former publications of these agencies—*World Index of Scientific Translations and List of Translations Notified to the ETC, Transatom Bulletin*, and *Bulletin des Traductions*— have been merged into the new common publication, the *World Transindex*.

Each year, the *World Transindex* announces over 30,000 translations from Asiatic and East European languages into Western languages. Translations from Western languages into French are also announced. Translations from all scientific and technical fields are covered, and are arranged under subject headings derived from the COSATI system. *World Transindex* consists of a source index and an author index, and is cumulated quarterly and annually. The index is produced by means of the PASCAL system developed by CNRS, and the database is expected to become available for online access and selective dissemination of information. A microfiche edition of *World Transindex* is also under consideration.

17.3.3 The British Library Lending Division

In England, the British Library Lending Division (BLLD) has a large collection of translations (over 430,000 items in 1979), including translations received from NTIS, JPRS, and the NTC. These translations held at the BLLD are announced in a monthly bulletin entitled *BLLD Announcement Bulletin: A Guide to British Reports, Translations and Theses*. Copies of translations announced in this bulletin can be acquired from the BLLD.

Each year about 6000 books in the pure and applied sciences are translated. These translations are listed in UNESCO's annual publication *Index Translationum* (1950-). Translations of single articles, reports, and patents are more difficult to trace because most of these are not published through the regular trade channels. In England, ASLIB maintains the *Commonwealth Index of Unpublished Translations* on cards, and also a register of specialist translators. The card index has about 500,000 entries, and is growing at the rate of 10,000 entries a year. Nearly one-fourth of these come from private companies and other similar sources who do not publish translations and in most cases do not wish it to be known that the topic of the translation is of interest to them [5]. This card index is a location tool, and includes all translations held by the BLLD and those listed in the *Translations Register Index* of NTC. ASLIB does not supply translations, but refers inquirers to appropriate sources where the desired translations may be available. Each year ASLIB receives some 15,000 inquiries for information on translations.

17.3.4 Directories of Translators

Information on individual translators specializing in various languages and subjects, and on translation agencies and pools, may be obtained from the following directories:

International Directory of Translators and Interpreters (London: Pond Press, 1967).
Over 2000 translators throughout the world are listed and indexed by subject specialization and geographic location.

Frances E. Kaiser. *Translators and Translations: Services and Sources in Science and Technology*, 2nd edition (New York: Special Libraries Association, 1965).
Individual translators, translation pools, and bibliographies of translations are listed.

Professional Services Directory of the American Translators Association (Croton-on-Hudson, N.Y.: American Translators Association, 1976).

Patricia Millard. *Directory of Technical and Scientific Translators and Services* (London: Crosby Lockwood, 1968).
This directory lists translation services and about 300 individual translators in Great Britain, covering among them about 50 languages and almost all major scientific disciplines.

Stefan Congrat-Butlar. *Translation and Translators: An International Directory and Guide* (New York: R. R. Bowker, 1979).
This is an annotated list of translators' associations and translation centers, awards, grants, fellowships, prizes, codes of practice, and books and journals on translation; extensive listing of translators classified by language.

Despite the existence of several national and international translation centers and translation indexes, information about existing translations and those in progress is not disseminated widely. Grogan has pointed out that owing to this inadequacy of information on translations, scientists are not well informed about the material that is available in translation. There are at least two instances where the same journal is being translated by different agencies. *Moscow University Physics Bulletin* is being translated by Allerton Press, New York, and also by the Aztec School of Languages, West Acton, Massachusetts. Both Plenum Publishing Corporation and the NTIS are translating the Russian journal *Cybernetics* [6]. Such a duplication is surely an inexcusable waste of manpower and other resources.

Indexes of Translations

1. *ASM Translations Index* (Metals Park, Ohio: American Society for Metals, 1978–). Quarterly.
2. *BLLD Announcement Bulletin* (Boston Spa, England: British Library Lending Division, 1971–). Monthly.
   Formerly: *NLL Announcement Bulletin*.

3. *A Guide to Scientific and Technical Journals in Translation.* 2nd edition. Carl J. Himmelsbach and Grace E. Brociner (New York: Special Libraries Association, 1972).
4. *Index to Translations Selected by the American Mathematical Society* (Providence, R.I.: American Mathematical Society, 1966–). Irregular. Author and subject indexes to translations in *American Mathematical Society Translations* (Series 1 and Series 2, 1929–), and *Selected Translations in Mathematical Statistics and Probability* (Providence, R.I.: American Mathematical Society, 1961–).
5. *Index Translationum: Repertoire International des Traductions* (Paris: UNESCO, 1932–). Annual. (From 1932 to 1940, quarterly).
6. *Index Translationum. Cumulative Index to English Translations, 1948–1968* (Boston, Mass.: G. K. Hall, 1973). 2 v.
7. *Transdex: Bibliography and Index to the United States Joint Publications Research Service Translations* (New York: CCM Information Corporation, 1962–). Monthly.
8. *Translations Register Index* (Chicago: National Translations Center, John Crerar Library, 1967–). Monthly. Semiannual and annual cumulated indexes.
   Continues *Technical Translations* (Washington, D.C.: U.S. Department of Commerce, Office of Technical Services, 1959–1967), which in turn superseded *Translations Monthly* (Chicago: Special Libraries Associations, 1955–1958).
9. *World Transindex* (Delft, Netherlands: International Translations Center, 1979–). Monthly.
   Formerly: *World Index of Scientific Translations* (1967–1971); and *World Index of Scientific Translations and List of Translations Notified to ETC* (1972–1978).

Selected List of Translated Journals

Mathematics and Statistics

1. *Algebra and Logic* (New York: Plenum Publishing Corporation, 1968–). Bimonthly.
   Translation of *Algebra i Logika* of the Academy of Sciences of the USSR.
2. *Differential equations* (New York: Faraday Press, 1965–). Monthly.
   Translation of the Russian journal *Differentsialnye Uravneniya.*
3. *Lithuanian Mathematical Journal* (New York: Plenum Publishing Corporation, 1973–, v. 13–). Quarterly.
   Formerly: *Lithuanian Mathematical Transactions.* Translation of *Leituvos Mathematikos Rinkinys (Litovskii Matematicheskii Sbornik).*
4. *Mathematics of the USSR Izvestia* (Providence, R.I.: American Mathematical Society, 1967–). Bimonthly.
   Translation of *Izvestia Mathematicheskaya Seriya.*

5. *Proceedings of the Steklov Institute of Mathematics* (Providence, R.I.: American Mathematical Society). Irregular.
Translation of *Priblizhenie Funktsii i Operatorov.*
6. *Siberian Mathematical Journal* (New York: Consultants Bureau, 1969-). Bimonthly.
Translation of *Sibirskii Matematicheskii Zhurnal* of the Siberian Branch of the Academy of Sciences of the USSR, Novosibirsk.
7. *Soviet Mathematics Doklady* (Providence, R.I.: American Mathematical Society, 1960-). Bimonthly.
Translation of the Mathematics Section of *Doklady Akademii Nauk SSSR.*
8. *Theory of Probability and Its Applications* (Philadelphia, Penn.: Society for Industrial and Applied Mathematics, 1956-). Quarterly.
Translation of the Russian journal *Teriya Veroyatnostei i ee Primeneniya.*
9. *Transactions of the Moscow Mathematical Society* (Providence, R.I.: American Mathematical Society, ca. 1952-). Annual.
Translated from the Russian *Trudy Moskovskogo Mathematicheskoe Obshchestvo* by the American Mathematical Society and the London Mathematical Society.
10. *Ukrainian Mathematical Journal* (New York: Plenum Publishing Corporation, 1967-). Bimonthly.
Translation of *Ukrainskii Matematicheski Zhurnal.*

Computer Science and Automation

11. *Automation and Remote Control* (New York: Plenum Publishing Corporation, 1958-). Monthly.
Translation of *Avtomatika i Telemekhanika* of the Academy of Sciences of the USSR.
12. *Cybernetics* (New York: Plenum Publishing Corporation, 1965-). Bimonthly.
Translation of *Kibernetika.*
13. *Programming and Computer Software* (New York: Plenum Publishing Corporation, 1978, v. 4-). Bimonthly.
14. *Soviet Automatic Control* (Washington, D.C.: Scripta Publishing Corporation, 1968, v. 13-). Bimonthly.
Translation of *Avtomatika* of the Institute of Cybernetics, Kiev.

Physics and Astronomy

15. *Astrophysics* (New York: Consultants Bureau, 1965-). Quarterly.
Translation of *Astrofizika* of the Academy of Sciences of the Armenian SSR.
16. *Optics and Spectroscopy* (New York: American Institute of Physics, for the Optical Society of America, 1959-). Monthly.
Translation of *Optika i Spektroskopiya.*

17. *Soviet Astronomy* (New York: American Institute of Physics, 1957-). Bimonthly.
    Translation of *Astronomicheskii Zhurnal*.
18. *Soviet Astronomy Letters* (New York: American Institute of Physics, 1978-). Bimonthly.
    Translation of *Pis'ma v Astronomicheskii Zhurnal*. Companion journal to *Soviet Astronomy*; contains short articles reporting findings of recent research.
19. *Soviet Journal of Low Temperature Physics* (New York: American Institute of Physics, 1975-). Monthly.
    Translation of *Fizika Nizkikh Temperatur* published by the Ukranian Academy of Sciences; includes contributions from all republics of the USSR.
20. *Soviet Journal of Nuclear Physics* (New York: American Institute of Physics, 1965-). Monthly.
    Translation of *Yadernaya Fizika* of the Academy of Sciences of the USSR.
21. *Soviet Journal of Optical Technology* (New York: American Institute of Physics, for the Optical Society of America, 1966-). Monthly.
    Translation of *Optiko-Mekhanicheskaya Promyshlennost'*.
22. *Soviet Journal of Particles and Nuclei* (New York: American Institute of Physics, 1972-). Bimonthly.
    Translation of *Fizika Elementarnykh Chastits i Atomnogo Yadra*.
23. *Soviet Journal of Plasma Physics* (New York: American Institute of Physics, 1975-). Bimonthly.
    Translation of *Fizika Plazmy*.
24. *Soviet Journal of Quantum Electronics* (New York: American Institute of Physics, 1971-). Monthly.
    Translation of the Russian journal *Kvantovaya Elektronika*.
25. *Soviet Physics Acoustics* (New York: American Institute of Physics, 1955-). Bimonthly.
    Translation of *Akusticheskii Zhurnal* of the Academy of Sciences of the USSR.
26. *Soviet Physics Crystallography* (New York: American Institute of Physics, 1957-). Bimonthly.
    Translation of *Kristallografiya* of the Academy of Sciences of the USSR.
27. *Soviet Physics Doklady* (New York: American Institute of Physics, 1956-). Monthly.
    Translation of the Physics section of *Doklady Akademii Nauk SSSR (Proceedings of the Academy of Sciences of the USSR)*.
28. *Soviet Physics Journal* (New York: Consultants Bureau, 1965-). Monthly.
    Translation of the Russian journal *Izvestia VUZ Fizika* published by the Higher Education Institutes of the USSR.
29. *Soviet Physics JETP* (New York: American Institute of Physics). Monthly.
    Translation of *Zhurnal Eksperimental'noi i Teoreticheskoi Fiziki* of the Academy of Sciences of the USSR.
30. *JETP Letters* (New York: American Institute of Physics, 1965-). Semimonthly.

Selected List of Translated Journals

Translation of *Pis'ma v Zhurnal Eksperimental'noi i Teoreticheskoi Fiziki* of the Academy of Sciences of the USSR.

31. *Soviet Physics Semiconductors* (New York: American Institute of Physics, 1967-). Monthly.
    Translation of *Fizika i Tekhnika Poluprovodnikov* of the Academy of Sciences of the USSR.
32. *Soviet Physics Solid State* (New York: American Institute of Physics, 1959-). Monthly.
    Translation of *Fizika Tverdogo Tela* (Leningrad) of the Academy of Sciences of the USSR.
33. *Soviet Physics Technical Physics* (New York: American Institute of Physics, 1956-). Monthly.
    Translation of *Zhurnal Tekhnicheskoi Fiziki* of the Academy of Sciences of the USSR.
34. *Soviet Physics Uspekhi* (New York: American Institute of Physics, 1958-). Monthly.
    Translation of *Uspekhi Fizicheskikh Nauk* of the Academy of Sciences of the USSR.
35. *Soviet Technical Physics Letters* (New York: American Institute of Physics, 1975-). Monthly.
    Translation of *Pis'ma v Zhurnal Tekhnicheskoi Fiziki* of the Academy of Sciences of the USSR. Contains articles of 1000 words or less providing an overview of new research findings that will eventually be published in *Soviet Physics Technical Physics*.
36. *Ukrainian Physics Journal* (New Delhi, India: Amerind Publishing Company).

Chemistry, Chemical Engineering, and Technology

37. *Biochemistry* (New York: Consultants Bureau, 1956-). Monthly.
    Translation of *Biokhimiya* of the Academy of Sciences of the USSR.
38. *Bulletin of the Academy of Sciences of the USSR. Division of Chemical Sciences* (New York: Consultants Bureau, 1952-). Monthly.
39. *Doklady Chemical Technology* (New York: Consultants Bureau, 1956-). Semiannual.
    Proceedings of the Academy of Sciences of the USSR. Chemical Technology Section.
40. *Doklady Chemistry* (New York: Consultants Bureau, ca. 1973-). Monthly.
    Proceedings of the Academy of Sciences of the USSR. Chemistry Section.
41. *Doklady Physical Chemistry* (New York: Consultants Bureau, 1973-). Bimonthly.
    Proceedings of the Academy of Sciences of the USSR. Physical Chemistry Section.
42. *Journal of Analytical Chemistry of the USSR* (New York: Consultants Bureau, ca. 1973-). Semimonthly.

Translation of *Zhurnal Analiticheskoi Khimii* of the Academy of Sciences of the USSR.
43. *Journal of General Chemistry of the USSR* (New York: Consultants Bureau). Semimonthly.
44. *Journal of Organic Chemistry of the USSR* (New York: Consultants Bureau, 1965-). Semimonthly.
Translation of *Zhurnal Organicheskoi Khimii* of the Academy of Sciences of the USSR.
45. *Polymer Mechanics* (New York: Consultants Bureau, 1965-). Monthly.
Translation of *Mekhanika Polimerov* of the Academy of Sciences of the Latvian SSR.
46. *Russian Journal of Inorganic Chemistry* (London: The Chemical Society, 1959-).
47. *Russian Journal of Physical Chemistry* (London: The Chemical Society, 1959-). Monthly.
Translation of *Zhurnal Fizicheskoi Khimii* of the Academy of Sciences of the USSR.
48. *Soviet Journal of Bioorganic Chemistry* (New York: Plenum Publishing Corporation). Monthly (v. 4, 1978).
Translation of *Bioorganicheskaya Khimiya* of the Academy of Sciences of the USSR.
49. *Soviet Journal of Coordination Chemistry* (New York: Plenum Publishing Corporation). Monthly (v. 4, 1978).
Translation of *Koordinatsionnaya Khimiya* of the Academy of Sciences of the USSR.
50. *Soviet Journal of Glass Physics and Chemistry* (New York: Plenum Publishing Corporation). Bimonthly (v. 4, 1978).

Earth Sciences, Geology, and Metallurgy

51. *Doklady-Earth Sciences Section* (Washington, D.C.: American Geological Institute, 1959-). Bimonthly.
Translation of *Doklady Akademii Nauk SSSR* of the Academy of Sciences of the USSR.
52. *Geochemistry International* (Washington, D.C.: American Geological Institute, 1964-). Bimonthly.
Translation of the Russian journal *Geokhimiya*.
53. *Geomagnetism and Aeronomy* (Washington, D.C.: American Geophysical Union, 1961-). Bimonthly.
Translation of *Geomagnetizm i Aeronomiya* of the Academy of Sciences of the USSR.
54. *Geotectonics* (Washington, D.C.: American Geophysical Union, 1967-). Bimonthly.
Translation of *Geotektonika* of the Academy of Sciences of the USSR.
55. *Metal Science and Heat Treatment* (New York: Consltants Bureau, 1959-). Bimonthly.

Translation of *Metallovedenie i Termicheskaya Obrabotka Metallov.* Cosponsors: American Institute of Mining, Metallurgical, and Petroleum Engineers, and American Society for Testing and Materials.
56. *Moscow University Geology* (New York: Allerton Press, 1974-). Bimonthly.
Translation of the Russian journal *Moskovskogo Gosudarstvennogo Universiteta Seriya IV: Geologiya.*
57. *Physics of Metals and Metallography* (Oxford: Pergamon Press, 1957-). Monthly.
Translation of the Russian journal *Fizika Metallov i Metallovedenie.*
58. *Russian Metallurgy* (London: Scientific Information Consultants, 1962-). Bimonthly.
Translation of the Russian journal *Metally.*
59. *Soviet Geology and Geophysics* (New York: Allerton Press, 1974-). Bimonthly.
Formed by merger of *Soviet Geology* and *Soviet Geophysics.* Translation of *Geologiya i Geofizika.*
60. *Soviet Journal of Non-Ferrous Metals* (New York: Primary Sources, 1960-). Monthly.
Translation of the Russian journal *Tsyetnye Metally.*
61. *Soviet Mining Science* (New York: Consultants Bureau). Bimonthly.
Translation of *Fizikotekhnicheskiye Problemy Razrabotki Poleznykh Iskopayemykh* of the Academy of Sciences of the USSR, Siberian Branch.
62. *Soviet Powder Metallurgy and Metal Ceramics* (New York: Consultants Bureau, 1962-). Monthly.
Translation of *Poroshkovaya Metallurgiya* of the Academy of Sciences of the Ukrainian SSR, Material Sciences Institute.

Aviation and Aerospace

63. *Academy of Sciences of the USSR. Bulletin (Izvestia). Atmospheric and Oceanic Physics Series* (Washington, D.C.: American Geophysical Union, 1965-). Monthly.
English edition of *Izvestiya Akademii Nauk USSR: Seriya Fizika, Atmosfery i Okeana.*
64. *Soviet Aeronautics* (New York: Allerton Press, 1966-, v. 9-). Quarterly.
Translation of the Russian journal *Izvestiya VUZ Aviatsionnaya Tekhnika.*

Civil Engineering

65. *Hydrotechnical Construction* (New York: American Society of Civil Engineers, 1967-). Monthly.
Translation of the Russian journal *Gidrotekhnicheskoi Stroitel'stvo.*
66. *Soil Mechanics and Foundation Engineering* (New York: Consultants Bureau, 1964-). Monthly.
Translation of the Russian journal *Osnovaniya Fundamenty i Mekhanika Gruntov.*

Mechanical Engineering

67. *Fluid Dynamics* (New York: Consultants Bureau, 1966-). Bimonthly.
Translation of *Izvestia Akademii Nauk SSSR: Mekhanika a Zhidkosti i Gaza* of the Academy of Sciences of the USSR.
68. *Journal of Applied Mathematics and Mechanics* (Oxford: Pergamon Press, 1958-). Bimonthly.
Translation of the Russian journal *Prikladnayia Mathematika i Mekhanika*.
69. *Magnetohydrodynamics* (New York: Consultants Bureau, 1965-). Quarterly.
Translation of *Magnitnaya Gidrodinamika* of the Academy of Sciences of the Latvian SSR.
70. *Power Engineering* (New York: Allerton Press, 1974-). Bimonthly.
Translation of *Izvestia Akademii Nauk SSSR: Energetika i Transport* of the Academy of Sciences of the USSR.
71. *Soviet Applied Mechanics* (New York: Consultants Bureau, 1966-). Monthly.
Translation of *Prikladnaya Mekhanika* of the Academy of Sciences of the USSR.
72. *Thermal Engineering* (Oxford: Pergamon Press, 1964-). Monthly.
Translation of the Russian journal *Teploenergetika*.

Electrical Engineering and Electronics

73. *Electric Technology USSR* (Oxford: Pergamon Press, 1957-). Quarterly.
Translation of *Elektrichestvo* of the Academy of Sciences of the USSR.
74. *Electronics and Communications in Japan* (Washington, D.C.: Scripta Publishing Corporation, 1970-, v. 53-). Monthly.
English edition of the *Transactions of the Institute of Electronics and Communication Engineers of Japan*.
75. *Soviet Microelectronics* (New York: Consultants Bureau, 1974-, v. 3-). Bimonthly.
Translation of the Russian journal *Mikroelektronika*.
76. *Radio Engineering and Electronic Physics* (Washington, D.C.: Scripta Publishing Co., 1961-, v. 6-). Monthly.
Translation of the Russian journal *Radiotekhnika i Elektronika* of the Academy of Sciences of the USSR.
77. *Soviet Electrical Engineering* (New York: Allerton Press, 1965-). Monthly.
Translation of the Russian journal *Elektrotekhnika*.
78. *Soviet Journal of Quantum Electronics* (New York: American Institute of Physics, 1971-). Monthly.
Translation of the Russian journal *Kvantovaya Elektronika*.
79. *Soviet Power Engineering* (Austin, Tex.: Ralph McElroy Co., 1972-). Monthly.
Translation of the Russian journal *Elektricheskie Stantsii*.

80. *Telecommunications and Radio Engineering* (Washington, D.C.: Scripta Publishing Co., 1963-). Monthly.
Translation of the journals *Elektrosvyaz* and *Radiotekhnika*.

Biology

81. *Applied Biochemistry and Microbiology* (New York: Plenum Publishing Corporation, 1965-). Bimonthly.
Translation of *Prikladnaya Biokhimiya i Mikrobiologiya* of the Academy of Sciences of the USSR.
82. *Biology Bulletin of the Academy of Sciences of the USSR* (New York: Plenum Publishing Corporation, ca. 1974-). Bimonthly.
Translation of the Russian journal *Izvestiya Akademii Nauk SSSR: Seriya Biologicheskaya*.
83. *Cytology and Genetics* (New York: Allerton Press, 1974-). Bimonthly.
Translation of the Russian journal *Tsitologiya i Genetika*.
84. *Doklady Biological Sciences* (New York: Consultants Bureau, 1973-, v. 208-). Bimonthly.
Translation of *Doklady Akademii Nauk SSSR* (Biology Section).
85. *Journal of Evolutionary Biochemistry and Physiology* (New York: Plenum Publishing Corporation, 1969-). Bimonthly.
Translation of *Zhurnal Evolutsionnoi Biokhimii i Fiziologii* of the Academy of Sciences of the USSR.
86. *Moscow University Biological Sciences Bulletin* (New York: Allerton Press, 1974-). Bimonthly.
Translation of the Russian journal *Moskovskogo Gosudarstvennogo Universiteta Vestnik Biologiya*.
87. *Soviet Journal of Marine Biology* (New York: Plenum Publishing Corporation, 1975-). Bimonthly.
Translation of the Russian journal *Biologiya Morya*.

Environment

88. *Water Resources* (New York: Plenum Publishing Corporation, 1975-). Bimonthly.
Translation of the Russian journal *Vodnye Resursy*.

Energy

89. *Annals of Nuclear Energy* (Oxford: Pergamon Press, 1959-). Monthly.
Contains translations from the Russian journal *Atomnaya Energiya*. Title until 1975: *Annals of Nuclear Science and Engineering*.
90. *Applied Solar Energy* (New York: Allerton Press, 1965-). Bimonthly.
Translation of the Russian journal *Geliotekhnika*.
91. *Soviet Atomic Energy* (New York: Consultants Bureau, 1973-). Monthly.
Translation of *Atomnaya Energiya* of the Academy of Sciences of the USSR.

References

1. William N. Locke, Translation by machine, *Scientific American*, 194(1):29–33 (January 1956).
2. D. N. Wood, The foreign language problem facing scientists and technologists in the U.K. Report of a recent survey, *Journal of Documentation*, 23(2):117–130 (June 1967).
3. *CAS Today: Facts and Figures about Chemical Abstracts Service* (Columbus, Ohio: Chemical Abstracts Service, 1980), p. 8.
4. Ritchie Calder, The nature and function of science, *ASLIB Proceedings*, 10(7):161–170 (July 1958).
5. Translations in the U.K., *ASLIB Proceedings*, 25(7):264–267 (July 1973).
6. Denis J. Grogan, *Science and Technology: An Introduction to the Literature*, 3rd edition (London: Clive Bingley, 1976), p. 293.

18

BIBLIOGRAPHIC CONTROL OF SCIENTIFIC AND TECHNICAL LITERATURE

18.1 Proliferation of Literature

Proliferation of scientific literature originating from many countries in numerous languages, increase in the number of scientists generating and using literature, and the need for rapid access to the most recent literature—all these factors have emphasized the need for a bibliographic apparatus to facilitate the identification, selection, and acquisition of scientific literature cutting across the diversities of format, language, and national origin. The bibliographic apparatus should be capable of providing access to scientific literature selectively (excluding all nonpertinent documents), comprehensively (including all pertinent documents), speedily, and economically.

The biggest deterrents to the bibliographic control of scientific and technical literature are its enormous volume and the alarming rate at which it appears to be growing. At the first annual lecture of the Science of Science Foundation delivered at the Royal Institution, London, on March 25, 1965, Derek J. de Solla Price said [1]:

> Science increases exponentially, at a compound interest of about 7% per annum, thus doubling in size every 10-15 years, growing by a factor of ten at least every half-century, and by something like a factor of a million in 300 years which separates us from the seventeenth century invention of the scientific paper when this process began.
>
> ... this is an alarming rate, much faster than any population explosion, much faster than growth rates of industry; those are standing still in comparison.

The problem caused by the growing volume of scientific and technical literature has been highlighted by J. C. R. Licklider thus [2]:

> To simplify back-of-the-envelope calculations, let us take the figures, 10^{13} characters and (say) 12 years, at face value; let us assume that one one-thousandth of all science and technology constitutes a field of

specialization; and let us consider the plight of a scientist who reads 3000 characters a minute, which is a rate more appropriate for novels than for journal articles. Suppose that he gathers together the literature of his field of specialization (10^{10} characters) and begins now to read it. He reads 13 hours a day, 365 days a year. At the end of 12 years, he sets down the last volume with a great sigh of relief—only to discover that in the interim another 10^{10} characters were published in his field. He is deterred by the realization that not only the volume but the rate of publication has doubled.

Journal literature constitutes the largest component of the totality of scientific literature. The proliferation of the primary journal has been discussed in Chapter 3. Other forms of literature have also been increasing. According to a recent survey by King Research, Inc., conducted for the National Science Foundation [3], scientific and technical books in the United States have registered the greatest rise, from 3379 titles in 1960 to 14,442 titles in 1974 (327% increase). Scholarly journal articles have risen from 106,000 in 1960 to 151,000 in 1974 (42% increase). During the same period, the costs attributed to scientific and technical books increased from $600 million to $2.1 billion (over 250% increase), and the costs associated with scientific and technical journals rose from $1.3 billion to $5.6 billion (330% increase). In the United States, patent applications represent a very large component of scientific literature, next only to journal literature. In 1974, approximately 108,000 patent applications were filed; this represents an increase of 28% over the 1960 figure of 84,500 applications. But not all the patent applications are eventually granted patents. In 1960, the number of patents issued was 50,000; in 1974 this number rose to 79,900, showing an increase of 60% during this period.

Technical reports represent the third largest category of scientific and technical literature. The number of technical reports processed by NTIS rose from 14,000 in 1965 to 61,000 in 1975, registering an annual mean growth rate of 15.8%. These numbers do not include the report literature processed by the U.S. Government Printing Office. Doctoral dissertations have also been steadily increasing, as seen in Chapter 5. Data on the coverage of secondary services gathered by the National Federation of Abstracting and Indexing Services (NFAIS) from its 40 member services showed a net increase of 145% during the 15-year period 1960-1974.

The total resources expended on scientific and technical communication in the United States have been growing more rapidly than the GNP. In the 15-year period 1960-1974, the GNP grew by 177%, whereas the total resources spent on scientific and technical communication grew from $2 billion in 1960 to $8.5 billion in 1974, registering an increase of 320%. The expenditure on scientific and technical communication in 1975 was estimated at $9.4 billion. This figure includes the costs incurred by authors, publishers, libraries, secondary services, and users in the production and use of scientific and technical books, journals, reports, and other forms of publications. The ratio of scientific and technical communication expenditure to the GNP has also been steadily increasing. In 1960, scientific and technical communication expenditure accounted for 0.4% of the GNP; it gradually increased to 0.6% in 1974.

At least one probable cause for this enormous growth in scientific and technical literature has been the increase in the number of scientists and engineers engaged in R&D as well as in publishing. The number of scientists and engineers in the United States rose by 3.8% per annum from 1960 to 1975. Price estimates that there are now approximately 400,000 scientific and technical authors, of whom about one-half are authors of scholarly publications. Price contends that seven of every eight scientists who had ever been alive are alive today [1].

The growth of scientific and technical literature has rendered impossible the mastery of the literature of even one branch of science such as physics or biology by an individual scientist. Scientists have reacted to this problem by resorting to narrow specialization within a branch of science. But the solution of problems in science and engineering calls for the application of ideas from a larger store of knowledge than an individual can possess at his command. With increasing specialization, the processes of storing, organizing, and retrieving information have become group activities rather than individual efforts. In the group approach, information storage and retrieval have been delegated to externalized (i.e., nonneural) bibliographic control systems. Bibliographies (both retrospective and current) and abstracting and indexing services are the most important tools of bibliographic control that help users of scientific literature to identify and select documents pertinent to their interests from the totality of scientific literature. These three major bibliographic control tools will be discussed in the following sections.

18.2 Bibliographies

The distinction between a current bibliography and a retrospective bibliography is rather temporary and pertains to the time span covered by the bibliography. Retrospective bibliographies describe material published in the past—sometimes including documents published a century or two ago—and are intended to aid comprehensive literature surveys. Current bibliographies include material published recently, usually within a short time span—a month or a few months— and are intended to aid selection of a few recent documents bearing on a given subject. Very often, current bibliographies are serial publications appearing at specified intervals, each issue covering the material published since the publication of the preceding issue. The periodical issues are then cumulated and made available as a retrospective bibliography covering a longer time span. It is also a common practice to prepare indexes to a retrospective bibliography in order to provide alternative or additional approaches to the documents, complementing the arrangement of entries in the main part of the bibliography. In this sense, then, most current serial bibliographies that are cumulated over a time span longer than that covered by the individual issues eventually become retrospective bibliographies. An example of this pehnomenon can be seen in the *Bibliography of the History of Medicine*, published annually since 1965 by the National Library of Medicine. The majority of citations are derived from *Index Medicus* and the *Current Catalog* of the National Library of Medicine. The annual issues of this bibliography are then cumulated every five years. So far two cumulated editions have been issued, covering the years 1964-1969 and 1970-1974.

18.2.1 Current Bibliographies

Besides regular bibliographies such as the *Bibliography of the History of Medicine* noted above, acquisition lists prepared by libraries and information centers, and reviews and announcements in journals, are also useful bibliographic sources for keeping track of recent publications. *New Technical Books* published by the New York Public Library and *ASLIB Book List* (1935-) are two notable examples. *British Scientific and Technical Books*, published since 1956, is compiled on the basis of the *ASLIB Book List*. Annotated lists of additions to the library of the Institution of Mechanical Engineers, London, are published as a regular feature in the monthly journal the *Chartered Mechanical Engineer*.

Thousands of primary journals publish reviews and announcements of recently published scientific and technical books. Notable among these are *Science, Nature, American Scientist, Physics Today, Scientific American*, and *Library Journal*, to name only a few. *Choice* is primarily a book-review journal, emphasizing college-level publications. *Library Journal* publishes an annual special feature on scientific, technical, and medical books, and also periodically announces new and forthcoming books. Announcements of new books and forthcoming books may also be seen in *Publishers' Weekly* and the *Weekly Record*. Publishers' announcements and catalogs are also useful sources of information on new and forthcoming books. Some abstracting and indexing services such as *Chemical Abstracts* and *Computing Reviews* also include annotations, abstracts, or reviews of books. *Science Books and Films* (formerly, *Science Books: A Quarterly Review*), published by the American Association for the Advancement of Science (AAAS) is yet another source of reviews of science books.

Reviews of scientific and technical books scattered in thousands of journals may be located through *Technical Book Review Index* (1935-) compiled and

18.2 Bibliographies

edited for the Special Libraries Association in the Science and Technology Department of the Carnegie Library of Pittsburgh. It gives extracts from published book reviews and serves as an index to book reviews in some 2500 journals. *Book Review Index* (Gale Research Co.), *Book Review Digest* (H. W. Wilson Co.), and *Index to Book Review Citations* (H. W. Wilson Co.) are similar indexes to book reviews, but the scope of these is far more general than that of the *Technical Book Review Index*. *Index to Book Reviews in the Sciences* is a new monthly index published by the Institute for Scientific Information, Philadelphia. It covers nearly 35,000 reviews of recently published books each year.

Standard lists of books are compiled and published periodically by large libraries, commercial publishing companies, and societies. These are usually designed for particular types of libraries. The following are some examples of standard book lists:

A Basic Collection for Scientific and Technical Libraries compiled by E. B. Lunsford and T. J. Kopkin (Special Libraries Association, 1975), has 2400 annotated entries arranged by subject.

Science and Technology: A Purchase Guide for Branch and Small Public Libraries, published annually by the Carnegie Library of Pittsburgh, Science and Technology Department, is a classified and annotated list of titles selected for the student and nonspecialist adult reader. The list is compiled from the books received in the library during the preceding year.

British Scientific and Technical Reference Books: A Select and Annotated List (1976), with 250 entries, is published by the British Council, London, keeping in view the particular needs of overseas libraries.

Special mention should be made of the *McGraw-Hill Basic Bibliography of Science and Technology* (1966). This is a companion volume to the *McGraw-Hill Encyclopedia of Science and Technology,* and contains lists of books on some 7000 subjects corresponding to the articles in the *Encyclopedia.* The books listed are aimed at students and general readers.

Bibliographic series issued periodically by libraries, societies, and research institutions are excellent sources of information on current publications. *Science Tracer Bullets* issued by the Library of Congress, Science and Technology Division, and the *Bibliographical Series* of the Iron and Steel Institute in England are examples of current bibliographies of this type. A number of trade bibliographies, notably those published by R. R. Bowker Company and the H. W. Wilson Company, are especially useful for bibliographic verification and acquisition. These include such well-known tools as *Cumulative Book Index, Books in Print, Scientific and Technical Books in Print,* and *American Scientific Books.*

Reference Sources (Ann Arbor, Mich.: Pierian Press, 1977-) edited by Linda Marks is an annual index to reviews of reference works of various kinds in all fields. The first volume (1977) contained 4500 references to reviews of over 3500 reference titles. Excerpts are also quoted from reviews.

Reference Services Reviews (Ann Arbor, Mich.: Pierian Press, 1973-) is a quarterly review journal containing reviews of a wide variety of reference works and current surveys of reference works in different subjects. The following are some recent examples of current surveys:

Arthur Antony, Current survey of reference sources in chemistry, *Reference Services Reviews*, 7(2):13-24 (June 1979).

Elisabeth B. Davis, Current survey of reference sources in botany, *Reference Services Reviews*, 7(1):7-12 (March 1979).

Sarojini Balachandran, State of the art survey of reference sources in engineering, *Reference Services Reviews*, 6(4):25-28 (October 1978).

Jack Weigel, State of the art survey of reference sources in physics and astronomy, *Reference Services Reviews*, 6(3):21-24 (September 1978).

Berle Reiter, State of the art survey of reference sources in mathematics and statistics, *Reference Services Reviews*, 6(3):17-20 (September 1978).

Another source of reviews of current reference works is the quarterly journal *Reference Book Review* (Columbia, S.C.: Cameron Northouse, 1976-). *American Reference Books Annual* (Littleton, Colo.: Libraries Unlimited, 1970-) contains descriptive and evaluative notes on reference books published during the preceding years. References to selected reviews are also included. A cumulative index to the first five volumes was published in 1975. Journals and other publications of societies are described in *Scientific, Engineering, and Medical Societies Publications in Print, 1976-1977,* compiled by James M. Kyed and James M. Matarazzo (New York: R. R. Bowker, 1976).

18.2.2 Retrospective Bibliographies

An early bibliography of articles in the publications of scientific societies during the eighteenth century was the *Reportorium Commentationum a Societatibus Litterariis Editorum* . . . compiled by Jeremias David Reuss (Gottingen: Henricum Dieterich, 1801-1821, reprinted in 1961 by B. Franklin, New York). This was a 16-volume work: Volume 1, natural history; volume 2, botany and mineralogy; volume 3, chemistry; volume 4, physics; and volumes 10-16, science and medicine.

The National Library of Medicine has been responsible for a series of monumental retrospective bibliographies covering early medical literature including incunabula. *A Catalogue of Sixteenth Century Printed Books in the National*

18.2 Bibliographies

Library of Medicine, 1967, compiled by Richard L. Darling, contains descriptions of over 4800 imprints. Another bibliography compiled by Dorothy M. Schullian, entitled *A Catalogue of Incunabula and Manuscripts in the Army Medical Library* was published in 1950. A supplement to both these bibliographies was issued in 1972: *A Catalogue of Incunabula and Sixteenth Century Printed Books in the National Library of Medicine. First Supplement, 1972.* This supplement, compiled by Peter Krivatsy, lists 27 incunabula and 272 books published during the sixteenth century.

Another retrospective bibliography covering early medical literature is a contribution from the Department of the History of Science and Medicine, Yale University: *American Medical Bibliography, 1639-1783,* compiled by Francisco Guerra (New York: Lathrop C. Harper, 1962). This is "A chronological catalogue, and critical bibliographical study of books, pamphlets, broadsides, and articles in periodical publications relating to the medical sciences—medicine, surgery, pharmacy, dentistry, and veterinary medicine—printed in the present territory of the United States of America during British Dominion and the Revolutionary War" (subtitle). Material in this bibliography is arranged in three groups: (1) books, pamphlets, and broadsides; (2) almanacs; and (3) periodicals and magazines. The first two parts are arranged in chronological order with an alphabetical index of authors and titles. The periodicals and magazines section has been arranged alphabetically and is followed by a chronological conspectus.

A tremendous amount of studied effort by the National Research Council, the State Department, the New York Public Library, a special committee of book publishers, and many other groups and individuals resulted in what has been hailed as the first comprehensive national bibliography of American scientific literature: *Scientific, Medical and Technical Books, Published in the United States of America 1930-1944: A Selected List of Titles in Print with Annotations* [4]. This bibliography, edited by R. R. Hawkins, and published in 1946, listed 6413 titles published in the years 1930-1944. Two supplements were issued. The first supplement, issued in 1950, covered literature published during the period 1945-1948; literature published during 1949-1952 was covered in the second supplement issued in 1953. A second edition of the entire work, also edited by Hawkins, under the direction of the National Academy of Sciences-National Research Council's Committee on Bibliography of American Scientific and Technical Books, was published in 1958. "The purpose of this bibliographic series is to supply descriptions of the outstanding scientific, medical and technical books written by citizens of Canada and the United States of America, published in the United States of America, and available for both domestic and foreign distribution" (from the preface to the second edition). The entries in this bibliography are arranged under subject headings. A directory of publishers, an author index, and a subject index are also included.

A somewhat unusual bibliographic tool is a composite index to a large number of books: *Index to All Books on the Physical Sciences in English, 1967 through January 1974* (Rockville, Md.: Marc 2 Research, Inc., 1974). This is an alphabetical subject index of over 200,000 terms providing subject access to about 22,000 monographs published during the years 1967-1974.

Published book catalogs of large libraries serve as valuable retrospective bibliographies. Printed catalogs of the John Crerar Library, Chicago; the Library of the Academy of Natural Sciences of Philadelphia; the library of the Museum of Comparative Zoology, Harvard University; and the Engineering Societies Library, New York, are representative examples. All these and many similar catalogs of large libraries and special collections are published by G. K. Hall & Company of Boston, Mass. The John Crerar Library catalog consists of the following three parts:

1. *Author-Title Catalog, The John Crerar Library (Chicago)*, 1967. 599,000 cards. 28,554 pages. 35 volumes.
2. *Classified Subject Catalog, The John Crerar Library (Chicago)*, 1967. 730,000 cards. 33,167 pages. 42 volumes, including subject index.
3. *Subject Index to the Classified Subject Catalog, 1967.* 47,000 entries. 610 pages. 1 volume.

The John Crerar Library is one of the major scientific, technical, and medical libraries in the world, with a collection of some 1,100,000 volumes and pamphlets including current and historical research materials in the pure and applied sciences. The collections, which include the holdings of the Illinois Institute of Technology, are of research strength in the basic sciences, such as physics, chemistry, and biology; medicine, including anatomy, physiology, biochemistry, and pharmacology; agriculture, especially agricultural engineering and chemicals; and technology, including all branches of engineering.

The Gray Herbarium Index (Boston, Mass.: G. K. Hall, 1968) is a reproduction of a special index maintained at the Harvard University. The index contains approximately 259,000 cards with single or multiple entries devoted to names and literature citations of newly described or established vascular plants of the western hemisphere. Literature from 1886 onwards has been covered. The printed index is a massive publication of 8121 pages in 10 volumes. These two examples illustrate the scope and magnitude of published library catalogs that serve as comprehensive retrospective bibliographies.

Printed catalogs of private collections, typified by the *Catalog of Botanical Books in the Collection of Rachel McMasters Miller Hunt* (Carnegie-Mellon University, Pittsburgh, 1958-1961), are also useful as retrospective bibliographies.

18.2 Bibliographies

Other types of retrospective bibliographies include catalogs of books displayed in exhibitions and biobibliographies of scientists. *One Hundred Books Famous in Science,* based on an exhibition held at the Grolier Club, compiled by H. D. Horblitt (New York: Grolier Club, 1964) is an example of book exhibition catalogs. *A Bibliographical Checklist and Index to the Published Writings of Albert Einstein,* compiled by Nell Boni and others (Paterson, N.J.: Pageant Books, 1960), and *A Bibliography of the Honourable Robert Boyle,* compiled by J. H. Fulton (Oxford: Clarendon Press, 1961), are typical of biobibliographies.

18.2.3 Special Subject Bibliographies

Most of the retrospective bibliographies discussed above are very broad in scope and encompass several or all branches of science, medicine, engineering, and technology. Numerous specialized bibliographies, limited in scope to one scientific discipline or subdiscipline, provide access to literature in specific subject areas. Only a few representative examples are listed below to indicate the scope of such specialized bibliographies.

A Select Bibliography of Chemistry, 1492-1897, by Henry Carrington Bolton (Washington, D.C.: The Smithsonian Institution, 1893). A supplement was published in 1899 to include the works omitted in the main volume and those published during the year 1897. A second supplement was brought out in 1904 to cover the books published during the years 1898-1902, both years inclusive. The bibliography is organized into eight sections: (1) bibliography, (2) dictionaries, (3) history, (4) biography, (5) chemistry, pure and applied, (6) alchemical literature in the 19th century, (7) periodicals, and (8) academic dissertations. "No attempt has been made to *index* books and periodicals, as this is accomplished in the *International Catalogue of Scientific Literature,* directed by the Royal Society, London, and that undertaking is not duplicated in the present work" (from the preface to the second supplement).

A Bibliography of Birds with Special Reference to Anatomy, Behavior, Biochemistry, Embryology, Pathology, Physiology, Genetics, Ecology, Aviculture, Economic Ornithology, Poultry Culture, Evolution, and Related Subjects, by Reuben Myron Strong (Chicago: Field Museum of Natural History, 1939-1946), with 25,000 entries in three volumes. Volumes 1 and 2 comprise the author catalog; the third volume, published in 1946, is a subject index. The library in which the publication was found during verification is indicated at the end of each entry.

A Bibliography of Meteorites, by H. S. Brown (Chicago University Press, 1953). The earliest entry in this bibliography dates from 1491.

Bibliography of Aeronautics (Washington, D.C.: The Smithsonian Institution, 1910). Published as Volume 55 of the Smithsonian Miscellaneous Collections, this bibliography covered the material published prior to 1909. On

July 1, 1918, the United States Congress approved publication of a further series of bibliographies by the National Advisory Committee for Aeronautics. *Bibliography of Aeronautics, 1909-1916,* covering the literature published from July 1, 1909 to December 31, 1916 was brought out in 1921. A further volume covering the literature of the period 1917-1919 was issued in 1922; this was followed by a series of volumes issued by the National Advisory Committee for Aeronautics (which subsequently became the present National Aeronautics and Space Administration) until 1932. In all these volumes, references to the publications of all nations have been included in their original languages. The arrangement is in dictionary form with author and subject entries interfiled in one alphabetical sequence.

18.2.4   Bibliographies of Bibliographies

Besides numerous separately published bibliographies and bibliographic serials of the type described above, a very large number of "hidden" bibliographies appear as parts of other documents such as books, dissertations, technical reports, and review articles. Because of the very large number and diversity of sources of bibliographies, some bibliographic control of bibliographies themselves is necessary. Compilation of bibliographies is a tedious and time-consuming endeavor, calling for a great deal of effort. In order to avoid unintended duplication of bibliographic effort, it is important for science bibliographers and users of scientific literature to be able to identify and use existing bibliographies. *A World Bibliography of Bibliographies and of Bibliographical Catalogues, Calendars, Abstracts, Digests, Indexes and the Like* by Theodore Besterman is a monumental work in five volumes in which 117,187 bibliographies are listed under 15,829 subject headings in an alphabetical dictionary arrangement. The fourth edition of this retrospective bibliography of bibliographies was published in 1965-1966 by Societas Bibliographica, Lausanne. A decennial supplement to this work was published in 1977: *A Word Bibliography of Bibliographies,* by Alice F. Toomey (Totowa, N.J.: Rowman & Littlefield, 1977), two volumes, 1166 pages.

Another retrospective bibliographic guide by Besterman is *Early Printed Books to the End of the Sixteenth Century: A Bibliography of Bibliographies,* 2nd edition (Lausanne: Societas Bibliographica, 1961). Both these works are of very broad scope, and cover bibliographies in all subjects. The following retrospective bibliographies of bibliographies are available for tracing special subject bibliographies:

World Bibliography of Agricultural Bibliographies. R. Lauche (Munich: BLV, 1964).

Bibliography of Scientific and Technical Bibliographies. M. Bloomfield and others (Culver City, Calif.: Hughes Aircraft Co., 1968). Available as

AD 676 797 and AD 676 798 from NTIS. In the first part of this guide, 6677 bibliographies are listed alphabetically by author/compiler/corporate source. The second part is a subject index to the bibliographies listed in the first part.

Botanical Bibliographies: A Guide to Bibliographic Materials Applicable to Botany (Minneapolis, Minn.: Burgess, 1970).

Physical Sciences: A Bibliography of Bibliographies. Theodore Besterman (Totowa, N.J.: Rowman & Littlefield, 1971). 2 v.

The above mentioned bibliographic guides are specially useful for locating older bibliographies. For identifying current bibliographies and "hidden" bibliographies which appear as parts of other documents, the *Bibliographic Index* (New York: H. W. Wilson Co., 1945 -) is a very useful tool. First published in 1945 to cover the period 1937-1942, this index is now published three times a year and is cumulated annually. Bibliographies of more than 50 items appearing in journal articles and other sources are indexed in the *Bibliographic Index*. Other abstracting and indexing services such as *Chemical Abstracts* and *Physics Abstracts* also cover bibliographies; the latter has a separate index to bibliographies. Mention should be made of a series of ten bibliographic essays published in a special issue of *Library Trends,* 15(4), April 1967. These essays describe publications in many branches of science, engineering, and medicine, published in many countries including England, Germany, France, and the USSR.

18.3 Abstracting Services

Abstracts, summaries, and digests of documents have always been used as substitutes for full documents by scholars, scientists, and busy executives. According to Francis Witty, a device similar in function to an abstract was first used "on some of the clay envelopes enclosing Mesopotamian cuneiform documents of the early second millennium B.C. The idea of the envelope, of course, was to preserve the document from tampering; but to avoid having to break the solid cover, the document would either be written in full on the outside with the necessary signature seals, or it would be abstracted on the envelope accompanied likewise by the seals" [5]. Borko and Bernier have pointed out that written or oral abstracts of documents were used not only by scholars and scientists, but also by statesmen, rulers, and religious leaders in ancient and medieval times for private communication and current awareness [6]. In the seventeenth century, the use of abstracts changed from a means of private communication to a system of public dissemination of information, but the function of the abstract remained the same: to facilitate the identification and selection of documents pertinent to one's field of inquiry, and in some cases, to be used as a substitute for the full document. A very interesting account of the emergence and development of abstracts journals during the period 1790-1920 has been provided by Bruce Manzer [7].

During the nineteenth century, many specialized abstracting journals began. Notable among these are: *Pharmaceutisches Central-Blatt* (1830-), *Engineering Index* (1884-), and *Science Abstracts* (1898-). The United States Patent Office started its weekly *Official Gazette* in 1872. The number of abstracting services has continued to increase in the twentieth century. A few well-known examples are: *Chemical Abstracts* (1907-), *Biological Abstracts* (1926-), *Mathematical Reviews* (1940-), *Bulletin Signaletique* (1940-), *Excerpta Medica* (1947-), *Applied Mechanics Reviews* (1948-), *Analytical Abstracts* (1954-), *Referativnyi Zhurnal* (1954-), *Excerpta Botanica* (1959-), *International Aerospace Abstracts* (1961-), and *Astronomy and Astrophysics Abstracts* (1969-). Both *Bulletin Signaletique* and *Referativnyi Zhurnal* are large, omnibus abstracting services published respectively in France and the USSR.

In recent years, both the number of abstracting services and the number of items covered in each of them have been increasing. *Chemical Abstracts* provides a picture of continued growth from its inception in 1907. The rate of growth of *Chemical Abstracts* has been steadily increasing, as can be seen from Table 18.1. It took nearly 32 years (from 1907 to 1938) to produce the first million abstracts. The second million was reached in 1956 (after 18 years). The third million was reached after 7 years in 1963, and the fourth million in 1968 after 5 years. The five millionth abstract was reached in 1971, after only three years. *Chemical Abstracts* now cites nearly one half million documents each year, drawn from nearly 13,000 serials and other publications.

Because of their extensive coverage and wide scope, large abstracting services such as *Physics Abstracts, Chemical Abstracts,* and *Biological Abstracts* have become too "general" to meet the needs of individual scientists specializing in a very narrow branch of science. A noticeable trend in the last few decades has been the emergence of specialized abstracting services covering narrow subject areas: *Helminthological Abstracts* (1932-), *Apicultural Abstracts* (1950-), *Rheology Abstracts* (1958-), *Electroanalytical Abstracts* (1963-), *Journal of Current Laser Abstracts* (1967-), *Nucleic Acid Abstracts* (1971-), *Amino Acid Peptide and Protein Abstracts* (1972-), and so on. Abstracting services devoted to technical reports are also of recent origin. *Government Reports Announcements and Index* of the National Technical Information Service started as the *Bibliography of Scientific and Industrial Reports* in January 1946. *Scientific and Technical Aerospace Reports* of the National Aeronautics and Space Administration began in 1963.

Growing concern with sociotechnological issues such as environmental pollution and energy conservation has spurred an enormous output of literature in these areas in recent decades. Inevitably, this increase in the volume of

18.4 Indexing Services

Table 18.1 Growth of *Chemical Abstracts*

Year	Number of abstracts and citations	Cumulative total number of documents cited
1907	11,847	11,847
1915	18,981	175,622
1920	19,326	256,122
1925	27,097	379,726
1930	55,146	586,029
1940	53,680	1,206,377
1950	59,098	1,662,559
1960	134,255	2,613,069
1970	309,742	4,907,588
1973	356,549	5,993,290
1974	375,663	6,368,953
1975	454,245	6,823,198
1976	458,508	7,281,706
1977	478,225	7,759,931
1978	498,559	8,258,490
1979	515,741	8,774,231

From Ref. 8.

interdisciplinary literature is accompanied by the emergence of new abstracting services. Notable among these are: *Pollution Abstracts* (1970-), *Air Pollution Abstracts* (1970-), *Water Pollution Abstracts* (1972-), and *Abstracts on Health Effects of Environmental Pollutants* (1972-).

18.4 Indexing Services

The origin of alphabetical subject indexing has been traced back to antiquity. According to Francis Witty, the earliest approach to an alphabetical subject index appears in an anonymous work of the fifth century, the *Apothegmata*, a list of sayings of various Greek fathers on certain theological topics [5]. Although alphabetical indexing was used throughout the Middle Ages, mainly in works on theology, philosophy and law, indexing of scientific literature did not gain momentum until the eighteenth century when the scientific journal became firmly established as the favored medium for the dissemination of scientific information.

The Royal Society of London was responsible for two monumental indexes to the scientific literature of the nineteenth century and early twentieth century.

The *Catalogue of Scientific Papers,* published in 19 volumes between 1866 and 1925, is an author index of articles in over 1500 scientific periodicals of the nineteenth century. Some European and American periodicals were also covered. A subject index was published between 1908 and 1914, but this covers only mathematics, mechanics, and physics. Because of various problems, the project had to be discontinued after publishing only three of the projected 17 volumes [9]. Between the years 1902 and 1921, the Royal Society published the *International Catalogue of Scientific Literature* in 14 volumes covering the scientific literature published during the period 1901-1914. This was an author and subject index to books and journal articles in all branches of science. The second monumental effort was also given up because of numerous organizational and economic problems, and because of the disruption caused by the First World War.

Zoological Record (1864-) and *Index Medicus* (1879-) are among the notable indexing services that began during the 19th century. The following are some of the numerous indexing services that began during the 20th century: *Applied Science and Technology Index* (supersedes in part *Industrial Arts Index,* 1958-), *Biological and Agricultural Index* (started in 1916 as *Agricultural Index*), *Bibliography of Agriculture* (1942-), *Air University Library Index to Military Periodicals* (1949-), *Science Citation Index* (1961-), *British Technology Index* (1962-), and *Pandex: Current Index to Scientific and Technical Literature* (1967-).

Extensive listings of abstracting and indexing services can be found in the following publications:

Abstracts and Indexes in Science and Technology: A Descriptive Guide. Dolores B. Owen and Marguerite M. Hanchey (Metuchen, N.J.: Scarecrow Press, 1974). 154p.

International Federation for Documentation. *Abstracting Services.* 2nd edition (The Hague, Netherlands: FID, 1969). 284p. v. 1: *Science, Technology, Medicine, Agriculture.*

An Introduction to Periodicals Bibliography. Paul E. Vesenyi (Ann Arbor, Mich.: Pierian Press, 1974). 382p.
Includes an annotated list of selected periodical bibliographies (and abstracting and indexing services).

Ulrich's International Periodicals Directory (New York: R. R. Bowker, 1932-).

U.S. Library of Congress. Science and Technology Division. *A Guide to the World's Abstracting and Indexing Services in Science and Technology* (Boston, Mass.: Gregg Press, 1972, reprint of 1963 edition). 183p.

18.5 Characteristics of Abstracting and Indexing Services

The *Guide to the World's Abstracting and Indexing Services in Science and Technology* was prepared by the National Federation of Abstracting and Indexing Services in 1963, and listed 1885 titles originating from some 40 countries. This directory is now out of date, and a new directory is in the process of being published by NFAIS in collaboration with the International Federation for Documentation (FID) and the National Science Foundation. This new directory is expected to contain descriptions of approximately 2500 abstracting and indexing services.

18.5 Characteristics of Abstracting and Indexing Services

18.5.1 Scope, Coverage, and Speed

The major function of abstracting and indexing services is twofold: (1) to facilitate retrospective literature searches, and (2) to satisfy the current awareness needs of scientists. In order to facilitate comprehensive retrospective searches, a secondary journal should have the widest possible coverage; in addition, the abstracting journal is required to provide indexes to the abstracts for easy and rapid location of the desired abstracts. For a secondary journal to be an efficient current awareness tool, it is essential that the journal be produced with minimum delay so that scientists can use it to identify current literature of interest to them. It is apparent that comprehensiveness of coverage and speed of production are mutually incompatible characteristics, and that any attempt to improve one of them will tend to have an adverse effect on the other. Publishers of secondary services have attempted to tackle this anomalous situation primarily in two ways: (1) by reducing the scope and coverage of the secondary service, and (2) by producing separate products to meet the retrospective and current awareness needs.

The scope of a secondary service refers to the subject area that it attempts to cover; its coverage refers to the exhaustiveness with which primary literature within the chosen subject area is monitored. Expansion of the scope and coverage of a secondary service will improve its efficiency as a retrospective search device, but only at the cost of its efficiency as a current awareness tool, because of the greater delay involved in its production. In view of the enormous volume of scientific literature currently published, secondary services encompassing all branches of science and technology are becoming a rarity. *Applied Science and Technology Index* and *British Technology Index* are examples of indexing services that attempt to cover all or most branches of science and technology. The coverage of these services is rather limited. The *Applied Science and Technology Index* covers fewer than 250 journals published in English, mostly in the United States. These journals are chosen by the subscribers themselves in a poll conducted periodically by the H. W. Wilson Company. The *British Technology Index* covers a similar number of journals with a heavy emphasis on British publications. Considering the wide scope of these publications, their coverage is very limited. This is not to say that these two indexing services are of limited usefulness; because of the balanced and careful selection of the primary journals that they monitor, both these indexing services are used very heavily in libraries.

A relatively recent multidisciplinary index with a broader coverage is *Pandex: Current Index to Scientific and Technical Literature* (New York: CCM Information Services, 1967-), published biweekly, with quarterly and annual cumulations. This is a computer-generated subject and author index and claims to cover 2200 journals, 5000 patents, 40,000 research reports, and 6000 books annually. The index is also available on microfilm and microfiche.

Omnibus abstracting services such as *Referativnyi Zhurnal* and *Bulletin Signaletique* are not single publications, but are families of large numbers of separate sections, each section being devoted to a specific branch of science or engineering. *Referativnyi Zhurnal* is published in over 65 sections, and *Bulletin Signaletique* in some 50 sections; the sections are available separately for sale.

When *Chemical Abstracts* began publication in 1907, it was a semimonthly publication, and each issue had fewer than a thousand abstracts. It was relatively easy for chemists to scan the semimonthly issues of *Chemical Abstracts* for current awareness. With increasing volume of chemical literature, the semimonthly issues became unmanageably large, and beginning from January 1967, the semimonthly issues had to be split into two weekly issues with the first 34 sections being published as odd-numbered weekly issues, and sections 35 to 80 as even-numbered issues every alternate week.

Separation of retrospective and current awareness functions of secondary services is a phenomenon that has become increasingly common in the last few decades. In 1960, when Chemical Abstracts Service was processing about 150,000 abstracts annually, it started publishing a separate biweekly alerting service called *Chemical Titles* to take over the current awareness function, leaving *Chemical Abstracts* to serve primarily as a tool for comprehensive retrospective literature search. *Chemical Titles* is basically a computer-generated keyword-in-context (KWIC) index to the titles of articles in some 700 journals in chemistry. Because of its smaller coverage and its amenability to computerized production with minimal human indexing effort, *Chemical Titles* appears with a much smaller timelag than the abstracting service, and is thus more useful as a current awareness aid.

Faced with a rapidly growing volume of primary literature to be monitored, *Referativnyi Zhurnal* responded similarly in 1966 by starting a separate series of current awareness journals. As the number of abstracts processed annually approached one million in 1966, the delay of six to eight months in the

18.5 Characteristics of Abstracting and Indexing Services

publication of *Referativnyi Zhurnal* necessitated a separate series of current awareness publications entitled *Ekspress Informatsii*. This is a series of over 70 separate sections that appear with an average delay of about 8 weeks. In general, indexing services are faster than abstracting services. The average timelags of indexing in the *Applied Science and Technology Index* and the *British Technology Index* are 11 weeks and 7 weeks, respectively.

The International Information System for the Physics and Engineering Communities (INSPEC) of the Institution of Electrical Engineers, London, has also been producing separate sets of products for current awareness function and for retrospective search function. INSPEC is now publishing three current awareness services: *Current Papers in Physics*, *Current Papers in Electrical and Electronics Engineering*, and *Current Papers in Computers and Control*. The three abstracting services of INSPEC, namely, *Physics Abstracts* (1898-), *Electrical and Electronics Abstracts* (1898-), and *Computer and Control Abstracts* (1966-), collectively called *Science Abstracts*, serve as primarily retrospective search tools. INSPEC also produces SDI services which are obtainable on a subscription basis.

18.5.2 Overlap and Gap in Coverage

It is estimated that there are about 3000 abstracting and indexing services published throughout the world; of these, about 2500 cover scientific and technical literature. These secondary services process about 8 million references annually. The estimated annual output of scientific and technical articles is about 2 million items. Nevertheless, all of these documents are not covered by the abstracting and indexing services. In practice, large numbers of documents are covered by several secondary services, and an indeterminate number of documents are not covered by any secondary journal. This twin problem of overlap and omission in the coverage of secondary services was observed by S. C. Bradford in 1937 [10]. He noted that nearly two-thirds of scientific and technical papers were not covered in indexing and abstracting periodicals, and the remaining one-third of the papers were covered in several secondary periodicals. In a statistical test on abstracting journals carried out for ASLIB, it was discovered that 46% of a sample of 1634 references were abstracted more than once, and 27% were not covered by any abstracting journal [11]. In a further study of 3420 references by John Martyn, it was discovered that 47% of the references were covered by more than one service, 32% were covered in one service, and 21% were not covered by any of the six abstracting services studied [12]. In the same study, it was also found that 22% of the literature on biological control of insects, pests, and weeds was covered in four abstracting journals.

One possible reason for such overlap in abstracting could be that each abstracting journal prepares a slanted abstract keeping in view the particular needs of its clientele. The Commonwealth Agricultural Bureaux often prepares three or four slanted abstracts of the same paper emphasizing different aspects, for publication in its various abstracting journals. But this intentional multiple abstracting is a very rare phenomenon, and is not a significant factor in the overlapping coverage of abstracting services. in the ASLIB study mentioned above, John Martyn found little evidence of genuine slanting of abstracts. Most of the secondary services use the author's abstract that appears with the journal article. The author's abstract is usually edited to suit the needs of the secondary service. The American Society of Civil Engineers receives 99% of its abstracts from authors for the *ASCE Publication Abstracts* [13]. Original abstracting is done only when the author abstract is either unavailable or unsatisfactory. Hence, overlap in abstracting may be attributed to overlap in the lists of journal titles monitored by the secondary services.

Another investigation was undertaken by Wood, Flanagan, and Kennedy to study the overlap of journal titles monitored and journal articles actually abstracted by BioSciences Information Service of Biological Abstracts (BIOSIS), Chemical Abstracts Service, and Engineering Index, Inc. Of the 14,592 different journals monitored, 1% were monitored by all the three services, 27% were monitored by two of the three services, and 72% were monitored by only one of the three services [14]. In a sample of 29,182 articles drawn from the journals covered by all these three services, a maximum of only 822 articles might have been abstracted in all the three abstracting services [15]. It was felt that the number of occurrences of all the three services having selected the same article for abstracting and indexing was not great enough to warrant concern.

A massive investigation was recently conducted by NFAIS with financial support from the National Science Foundation, to study the overalp in the coverage of 14 major science abstracting services [16]. Of the approximately 26,000 journals covered by at least one of the 14 services studied, one or more articles from the 1973 issues of 5466 journals had been covered by two or more services. The articles selected by the services were not necessarily the same. The 5466 journals were then studied to estimate the actual overlap at the individual article level. It was found that 23.4% of those articles abstracted or indexed by any of the secondary services from the 5466 journals were covered by at least two of the services. The remainder of the 26,000 journals had no article overlap whatsoever.

Multiple coverage of the same paper in several secondary services is in itself not an undesirable process; it ensures that articles of even peripheral interest are not overlooked by scientists using any of the secondary services. What is undesirable is the avoidable duplication of intellectual effort involved in repeated indexing and abstracting of the same documents by several secondary services. A solution to this situation lies in the use of authors' abstracts and

18.5 Characteristics of Abstracting and Indexing Services

and exchange of surrogates by secondary services. While most secondary services use authors' abstracts whenever these are available, exchange of surrogates is not practised widely because of administrative and technical problems and lack of standardization of bibliographic description of documents and exchange of bibliographic data.

In at least two instances, extensive duplication of coverage in secondary services has triggered major changes. In 1967, it was noticed that there was about 80% overlap in the coverage of the *Review of Metals Literature* of the American Society for Metals and the *Metallurgical Abstracts* of the Institute of Metals (London). The two societies then integrated their secondary services and started a new set of publications entitled *Metals Abstracts* and *Metals Abstracts Index*, published jointly by the two societies since 1968. More recently, *Nuclear Science Abstracts* was discontinued in June 1976 in view of the substantial duplication of its contents in *Atomindex* published by the International Nuclear Information System.

The problem of omission of coverage is more serious than that of overlap in the coverage of secondary services. In the ASLIB study mentioned earlier, 21% of the 3420 references were not covered by any of the abstracting journals studied. According to one estimate made in 1962, less than a quarter of the published literature in biology was actually abstracted and indexed in secondary journals. There is reason to believe that this situation has improved in recent years.

The coverage of secondary services is often influenced by geographic and linguistic predilections. For example, *Applied Science and Technology Index* covers only English language journals, mostly those published in the United States. The time and expense required for indexing and abstracting foreign language material are deterrents to the coverage of such material in English language secondary services. Also, expanding the coverage of secondary services tends to increase production delays and costs, which in turn tends to decrease the number of subscriptions. Thus, owing to economic limitations as well as geographic and linguistic barriers, an indeterminate quantity of primary literature remains unnoticed by secondary services.

18.5.3 Indexes to Abstracts

A collection of abstracts can be only as good as its indexes. Most abstracting services arrange the abstracts under broad subject categories or according to a subject classification scheme. In order to provide a subject approach to abstracts at a more specific level than the broad subject categories under which the abstracts are usually grouped, a specific subject index becomes necessary. Besides a specific subject approach, additional access points (e.g., by author's name, patent number, or report number) are also required to facilitate identification of abstracts pertinent to a given inquiry. In *Chemical Abstracts,* the abstracts are grouped under 80 broad sections (e.g., general biochemistry, enzymes, immunochemistry, physical organic chemistry, general physical chemistry, electric phenomena, and inorganic analytical chemistry). These broad section headings are not further subdivided into more specific subheadings, with the result that in each of these sections hundreds of abstracts may be found in the weekly issues of *Chemical Abstracts.* Each weekly issue of *Chemical Abstracts* carries a keyword index that provides a specific subject access to the abstracts; three additional indexes—a numerical patent index, a patent concordance, and an author index—provide alternative approaches to the abstracts.

A subject index to the abstracts becomes even more essential in those secondary services that use a classification scheme such as the Universal Decimal Classification for arrangement of the abstract entries. Mention should be made of *Physics Abstracts* of INSPEC in which abstracts are arranged according to a special classification scheme for physics and astronomy. Each semimonthly issue of *Physics Abstracts* has the following indexes:

Subject index

Author index

Bibliography index

Book index

Conference index

Corporate author index

The subject index in the semimonthly issues of *Physics Abstracts* is an abbreviated alphabetical index to the classes in the classification scheme. A detailed cumulated subject index to the individual abstracts is published separately twice a year covering the periods January-June and July-December.

While indexing of abstracts is necessary in order to provide alternative and additional access points to the abstracts in an abstracting service, periodical cumulation of the indexes is also important to facilitate retrospective literature searches over extended periods of time. One of the recommendations of the Royal Society Scientific Information Conference, 1948, was that abstract publications should issue a consolidated subject index at least once in ten years. At the end of each volume in June and December each year, *Chemical Abstracts* provides the following volume indexes:

General subject index

Chemical substance index

Numerical patent index

18.5 Characteristics of Abstracting and Indexing Services

Patent concordance

Author index

Formula index

The patent indexes and author indexes are cumulations of those appearing in the weekly issues. The subject indexes are not simple cumulations of the weekly keyword indexes. The general subject index and the specific substance index appearing at the end of each volume are more complete and systematically prepared indexes based on a controlled vocabulary. A full description of the volume indexes can be found in the *Chemical Abstracts Index Guide* and its supplements.

For retrospective searches spanning several volumes of *Chemical Abstracts,* the collective indexes are useful. From 1907 until 1956, these were decennial indexes. Starting from 1957, the collective indexes are published quinquennially. The eighth collective index to *Chemical Abstracts* covering the years 1967-1971 is a massive publication of over 75,000 pages in 35 volumes. The ninth collective index covering the period 1972-1976, which was completed in August 1978, is believed to be the largest collective index published by Chemical Abstracts Service. It contains 21 million entries for 2,024,013 papers, patents, and other documents cited in *Chemical Abstracts.* The chemical substance index alone has 7.4 million entries and occupies more than 40,000 pages. The index was compiled and photocomposed by computer from computer-readable files of the semiannual volume indexes. The entire index, occupying 96,000 pages in 57 volumes, was completed in a record time of less than 20 months. The tenth collective index for the years 1977-1981 is expected to be one-third again as large as the ninth collective index [17].

The frequency of cumulation of indexes varies considerably among individual abstracting services. Annual cumulations are most common. Cumulations covering a longer time span (5 or 10 years) are more useful for retrospective searches, as they minimize the number of separate indexes to be scanned. In the case of most abstracting services, only the indexes are cumulated at various intervals. *Engineering Index* is a unique example in which the entire publication including the abstracts, is cumulated annually. The *Engineering Index Monthly* issues may be discarded when the annual cumulated edition is received.

Some indexing services (e.g., the *Applied Science and Technology Index* and the *British Technology Index*) also cumulate periodically. The monthly issues of these indexes may be used as current awareness tools, and these may be discarded upon arrival of the annual cumulated volumes which are more useful for retrospective search.

Various types of indexes are provided in abstracting services, depending upon the nature of the material abstracted. Abstracts of technical reports call for corporate source index and report/accession number index, as in the case of the abstracting services of NASA and NTIS. The *Official Gazette* of the U.S. Patent Office provides a subject matter index and an index to inventors and assignees of patents. *Biological Abstracts* contains a number of specialized indexes particularly suitable for biological literature. These include: a generic and a biosystematic index, which are especially useful for searching by taxonomic categories, a concept index (formerly called Computer Rearrangement of Subject Specialities or CROSS index), and a KWIC index to document titles augmented with appropriate additional keywords.

The depth and quality of indexes vary widely in abstracting services. In his tests on abstract journals, John Martyn observed many inconsistencies in the subject indexes to abstracts, and concluded: ". . . if 80% of the coverage of the particular topic on which the information is required be available from one abstracting and indexing service, the searcher is unlikely to be able to find more than three-quarters of this via the subject indexes, and he is unlikely to be able to find more than half without the exercise of considerable ingenuity or a good knowledge of the subject" [18].

18.6 Guides to Literature

In recent years, the volume of secondary literature (abstracting and indexing services, and reference works of various kinds discussed in earlier chapters) has been rapidly growing to keep pace with the growth of primary literature. A stage has now been reached when "tertiary" bibliographic devices have become necessary to keep track of the growing volume of secondary literature. In response to this need, a number of guides to the literature of science and technology are being published. An illustrative list of these guides is included at the end of this chapter. These guides are not bibliographies or indexing services, but are tertiary bibliographic devices that list and describe secondary sources. They identify the surrogated secondary sources (e.g., bibliographies and indexes) that may be useful for current awareness or retrospective searching, and also repackaged and condensed secondary sources (e.g., directories, reviews, encyclopedias) that are useful for answering specific reference questions and for background reading.

Two types of guides to literature may be identified: the "expository" type, and the "inventory" type. In the expository type of guide, the emphasis is on exposition of various forms of literature and search procedures rather than on comprehensive listing of individual works. Bernard Houghton's *Mechanical Engineering: The Sources of Information* (London: Clive Bingley), is not intended as a "bibliography of mechanical engineering literature or as a listing of titles, but as a map to help the engineer find his way through the varying forms of literature" (from the Introduction, p. 7). Some guides of this type

18.6 Guides to Literature

also include chapters on principles and procedures of literature survey, hints on how to use the library, and occasionally practice exercises for students of scientific literature. R. T. Bottle's *The Use of Chemical Literature,* 2nd edition (London: Butterworths, 1969) and M. G. Mellon's *Chemical Publications: Their Nature and Use,* 4th edition (New York: McGraw-Hill, 1965) contain hints on the use of bibliographic tools in libraries and practical exercises useful to students of chemical literature. Solutions to the exercises are also provided in Bottle's book.

The inventory type of guide is designed as a working tool for the reference librarian and the bibliographer, and aims at comprehensiveness. An example of this type of guide is the *Sources of Information on the Rubber, Plastics and Allied Industries,* by E. R. Yescombe (Oxford: Pergamon Press, 1968) in which "every effort has been made to include all important sources" (p. ix).

Two classical guides of the inventory type used extensively in libraries are A. J. Walford's *Guide to Reference Materials,* 3rd edition (London: The Library Association, 1973), and Constance M. Winchell's *Guide to Reference Books,* 8th edition (Chicago: American Library Association, 1967). Eugene Paul Sheehy's *Guide to Reference Books,* 9th edition (Chicago: American Library Association, 1975) is an expanded and updated version of Winchell's *Guide* and its supplements. Although these are general guides to literature of all branches of knowledge, they are nevertheless extremely useful in approaching scientific and technical literature. Volume 1 of Walford's *Guide* is entirely dedicated to science and technology.

Guides to literature are published by professional societies and library associations as well as commercial publishers. *Literature of Chemical Technology,* Advances in Chemistry Series No. 78 (Washington, D.C.: American Chemical Society, 1968) is one of a series of guides to chemical literature published by the American Chemical Society. The American Society for Engineering Education (ASEE) has published a number of short guides to literature. *Guide to Literature on Metals and Metallurgical Engineering* (1970) and *Guide to Literature on Mining and Mineral Resources Engineering* (1972) are two examples of ASEE guides; these are short lists of reference books, secondary services, and bibliographies, without annotations. The Special Libraries Association has published several bibliographic guides including *Guide to Metallurgical Information* by Eleanor B. Gibson and Elizabeth W. Tapia (2nd edition, 1965).

How to Find Out About Chemical Industry by Russell Brown and G. A. Campbell (1969), and *How to Find Out in Iron and Steel,* by D. White (1970) are two publications in a series of guides to literature published by Pergamon Press. The Management Information Guides of Gale Research Company are a set of guides including such titles as *Textile Industry Information Resources* (Management Information Guide No. 4, 1964), and *Computer and Information Processing Information Sources* (Management Information Guide No. 15, 1969).

There is considerable variety in the scope of the guides to literature. Some attempt to cover all branches of science and technology. A. J. Walford's *Guide to Reference Materials* (Volume 1), mentioned earlier, and Ching Chih Chen's *Scientific and Technical Information Sources* (Cambridge, Mass.: MIT Press, 1977) are examples of this type. But most guides are usually limited to the literature of one discipline, e.g., *The Use of Physics Literature,* by Herbert Coblans (London: Butterworths, 1975).

Guides to literature may also appear as journal articles, conference proceedings, catalogs of holdings of individual libraries and special collections, encyclopedia articles, or chapters in handbooks. The following are a few examples to illustrate the diverse kinds of "hidden" guides to literature:

How to Obtain Information in Different Fields of Science and Technology: A User's Guide (AGARD LS-69, 1974) is a collection of papers presented at a NATO conference.

Current Reference Materials for the Physical Sciences is an article by Raphaella Kingsbury in *Special Libraries,* 63(9):394-399 (September 1972).

Environmental Update: A Review of Environmental Literature and Developments is a regular feature published in *Library Journal* in May each year.

Metallurgical Libraries and Literature, by Carol Mulvaney (*Encyclopedia of Library and Information Science,* 17:464-472, 1972) is one of a number of similar articles published in the *Encyclopedia.*

Literature of Chemistry and Chemical Technology is a two-part article in the *Kirk-Othmer Encyclopedia of Chemical Technology,* 2nd edition, v. 12, pp. 500-528.

State of the Art Survey of Reference Sources in Biology, by Elisabeth Davis, and State of the Art Survey of Reference Sources in Engineering by Sarojini Balachandran (*Reference Services Review,* October/December 1978) are typical of surveys that appear regularly in *Reference Services Review.*

The following bibliographies and guides are useful in locating guides to literature:

American Reference Books Annual (Littleton, Colo.: Libraries Unlimited, 1970-).

Reference Books in Paperback: An Annotated Guide, 2nd edition (Littleton, Colo.: Libraries Unlimited, 1976).

Basic Reference Books: Titles of Lasting Value Selected from American Reference Books Annual, 1970-1976 (Littleton, Colo.: Libraries Unlimited, 1976).

Readers Advisory Service: Selected Topical Booklists (New York: Science Associates International, 1973-).
This is a series of subject bibliographies and guides to literature prepared by professional societies, libraries, and research centers. Approximately 75 guides and bibliographies are issued annually.

General guides to literature including those by Walford and Sheehy, and reviews in journals (e.g., *Choice, Library Journal, RQ,* and *Reference Services Review*) are also important sources of information on guides to literature.

Science Tracer Bullets published by the Library of Congress on specific topics such as boundary layer flow, artificial kidneys (hemodialyzers), and harbor design are useful as guides to literature on specific topics. A list of *Science Tracer Bullets* is available from the Library of Congress, Science and Technology Division, Washington, DC 20540.

Library pathfinders (published by Addison-Wesley Publishing Co., Reading, MA 01867) are very short guides to literature on specific topics, and are intended to assist literature searchers in the beginning stages of the search process. Pathfinders were first conceived as bibliographic guides to aid literature searchers as a part of the Massachusetts Institute of Technology's Project INTREX [19].

18.7 Information Analysis Centers

The dissemination and bibliographic control of scientific and technical information in the United States depends on voluntary cooperation among numerous publishers, information systems, and abstracting and indexing services, in both the public and the private sectors. There are also very many specialized information centers, both government and private, engaged in the acquisition, processing, and dissemination of scientific and technical literature. Most of these centers specialize in one branch of science or in a narrow subdiscipline. The major functions of these information centers in general are as follows:

1. Production, maintenance, and distribution of machine-readable bibliographic databases (e.g., MEDLARS of the National Library of Medicine, COMPENDEX of Engineering Index, Inc.)
2. Provision of current awareness services (e.g., Automatic Subject Citation Alert of the Institute for Scientific Information)
3. Publication of abstracting and indexing services (e.g., *Chemical Abstracts, Index Medicus,* etc.)
4. Referral service (e.g., National Referral Center, Library of Congress)
5. On-demand bibliographic searches (such as those provided by MEDLARS and the NTIS)

A chain of information analysis centers is supported by the U.S. Department of Defense, the Department of Energy, and other federal agencies such as NASA. "An information analysis center is a formally structured organizational unit established for the purpose of acquiring, storing, retrieving, evaluating, analyzing, and synthesizing a body of information in a clearly defined specialized field or pertaining to a specified mission with the intent of compiling, digesting, repackaging, or otherwise organizing and presenting pertinent information in a form most authoritative, timely, and useful to a society of peers and management" [20].

A typical information analysis center is the Radiation Shielding Information Center functioning since 1962 at the Oak Ridge National Laboratory, Oak Ridge, Tennessee. This center is jointly sponsored by the erstwhile United States Atomic Energy Commission and the United States Defense Nuclear Agency. Staffed by research scientists, information specialists, and support personnel, the center serves as a national technology resource for collecting, organizing, evaluating, and disseminating information related to radiation from reactors, radioisotopes, weapons, accelerators, and outer space. It also collects, repackages, and disseminates digital computer programs for use in radiation shielding calculations. With the assistance of the computerized Storage and Retrieval Information System (SARIS), the center answers technical inquiries, publishes bibliographies and state-of-the-art reviews, and routinely disseminates selected information to fill individual needs indicated by customer profiles [21]. In 1974 there were about 27 information analysis centers similar in scope and functions to the Radiation Shielding Information Center, sponsored wholly or jointly by the former USAEC. The U.S. Department of Defense also has a chain of similar information analysis centers typified by the Defense Metals Information Center at the Batelle Memorial Institute, Columbus, Ohio, and the Military Entomology Information Service at the Walter Reed Army Medical Center, Washington, D.C. [22]. Descriptions of federally supported information analysis centers may be found in the *Directory of Federally Supported Information Analysis Centers,* 3rd edition (Washington, D.C.: Library of Congress, National Referral Center, 1974). The National Referral Center has also published a few other directories that are helpful in locating brief descriptions of other information centers [23-25]. Aluri and Yannarella have compiled a bibliography of the publications of selected information analysis centers [26].

18.8 Decentralized Bibliographic Control

As noted earlier, there is no one national system or agency in the United States that is solely responsible for the bibliographic control of scientific and technical literature. The government's concern for bibliographic control of scientific

18.8 Decentralized Bibliographic Control

and technical literature is shared by numerous agencies in the private sector including commercial enterprises, academic institutions, professional associations, and scholarly societies. There is also no government mandate that calls for cooperation or coordination within this loose assemblage of diverse agencies all engaged in the common task of disseminating scientific and technical information. However, there is a growing awareness of the need for cooperation, and the movement for voluntary cooperation among the agencies in both the public and the private sector is gaining momentum. The efforts of federations such as the Association of Scientific Information Dissemination Centers (ASIDIC) and the National Federation of Abstracting and Indexing Services (NFAIS) in this direction are noteworthy.

An obvious disadvantage of such a decentralized system is the watesful duplication of effort. But this is balanced by a number of advantages. The coexistence of a plurality of bibliographic agencies in both the private and the public sector promotes a competitive atmosphere in which every agency is forced to strive for better performance, speed, and economy in its operations. The net result to the users of scientific and technical information is the availability of a great variety of products and services at competitive prices. A striking example of the net advantage to the user is the progressive reduction in the cost of online bibliographic searching accompanied by the ever-expanding repertoire of databases made possible by the healthy competition among three commercial retailers of online bibliographic access. The companies involved are: Bibliographic Retrieval Services, Inc. (BRS), System Development Corporation (ORBIT System), and Lockheed Missiles and Space Company, Inc. (DIALOG System). The competition among companies that is inevitable in a pluralistic society and the advances in computer technology have made it possible for these companies to announce "the era of the $5 search" [27].

A highly centralized organization for bibliographic control seems to be more conducive to an aggressive dissemination and utilization of scientific and technical information. In the Soviet block countries, abstracting and indexing services are extensively and purposefully exploited as a primary means of propagating scientific and technical information. These services are aggressively disseminated not only to scientists and engineers, but also to factory foremen and graduate students. In his study of the diffusion of abstracting and indexing services, Irving Klempner observed [28]:

> Nowhere has the abstract been more fully adopted and more doggedly exploited in the diffusion of scientific and technical information than in communist countries. Whether indicative or informative, in card format or published in primary journals or in abstracting and indexing services, the abstract has been frequently used as the official medium for current awareness and information retrieval.

The diffusion and utilization of abstracting and indexing services in the United States appear to be much less extensive. Klempner's study indicated that in 1967, VINITI disseminated its *Referativnyi Zhurnal* (excluding *Ekspress Informatsii* and other related services) to approximately 375,000 Soviet recipients. During the same year in the United States only 12,255 recipients received the four federally sponsored secondary services: *Nuclear Science Abstracts, Scientific and Technical Aerospace Reports, Technical Abstracts Bulletin,* and *United States Government Research and Development Reports* [29]. However, such comparisons should be tempered by an appreciation of the widely divergent political ideologies and sociocultural environments, as well as methods of reporting statistics, obtaining in these two countries.

Selected List of Guides to Literature, Bibliographies, Catalogs, and Abstracting and Indexing Services

General Science and Engineering

Guides to Literature

1. Ching-Chih Chen. *Scientific and Technical Information Sources* (Cambridge, Mass.: MIT Press, 1977). 519p.
2. Denis J. Grogan. *Science and Technology: An Introduction to the Literature.* 3rd edition (London: Clive Bingley, 1976). 343p.
3. James M. Doyle and George H. Grimes. *Reference Sources: A Systematic Approach* (Metuchen, N.J.: Scarecrow Press, 1976). 301p.
4. Bernard Houghton. *Technical Information Sources: A Guide to Patent Specifications, Standards, and Technical Report Literature.* 2nd edition (Hamden, Conn.: Shoe String Press, 1972). 119p.
5. Earl J. Lasworth. *Reference Sources in Science and Technology* (Metuchen, N. J.: Scarecrow Press, 1972). 305p.
6. Michael Karol. *Guide to Russian Reference Books.* Hoover Institution Bibliographical Series 32 (Stanford, Calif.: Hoover Institution on War, Revolution, and Peace, Stanford University, 1967). 384p.
7. Harold R. Malinowsky, Richard A. Gray, and Dorothy A. Gray. *Science and Engineering Literature: A Guide to Reference Sources.* 2nd edition (Littleton, Colo.: Libraries Unlimited, 1976). 368p.
8. K. W. Mildren and N. G. Meadows. *Use of Engineering Literature* (London: Butterworths, 1976). 621p.
9. Ellis Mount. *Guide to Basic Information Sources in Engineering* (New York: Jeffrey Norton, 1976). 196p.
10. North Atlantic Treaty Organization. Advisory Group for Aerospace Research and Development. *How to Obtain Information in Different Fields of Science and Technology: A User's Guide.* AGARD LS-69 (Neuilly-sur-Seine, France: AGARD, 1974). Available as AD 780 061 from NTIS.

Selected Lists

11. Eugene P. Sheehy. *Guide to Reference Books.* 9th edition (Chicago: American Library Association, 1976). 1015p.
12. Albert J. Walford. *Guide to Reference Material.* 3rd edition (London: The Library Association, 1973). v. 1: *Science and Technology.* 615p.
13. Sylvia G. Weiser. *Guide to the Literature of Engineering, Mathematics, and the Physical Sciences.* 3rd edition. Technical Memorandum TG 230-B3 (Silver Spring, Md.: Johns Hopkins University, Applied Physics Laboratory, 1972). 209p.

Bibliographies of Bibliographies

14. *Bibliographic Index: A Cumulative Bibliography of Bibliographies* (New York: H. W. Wilson, 1938-). Published three times a year; annual cumulation.
15. Theodore Besterman. *A World Bibliography of Bibliographies and of Bibliographical Catalogues, Calendars, Abstracts, Digests, Indexes, and the Like.* 4th edition (Lausanne: Societas Bibliographica, 1965-66). 5 v.
16. Alice F. Toomey. *A World Bibliography of Bibliographies, 1964-1974* (Totowa, N.J.: Rowman and Littlefield, 1977). 2 v.
    A decennial supplement to Theodore Besterman's *A World Bibliography of Bibliographies.*
17. Theodore Besterman. *Physical Sciences: A Bibliography of Bibliographies* (Totowa, N.J.: Rowman and Littlefield, 1971). 2 v.

Bibliographies and Catalogs

18. *Bibliographic Guide to Technology, 1975.* The Research Libraries of the New York Public Library, the Library of Congress, and the Engineering Societies Library, New York (Boston, Mass.: G. K. Hall, 1976).
19. *Classified Subject Catalog of the Engineering Societies Library, New York City* (Boston, Mass.: G. K. Hall, 1965). 12 v. 215,000 cards.
    Index to Classified Subject Catalog. 26,700 entries.
    First Supplement, 1964; *Tenth Supplement,* 1974.
20. The John Crerar Library, Chicago. *Author-Title Catalog* (Boston, Mass.: G. K. Hall, 1967). 35 v. 28,554p. 599,000 cards.
    Classified Subject Catalog. 1967. 42 v. 33,167p. 730,000 cards.
    Subject Index to the Classified Subject Catalog. 1967. 610p. 47,000 entries.
21. Northeastern University, Boston. Library. *A Selective Bibliography in Science and Engineering* (Boston, Mass.: G. K. Hall, 1964). 550p. 16,500 entries.
22. David M. Knight. *Natural Science Books in English 1600-1900* (London: B. T. Batsford, Ltd., 1972). 262p.
23. James M. Matarazzo and James M. Kyed. *Scientific, Engineering, and Medical Societies Publications in Print, 1976-1977.* 2nd edition (New York: R. R. Bowker, 1976). 509p.

24. *McGraw-Hill Basic Bibliography of Science and Technology.* Compiled and annotated by the editors of the *McGraw-Hill Encyclopedia of Science and Technology* (New York: McGraw-Hill, 1966). 738p.
25. Kenneth J. Rider. *History of Science and Technology: A Select Bibliography for Students.* 2nd edition (London: The Library Association, 1970). 75p.
26. Rutgers University, New Brunswick, N.J. Bureau of Information Science Research. *Bibliography of Research Relating to the Communication of Scientific and Technical Information* (New Brunswick, N.J.: Rutgers University, 1967). 732p.
27. *Scientific and Technical Books and Serials in Print, 1979.* 5th edition (New York: R. R. Bowker, 1978). Annual.
    Bibliography of approximately 66,800 books indexed by author, title, and some 12,000 subjects. Includes about 18,000 serials listed by subject and title. Directory of 2000 publishers. Ordering information included.

Abstracting and Indexing Services

28. *Applied Science and Technology Index* (New York: H. W. Wilson, 1958-). Monthly except July; quarterly and annual cumulations.
    One of the two indexes that replaced *Industrial Arts Index*, 1913-1958.
29. *Bulletin Signaletique* (Paris, France: Centre National de la Recherche Scientifique, 1940-). 33 sections.
    Title from 1940 to 1956: *Bulletin Analytique*.
30. *Catalogue of Scientific Papers, 1800-1900* (London: The Royal Society, 1867-1925). 23 v.
31. *International Catalogue of Scientific Literature* (London: The Royal Society, 1901-1914). 14 v.
32. *Current Contents* (Philadelphia, Penn.: Institute for Scientific Information, 1961-). Weekly.
    1. Agriculture, Biology and Environmental Sciences
    2. Engineering, Technology and Applied Sciences
    3. Clinical Practice
    4. Life Sciences
    5. Physical, Chemical and Earth Sciences
    6. Social and Behavioral Sciences
    7. Arts and Humanities
33. *Engineering Index* (New York: Engineering Index, Inc., 1892-). Monthly; annual cumulation.
34. *Pandex: Current Index to Scientific and Technical Literature* (New York: CCM Information Sciences, 1969-). Bimonthly.
35. *Referativnyi Zhurnal* (Moscow, USSR: VINITI and Academy of Sciences of the USSR, 1953-).
    About 70 subject series covering different branches of science, engineering, and technology. Various frequencies.

Selected Lists 313

36. *Science Citation Index* (Philadelphia, Penn.: Institute for Scientific Information, 1961-). Quarterly; annual cumulation.

Additional bibliographies and abstracting and indexing services are discussed in several other chapters (e.g., Conference Literature, Dissertations, Patents, Technical Reports, Bibliographical Literature).

Mathematics and Statistics

Guides to Literature

37. Nancy D. Anderson. *French Mathematical Seminars: A Union List* (Providence, R.I.: American Mathematical Society, 1978). 96p.
38. Elie M. Dick. *Current Information Sources in Mathematics: An Annotated Guide to Books and Periodicals, 1960-72* (Littleton, Colo.: Libraries Unlimited, 1973). 281p.
39. A. R. Dorling. *Use of Mathematical Literature* (London: Butterworths, 1977). 260p.
40. John E. Pemberton. *How to Find Out in Mathematics.* 2nd edition (Oxford: Pergamon Press, 1970).
    Sources are arranged by Dewey Decimal Classification number.
41. Barbara K. Schaefer. *Using the Mathematical Literature. A Practical Guide* (New York: Marcel Dekker, 1979). 160p.
42. Joseph Zaremba. *Mathematical Economics and Operations Research.* Economics Information Guide Series, v. 10 (Detroit, Mich.: Gale Research Co., 1978). 650p.

Bibliography of Bibliographies

43. H. O. Lancaster. *Bibliography of Statistical Bibliographies* (Edinburgh: Oliver & Boyd, 1968). 103p.

Bibliographies

44. *M. C. Publications, 1946-1971* (Amsterdam: Mathematical Centre, 1971). A bibliography of the publications of the Mathematical Centre, Amsterdam. Part I: Bibliography and author index, 190p. Part II: KWIC index, 121p.
45. William L. Schaaf. *A Bibliography of Recreational Mathematics* (Reston, Va.: National Council of Teachers of Mathematics, 1973). 175p.
46. Chia Kuei Tsao. *Bibliography of Mathematics Published in Communist China during the Period 1949-1960.* Contemporary Chinese Research Mathematics, v. 1 (Providence, R.I.: American Mathematical Society, 1961). 83p.
    Includes a list of 70 Chinese journals containing mathematical papers.
47. Herman O. A. Wold. *Bibliography on Time Series and Stochastic Processes. An International Team Project* (Cambridge, Mass.: MIT Press, 1965). 516p.

Abstracting and Indexing Services

48. *An Author and Permuted Title Index to Selected Statistical Journals.* Brian L. Joiner and others (Washington, D.C.: U.S. Government Printing Office, 1970). 506p. National Bureau of Standards Special Publication 321.
Journals indexed: *Annals of Mathematical Statistics* (1961-1969); *Biometrics* (1965-1969); *Biometrica* (1951-1969); *Journal of the American Statistical Association* (1956-1969); *Journal of the Royal Statistical Society, Series B* (1954-1969, #2); *South African Statistical Journal* (1967-1969, #2); and *Technometrics* (1959-1969). Bibliography: pp. 433-506.
49. *Current Mathematical Publications* (Providence, R.I.: American Mathematical Society, 1975-). Biweekly.
Continues: *Contents of Contemporary Mathematical Journals and New Publications,* 1969-1974.
50. *International Abstracts in Operations Research* (Amsterdam: North Holland Publishing Co., for the International Federation of Operational Research Societies, 1961-). Quarterly.
51. *Mathematical Reviews* (Providence, R.I.: American Mathematical Society, 1940-). Monthly.
52. *Index of Mathematical Papers* (Providence, R.I.: American Mathematical Society, 1970-).
Annual author and subject index to *Mathematical Reviews.*

Computer Science

Guides to Literature

53. ASLIB Computer Information Group. *World List of Computer Periodicals* (London: United Trade Press, 1977). 140p.
54. Ciel Michele Carter. *Guide to Reference Sources in the Computer Sciences* (New York: Macmillan Information, 1974). 237p.
55. Chester Morrill. *Computers and Data Processing Information Sources* (Detroit, Mich.: Gale Research Co., 1969). 275p.
56. Alan A. Prichard. *A Guide to Computer Literature: An Introductory Survey of the Sources of Information.* 2nd edition (Hamden, Conn.: Linnet Books, 1972). 194p.
57. Karen T. Quinn. *Guide to Literature on Computers* (Washington, D.C.: American Society for Engineering Education, 1970).

Bibliographies

58. A. H. Agajanian. *Computer Technology: Logic, Memory, and Microprocessors. A Bibliography* (New York: IFI/Plenum, 1978). 346p.
59. *International Computer Bibliography* (New York: Science Associates International, 1973). 700p.

Selected Lists

Nearly 6000 abstracts of books and technical reports on computer technology and applications (1960-1968) from 39 countries. Keyword subject and author indexes.
60. U.S. National Aeronautics and Space Administration. *COSMIC: A Catalog of Selected Computer Programs* (Washington, D.C.: U.S. Government Printing Office, 1977). 152p.

Abstracting and Indexing Services

61. *Computer Abstracts* (Jersey, British Channel Islands: Technical Information Co., 1957-). Monthly.
62. *Computer and Control Abstracts* (London: Institution of Electrical Engineers, INSPEC, 1966-). Monthly.
    Section C of *Science Abstracts*.
63. *Current Papers on Computers and Control* (London: Institution of Electrical Engineers, INSPEC, 1969-). Monthly.
64. *Quarterly Bibliography of Computers and Data Processing* (Phoenix, Ariz.: Applied Computer Research, 1979-).
    Annotated index to current computer literature arranged under more than 200 subject headings.

Physics and Astronomy

Guides to Literature

65. Stephen G. Brush. *Resources for the History of Physics* (Hanover, N.H.: The University Press of New England, 1972).
    Part I: Guide to Books and Audiovisual Materials. 86p.
    Part II: Guide to Original Works of Historical Importance and Their Translations into Other Languages. 90p.
66. Herbert Coblans. *Use of Physics Literature* (London: Butterworths, 1976). 290p.
67. Tom F. Connolly. *Solid State Physics Literature Guides* (New York: Plenum Publishing Corporation, 1970-).
    These are bibliographic guides largely based on papers received since 1960 at the Research Materials Information Center of the Oak Ridge National Laboratory. Permuted title subject index, author index, organization index, patent index, and report number index.
68. D. A. Kemp. *Astronomy and Astrophysics: A Bibliographical Guide* (Hamden, Conn.: Archon Books, 1970). 584p.
    A selective guide to books, articles, technical reports, catalogs, ephemeries, etc., arranged in 75 subject categories. Author and subject indexes.
69. Robert A. Seal. *A Guide to the Literature of Astronomy* (Littleton, Colo.: Libraries Unlimited, 1977).
    Updates Kemp's *Astronomy and Astrophysics*.

70. Robert H. Whitford. *Physics Literature: A Reference Manual.* 2nd edition (Metuchen, N.J.: Scarecrow Press, 1968). 272p.
71. B. Yates. *How to Find Out About Physics* (Oxford: Pergamon Press, 1965). 175p.
    "A guide to sources of information arranged by the Dewey Decimal Classification" (subtitle).

Bibliographies and Catalogs

72. *Dictionary Catalog of the Princeton University Plasma Physics Laboratory Library* (Boston, Mass.: G. K. Hall, 1970). 4 v. 2957p. 62,000 cards. *First Supplement.* 1973. 874p. 18,300 cards.
    Includes entries for journal articles.
73. G. C. Battle, Tom Connolly, and M. Keese. *Laser Window and Mirror Materials* (New York: IFI/Plenum, 1977). 290p. (Solid State Physics Literature Guides, v. 9).
74. Great Britain. Department of Industry. National Engineering Laboratory. *Heat Bibliography* (Edinburgh, England: HMSO, 1948-). Annual
75. United States. Naval Observatory. Library. *Catalog of the Naval Observatory Library,* Washington, D.C. (Boston, Mass.: G. K. Hall, 1976). 6 v.

Abstracting and Indexing Services

76. *Acoustics Abstracts* (London: Multi-Science Publishing Co., 1967-). Bimonthly.
    Formerly: *Acoustics and Ultrasonics Abstracts.*
77. *Astronomischer Jahresbericht* (Berlin: G. Reimer, 1899-1968). Annual. International classified bibliography of journal articles, books, reports, conference literature, etc., in astronomy, astrophysics, and related subjects. Author and subject indexes. Continued by *Astronomy and Astrophysical Abstracts,* 1969-.
78. *Astronomy and Astrophysical Abstracts* (Berlin: Springer-Verlag, 1969-). Semiannual.
    Abstracts are in English, French, or German, and are arranged in 108 categories. Author and subject indexes.
79. *Astrophysical Abstracts* (Glasgow: Gordon and Breach Science Publishers, 1969-).
80. *Current Papers in Physics* (London: Institution of Electrical Engineers, INSPEC, 1966-). Semimonthly.
81. *Physics Abstracts* (London: Institution of Electrical Engineers, INSPEC, 1898-). Semimonthly.
    To be superseded by *Physics Briefs.*
82. *Physics Briefs* (Fachinformationszentrum Energie Physik Mathematik, West Germany, and American Institute of Physics, New York). Semimonthly. Supersedes *Current Physics Index.*

Selected Lists

Chemistry, Chemical Engineering and Technology

Guides to Literature

83. A. Antony. *Guide to Basic Information Sources in Chemistry* (New York: John Wiley, Halsted Press, 1979). 219p.
84. Janet E. Ash and Ernest Hyde. *Chemical Information Systems* (New York: Halsted Press, 1975). 309p.
85. Robert T. Bottle. *The Use of Chemical Literature*. 2nd edition (London: Blackwell, 1969). 249p.
86. Charles R. Burman. *How to Find Out in Chemistry*. 2nd edition (Oxford: Pergamon Press, 1966). 226p.
87. Charles H. Davis and James E. Rush. *Information Retrieval and Documentation in Chemistry* (Westport, Conn.: Greenwood Press, 1974).
88. Clive Evers. *How to Find Out in Chemical Spectroscopy* (London: City of London Polytechnic Library and Learning Resources Service, 1975). 12p.
89. Melvin G. Mellon. *Chemical Publications: Their Nature and Use*. 4th edition (New York: McGraw-Hill, 1965). 324p.
90. Austin V. Signeur. *Guide to Gas Chromatography Literature* (New York: Plenum Press, 1964–1974). 3 v.
91. Henry M. Woodburn. *Using Chemical Literature: A Practical Guide* (New York: Marcel Dekker, 1974). 302p.
92. Russell Brown and G. A. Campbell. *How to Find Out About the Chemical Industry* (Oxford: Pergamon Press, 1969). 219p.
93. Kathleen Bourton. *Chemical and Processing Engineering Unit Operations. A Bibliographical Guide* (New York: IFI/Plenum, 1968). 534p.
94. Lawrence Franko. *The Petroleum Industry in Western Europe: A Guide to Information Sources* (New York: Garland STPM Press, 1975). 170p. Describes bibliographies, abstracts, indexes, clipping services, associations and organizations, conferences, books, reports, serials, directories, and company publications.
95. Sidney Green. *Guide to English Language Publications in Food Science and Technology*. 4th revised edition (London: Food Trade Press, 1975). 133p.
96. Theodore P. Peck. *Chemical Industries Information Sources* (Detroit, Mich.: Gale Research Co., 1978). 194p.
97. Theodore P. Peck. *Chemical Engineers' Guide to Information Sources* (Minneapolis, Minn.: University of Minnesota, Department of Chemical Engineering and Materials Science, 1973). 194p.
98. R. E. Maizell. *How to Find Chemical Information: A Guide for Practising Chemists, Teachers and Students* (New York: John Wiley, 1979). 384p.
99. Julian F. Smith. *Literature of Chemical Technology* (Washington, D.C.: American Chemical Society, 1968). 732p. (Advances in Chemistry Series, No. 78).

100. E. E. Yescombe. *Sources of Information on the Rubber, Plastics and Allied Industries* (Oxford: Pergamon Press, 1968). 253p.
101. *A Guide to Sources of Information in the Textile Industry* (London: ASLIB, Textile Group, 1970), 125p.
102. David M. Hall, Nicholas Achee, and Caroline C. Persons. *Guide to Literature on Textile Engineering and Textile Science* (Washington, D.C.: American Society for Engineering Education, 1972). 19p.
103. Joseph V. Kopycinski. *Textile Industry Information Sources* (Detroit: Mich.: Gale Research Co., 1964). 194p.
104. Valerie H. Ralston. *Textile Reference Sources: A Selective Bibliography* (Storrs, Conn.: Wilbur Cross Library, 1973).
105. Helen G. Sommar. *A Brief Guide to Sources of Fibre and Textile Information* (Washington, D.C.: Information Resources Press, 1973). 138p.
106. Theodore P. Peck. *Occupational Safety and Health: A Guide to Information Sources* (Detroit, Mich.: Gale Research Co., 1974). 261p.
107. Albert C. Vara. *Food and Beverage Industries: A Bibliography and Guidebook* (Detroit, Mich.: Gale Research Co., 1970). 215p.

Bibliographies and Catalogs

108. Donald T. Hawkins, Warren E. Falconer, and Neil Bartlett. *Noble Gas Compounds: A Bibliography* (New York: IFI/Plenum, 1978). 179p.
109. B. J. Haywood. *Thin Layer Chromatography: An Annotated Bibliography, 1964-1968* (Ann Arbor, Mich.: Ann Arbor Science Publishers, 1968). 284p.
110. D. Hunter. *The Literature of Paper Making, 1390-1800* (New York: Burt Franklin, 1925; reprinted 1971). 47p. (Bibliography and Reference Serial, No. 411).
111. William E. Tarrants. *A Selected Bibliography of Reference Materials in Safety Engineering* (Chicago: American Society of Safety Engineers, 1967). 152p.
112. Jack Weiner and Vera Pollock. *Nonsulfur Pulping* (Appleton, Wis.: Institute of Paper Chemistry, 1977). 163p. (Bibliographic Series, The Institute of Paper Chemistry, No. 275).
113. University of Pennsylvania. *Catalog of the Edgar F. Smith Memorial Collection in the History of Chemistry* (Boston, Mass.: G. K. Hall, 1960). 524p. 11,000 cards.
    The nucleus of the collection was assembled by Edgar F. Smith, Professor of Chemistry and Provost of the University of Pennsylvania. Miss Eva Armstrong, curator of the collection, continued to expand it after his death in 1927.

Abstracting and Indexing Services

114. *Abstract Bulletin of the Institute of Paper Chemistry* (Appleton, Wis.: Institute of Paper Chemistry, 1930-). Monthly.

Selected Lists

115. *Analytical Abstracts* (London: Society for Analytical Chemistry, 1954-). Monthly.
 Supersedes: *British Abstracts, Series C: Analysis and Apparatus* (1944-1953).
116. *Chemical Abstracts* (Columbus, Ohio: Chemical Abstracts Service, American Chemical Society, 1907-). Weekly.
117. *Chemical-Biological Activities. An Index to Current Literature on the Biological Activity of Chemical Substances* (Columbus, Ohio: Chemical Abstracts Service, American Chemical Society, 1965-). Semimonthly.
118. *Chemical Titles* (Columbus, Ohio: Chemical Abstracts Service, American Chemical Society, 1961-). Semimonthly.
119. *Current Abstracts of Chemistry and Index Chemicus* (Philadelphia, Penn.: Institute for Scientific Information, 1960-). Weekly.
120. *Current Chemical Papers* (London: Chemical Society, 1954-). Monthly.
121. *Current Contents: Physical and Chemical Sciences* (Philadelphia, Penn.: Institute for Scientific Information, 1961-). Weekly.
122. *Electroanalytical Abstracts.* (Basel: Birkhauser Verlag, 1963-). Bimonthly.
 Continues the abstracts section of the *Journal of Electroanalytical Chemistry.*
123. *Gas and Liquid Chromatography Abstracts* (Barking, Essex, England: Applied Science Publishers, 1958-). Quarterly.
 Formerly: *Gas Chromatography Abstracts.*
124. *Lead Abstracts. A Review of Recent Technical Literature on the Uses of Lead and Its Products* (London: Lead Development Association, 1958-). Monthly.
125. *Polymer Science and Technology: Patents* (POST-P), and *Polymer Science and Technology: Journals* (POST-J) (Columbus, Ohio: Chemical Abstracts Service, American Chemical Society, 1967-). Semimonthly.
 Since 1971, available only in computer-readable format.
126. *Rolston's Index to the Society of Plastics Industry Reinforced Plastics/ Composites Institute Conference Proceedings, 1970-1976.* Albert J. Rolston (Alexandria, Va.: R. Wilson Publishing Co., 1977). 178p.
127. *Theoretical Chemical Engineering Abstracts* (London: Technical Information Co., 1964-). Bimonthly.
128. *Zinc Abstracts. A Review of Recent Technical Literature on the Uses of Zinc and Its Products* (London: Zinc Development Association, 1943-). Bimonthly.

Earth Sciences, Mining and Metallurgy

Guides to Literature

129. Eleanor B. Gibson and Elizabeth W. Tapia. *Guide to Metallurgical Information.* 2nd edition (New York: Special Libraries Association, 1965). 85p. (SLA Bibliography No. 3).

130. Marjorie R. Hyslop. *A Brief Guide to Information Sources in Mining, Minerals, and Geosciences* (New York: Interscience Publishers, 1965). 599p.
131. Dederick C. Ward and Marjorie W. Wheeler. *Geologic Reference Sources: A Subject and Regional Bibliography of Publications* 2nd edition (Metuchen, N.J.: Scarecrow Press, 1972). 453p.
132. D. White. *How to Find Out in Iron and Steel* (Oxford: Pergamon Press, 1970). 184p.
133. Virginia Lee Wilcox. *Guide to Literature on Mining and Mineral Resources Engineering* (Washington, D.C.: American Society for Engineering Education, Engineering School Libraries Division, 1972). 22p.
134. David N. Wood. *Use of Earth Sciences Literature* (Reading, Mass.: Butterworths, 1973). 459p.

Bibliographies and Catalogs

135. Filmore C. F. Earney. *Researcher's Guide to Iron Ore: An Annotated Bibliography on the Economic Geography of Iron Ore* (Littleton, Colo.: Libraries Unlimited, 1974). 595p.
136. Subhash C. Malhotra. *Bibliography on Copper Smelting, 1940-1973* (White Pine, Mich.: White Pine Copper Co., 1973). 271p.
137. John D. Ridge. *Annotated Bibliographies of Mineral Deposits in Africa, Asia (Exclusive of the USSR) and Australasia* (Oxford: Pergamon Press, 1976). 545p.
138. *Publications of the Geological Survey* (Washington, D.C.: U.S. Geological Survey, 1979-). Annual.
   A cumulated list covering the period 1879-1961 has been published.
139. American Geographical Society, New York. *Author/Title, Subject, and Geographic Catalogs of the Glaciology Collection, Department of Exploration and Field Research* (Boston, Mass.: G. K. Hall, 1971). 3 v. 2332p.
   48,900 catalog cards of books, journals, reprint photographs, maps and unpublished reports from the mid-1800s in glaciology. Other fields covered are: geography, geophysics, geology, and earth science.
140. *Catalog of the Library of the Arctic Institute of North America, Montreal, Canada* (Boston, Mass.: G. K. Hall, 1968). 4 v. 3174p.
   First Supplement, 1971. 902p.
   Second Supplement, 1974. 2 v. 1588p.
   Catalog of some 122,200 cards describing 9000 volumes of books, 20,000 pamphlets and reprints, and 1200 titles of periodicals.
141. *Catalog of the Library of the Marine Biological Laboratory and the Woods Hole Oceanographic Institution, Woods Hole, Mass.* (Boston, Mass.: G. K. Hall, 1971). 12 v. 9339p.
   An author catalog of some 339,000 cards. A separate *Journal Catalog* was published in 1971 (418p. 8700 cards).

Selected Lists 321

142. U.S. Department of the Interior, Washington, D.C. *Catalog of the United States Geological Survey Library* (Boston, Mass.: G. K. Hall, 1964). 25 v. 19,855p. 416,000 cards.
    First Supplement, 1972. 11 v. 8481p. 178,000 cards.
    Second Supplement, 1975. 4 v. 2841p. 59,600 cards.
    Author, title, and subject indexes to more than half a million documents (mostly bound volumes), pamphlets, and some maps, covering geology, paleontology, petrology, mineralogy, ground and surface water, cartography, and mineral resources. Selective coverage in supporting subjects such as mathematics, physics, engineering, chemistry, soil science, botany, zoology, oceanography, and natural resources.
143. University of California, San Diego. *Catalog of the Scripps Institution of Oceanography Library* (Boston, Mass.: G. K. Hall, 1970-1973). 16 v. 250,000 cards.
    Subjects covered in the collection: oceanography, marine biology, marine technology, physics, chemistry, mathematics, atmospheric sciences, fisheries, geology, geophysics, and zoology. Imprint dates start from 1633.
144. Woods Hole Oceanographic Institution, Woods Hole, Mass. *Oceanographic Index* (Boston, Mass.: G. K. Hall, 1972). 15 v. 300,000 cards.
    A set of working guides to journal articles, monographs, and other literature on the marine sciences held in the library of the Marine Biological Laboratory and the library of the Woods Hole Oceanographic Institution. They are also cumulative indexes to the bibliographic-abstract sections of *Deep Sea Research*. Coverage emphasizes biological and physical oceanography and marine chemistry, geology, and meteorology. Many related topics such as limnology, terrestrial geology, basic chemistry, fisheries biology, malacology, algology, etc., are excluded.

Abstracting and Indexing Services

145. *Alloys Index* (Metals Park, Ohio: American Society for Metals, 1974-). Monthly.
146. *Bibliography and Index of Geology* (New York: Geological Society of America, 1933-). Monthly.
    Title of Volume 1: *Bibliographic Contributions of the Geological Society of America*. Published annually from 1933 to 1940. Title until 1968: *Bibliography and Index of Geology Exclusive of North America*.
147. *Ceramic Abstracts* (Columbus, Ohio: American Ceramic Society, 1922-). Monthly.
148. *Corrosion Abstracts* (Houston, Tex.: National Association of Corrosion Engineers, 1962-). Bimonthly.
149. *Corrosion Control Abstracts* (London: Scientific Information Consultants, Ltd., 1966-). Bimonthly.

"English translation of *Referativnyi Zhurnal: Korroziya i Zashchita ot Korrozii,* with additional abstracts from other sources" (subtitle).
150. *Geophysical Abstracts* (Norwich, England: Geo Abstracts, Ltd., 1977–). Bimonthly.
151. *Geotitles Weekly* (London: Geosystems, 1969–). Weekly.
152. *IMM Abstracts* (London: Institution of Mining and Metallurgy, 1950–). Bimonthly.
153. *Metal Finishing Abstracts* (Middlesex, England: Finishing Publications, Ltd., 1959–). Bimonthly.
154. *Metals Abstracts. Metals Abstracts Index* (London: Institute of Metals; Metals Park, Ohio: American Society for Metals, 1968–). Monthly. Formed by merging *ASM Review of Metal Literature* and *Metallurgical Abstracts.*
155. *Mineralogical Abstracts* (London: Mineralogical Society, 1959–). Quarterly.
156. *Rheology Abstracts* (London: Pergamon Press for the British Society of Rheology, 1958–). Quarterly.
157. *World Aluminum Abstracts* (Metals Park, Ohio: American Society for Metals, 1968–). Monthly.

Aviation, Aerospace, and Meteorology

Guide to Literature

158. Bernard M. Fry and Foster Mohrhardt. *A Guide to Information Sources in Space Science and Technology* (New York: Wiley-Interscience, 1963). 579p.

Bibliographies

159. *Bibliography of Aeronautics.* Paul Brockett (Washington, D.C.: Smithsonian Institution, 1910). 940p. (Smithsonian Miscellaneous Collection, v. 55).
    13,500 items arranged alphabetically by author or title. Continued by *Bibliography of Aeronautics,* U.S. National Advisory Committee on Aeronautics, 1909–1932 (Washington, D.C.: U.S. Government Printing Office, 1921–1936). 14 v.
160. U.S. Library of Congress. Science and Technology Division. Special Bibliographies Section. *Air Force Scientific Research Bibliography, 1950/56–1965.* G. Vernon Hooker (Washington, D.C.: U.S. Government Printing Office, 1961–1970). 8 v.
161. *Index to Parachuting, 1900–1975. An Annotated Bibliography.* Michael Horan (New York: Garland STPM Press, 1976). 200p.
    Over 2700 entries from American and foreign periodicals, including references from a number of disciplines that overlap with parachuting, such as medicine, psychology, aerospace engineering, etc.

Selected Lists

Abstracting and Indexing Services

162. *Aeronautical Engineering: A Special Bibliography with Indexes* (Washington, D.C.: U.S. National Aeronautics and Space Administration, 1970-).
    A continuing series of annotated bibliographies of unclassified technical reports and journal articles announced in *Scientific and Technical Aerospace Reports* and *International Aerospace Abstracts*. Annual cumulative index.
163. *Aerospace Medicine and Biology: A Continuing Bibliography with Indexes* (Washington, D.C.: U.S. National Aeronautics and Space Administration, 1952-).
    A continuing series of annotated bibliographies of unclassified reports and journal articles announced in *Scientific and Technical Aerospace Reports* and *International Aerospace Abstracts*. Annual index.
164. *Air University Library Index to Military Periodicals* (Alabama: Maxwell Air Force Base, 1949-). Quarterly; annual cumulation.
    Supersedes *Air University Periodicals Index*. A subject index to articles, news items, and editorials in 65 English language military and aeronautical periodicals.
165. *International Aerospace Abstracts* (New York: American Institute of Aeronautics and Astronautics, 1961-). Semimonthly.
166. *Meteorological and Geoastrophysical Abstracts* (Boston, Mass.: American Meteorological Society, 1950-). Monthly.
167. *Cumulated Bibliography and Index to Meteorological and Geoastrophysical Abstracts, 1950-1969.* American Meteorological Society (Boston, Mass.: G. K. Hall, 1972).
    Author sequence and UDC System sequence are available separately.
168. *NASA Patent Abstracts Bibliography* (Washington, D.C.: U.S. National Aeronautics and Space Administration, 1972-). Semiannual.
169. *Aeronautical Engineering Index* (New York: Institute of the Aeronautical Sciences, 1947-). Annual.
170. *Scientific and Technical Aerospace Reports* (Washington, D.C.: U.S. National Aeronautics and Space Administration, 1963-).

Civil Engineering

Guides to Literature

171. Howard B. Bentley. *Building Construction Information Sources* (Detroit, Mich.: Gale Research Co., 1964). 181p.
172. Jules B. Godel. *Sources of Construction Information: An Annotated Guide to Reports, Books, Periodicals, Standards, and Codes* (Metuchen, N.J.: Scarecrow Press, 1977). 673p.
173. Beverly Hickok. *Guide to Literature on Transportation Engineering* (Washington, D.C.: American Society for Engineering Education, 1970).

174. Rita McDonald. *Guide to Literature on Civil Engineering* (Washington, D.C.: American Society for Engineering Education, 1972). 21p.
175. Kenneth N. Metcalf. *Transportation Information Sources* (Detroit, Mich.: Gale Research Co., 1965). 307p.
176. Jean Paul Reid. *Bibliography of Bibliographies on Transportation and Related Fields* (Monticello, Ill.: Council of Planning Librarians, 1978). 70p.
177. Denison L. Smith. *How to Find Out in Architecture and Building. A Guide to Sources of Information* (Oxford: Pergamon Press, 1967). 232p.

Bibliography

178. Emanuel Benjamin Ocran, Jr. *Transportation Costs and Costing, 1917-1973. A Select Annotated Chronological Bibliography* (New York: Garland STPM Press, 1975). 751p.

Abstracting and Indexing Services

179. *ASCE Publications Abstracts* (New York: American Society of Civil Engineers). Bimonthly.
180. *Architectural Periodicals Index* (London: Royal Institute of British Architects, 1973-). Quarterly; last issue of each year is an annual cumulative issue.
181. *Transportation Master File, 1921-1971* (Washington, D.C.: United States Historical Documents Institute, 1971). Semiannual cumulative supplements.
182. *Transportation Research Abstracts* (Washington, D.C.: National Academy of Sciences, Transportation Research Board, 1931-1975). Monthly.
    Formerly: *Highway Research Abstracts*. Beginning with February 1976, merged with the abstracts section of *Transportation Research News*.
183. *Transportation Research News* (Washington, D.C.: National Academy of Sciences, Transportation Research Board, 1976-). Bimonthly.
    Includes an abstracts section that supersedes *Transportation Research Abstracts*, 1931-1975.

Mechanical Engineering

Guides to Literature

184. Bernard Houghton. *Mechanical Engineering: The Sources of Information* (Hamden, Conn.: Archon Books, 1970). 311p.

Selected Lists

Bibliographies and Catalogs

185. Automotive Bibliography. *American Libraries,* 4:481–484 (September 1973).
186. Thomas Crawshaw. *High Temperature Materials for Gas Turbines* (Leicester: GEC Power Engineering Limited, Library, 1976). 75p.
187. U.K. Ministry of Technology. National Engineering Laboratory. *Heat Bibliography* (Edinburgh: Her Majesty's Stationery Office, 1948/52–1968). 15 v.
188. Paul B. Cors. *Railroads* (Littleton, Colo.: Libraries Unlimited, 1975). 152p. (Spare Time Guides, No. 8).
    Annotated list of 259 books on all aspects of railroading, including locomotives, rolling stock, stations, etc., arranged under subject headings. Directory of publishers and author-title-subject index.
189. *The Automotive History Collection of the Detroit Public Library: A Simplified Guide to Its Holdings* (Boston, Mass.: G. K. Hall, 1966). 2 v. 1091p. 22,900 cards.
    Covers technical, biographical, financial, corporate, statistical, and social aspects of automotive history. The catalog consists of four sections: dictionary catalog of books, periodicals shelf-list, check-list of automobile catalogs, and descriptions of special collections.

Abstracting and Indexing Services

190. *Applied Mechanics Reviews* (New York: American Society of Mechanical Engineers, 1948–). Monthly.
191. *Current Bibliography on Science and Technology: Mechanical Engineering* (Tokyo, Japan: Center of Science and Technology, 1958–). Semimonthly.
192. *MIRA Automobile Abstracts* (Warwickshire, England: Motor Industry Research Association, 1955–). Monthly.
    Formerly: *Automobile Abstracts.*
193. *SAE Quarterly Abstracts* (New York: Society of Automotive Engineers, 1970–). Quarterly.

Electrical Engineering and Electronics

Guides to Literature

194. J. Burkett and P. Plumb. *How to Find Out in Electrical Engineering* (Oxford: Pergamon Press, 1967). 234p.
    Information sources are arranged by UDC class numbers.
195. Richard L. Funkhouser and others. *Guide to Literature of Electrical and Electronic Engineering* (Washington, D.C.: American Society for Engineering Education, 1970).

196. J. F. Chaney and T. M. Putnam. *Electronic Properties Research Retrieval Guide, 1972-1976: A Comprehensive Compilation of Scientific and Technical Literature.* Center for Information and Numerical Data Analysis and Synthesis (CINDAS). Purdue University, Lafayette, Ind. (New York: IFI/Plenum, 1979). 4 v.
197. Gretchen R. Randle. *Electronic Industries Information Sources* (Detroit, Mich.: Gale Research Co., 1968). 227p.

Abstracting and Indexing Services

198. *Current Papers in Electrical and Electronics Engineering* (London: Institution of Electrical Engineers, INSPEC, 1969-). Monthly.
199. *Electrical and Electronics Abstracts* (London: Institution of Electrical Engineers, INSPEC, 1898-). Monthly.
    Section B of *Science Abstracts.* Formerly: *Electrical Engineering Abstracts.*
200. *Electronics and Communications Abstracts* (London: Multi-Science Publishing Co., 1961-). Monthly.
201. *Electronics and Communications Abstracts Journal* (Riverdale, Md.: Cambridge Scientific Abstracts, Inc., 1966-). 10/year.
    "An abstract journal involving theory, design and applications of electronic devices and systems" (subtitle).

Biology and Agriculture

Guides to Literature

202. R. T. Bottle and H. V. Wyatt. *The Use of Biological Literature.* 2nd edition (Hamden, Conn.: Archon Books, 1971). 379p.
203. Thomas G. Kirk. *Library Research Guide to Biology* (Ann Arbor, Mich.: Pierian Press, 1978).
204. Elizabeth P. Roberts. *Guide to Literature in Agricultural Engineering* (Washington, D.C.: American Society for Engineering Education, 1971).
205. Winifred Sewell. *Guide to Drug Information* (Hamilton, Ill.: Drug Intelligence Publications, 1976). 218p.
206. Roger C. Smith and W. Malcolm Reid. *Guide to the Literature of the Life Sciences.* 8th edition (Minneapolis, Minn.: Burgess Publishing Co., 1972). 166p.
    Earlier editions were published under the title *Guide to the Literature of the Zoological Sciences.*
207. Lloyd H. Swift. *Botanical Bibliographies: A Guide to Bibliographic Materials Applicable to Botany* (Minneapolis, Minn.: Burgess Publishing Co., 1970). 804p.

Bibliographies and Catalogs

208. Academy of Natural Sciences, Philadelphia. *Catalog of the Library of the Academy of Natural Sciences of Philadelphia* (Boston, Mass.: G. K. Hall, 1972). 16 v.

Selected Lists

209. Paul S. Galtsoff. *Bibliography of Oysters and Other Marine Organisms Associated with Oyster Bottoms and Estuarine Ecology* (Boston, Mass.: G. K. Hall, 1972). 857p.
210. I. Gilboa and H. G. Dowling. *A Bibliography on Albinism in Amphibians and Reptiles, 1849-1972* (New York: Herpetological Information Search Systems, American Museum of Natural History, 1974). 11p.
211. Harvard University. Gray Herbarium. *Gray Herbarium Index* (Boston, Mass.: G. K. Hall, 1968). 10 v.
212. Harvard University. Museum of Comparative Zoology. *Catalogue of the Library of the Museum of Comparative Zoology, Harvard University* (Boston, Mass.: G. K. Hall, 1967). 8 v.
213. Charles W. Huver. *A Bibliography of the Genus Fundulus* (Boston, Mass.: G. K. Hall, 1973). 138p.
214. Pauline A. Keehn. *A Bibliography on Marine Atlases* (Washington, D.C.: American Meteorological Society, 1968). 184p. (Special Bibliographies on Oceanography, Contribution No. 6).
215. Massachusetts Horticultural Society, Boston. *Dictionary Catalog of the Library of the Massachusetts Horticultural Society* (Boston, Mass.: G. K. Hall, 1963). 3 v. First supplement issued in 1972.
216. McGill University. Blacker-Wood Library of Zoology and Ornithology. *Dictionary Catalogue of the Blacker-Wood Library of Zoology and Ornithology* (Boston, Mass.: G. K. Hall, 1966). 9 v.
217. New York Botanical Garden Library. *Catalog of the Manuscript and Archival Collections and Index to the Correspondence of John Torrey* (Boston, Mass.: G. K. Hall, 1973). 473p.
    Describes 180,000 items including research notes, papers, diaries, unpublished manuscripts, and personal, administrative, and scientific correspondence.
218. Kenneth O. Rachie. *The Millets and Minor Cereals: A Bibliography of the World Literature on Millets Pre-1930 and 1964-1969; and of All Literature on Other Minor Cereals* (Metuchen, N.J.: Scarecrow Press, 1974). 225p.
219. Miloslav Rechigal, Jr. *World Food Problem: A Selective Bibliography of Reviews* (Cleveland, Ohio: Chemical Rubber Company Press, 1975). 211p.
220. Royal Botanic Gardens Library, Kew, England. *Author and Classified Catalogues of the Royal Botanic Gardens Library* (Boston, Mass.: G. K. Hall, 1974). Author catalog, 5 v.; classified catalog, 4 v.
    The collections of the Royal Botanic Gardens (founded in 1852) encompass plant taxonomy and distribution, economic botany, botanical travel and exploration, and plant cytology, physiology, and biochemistry. Much of the material is arranged by the Dewey Decimal Classification. For systematic works, the Bentham and Hooker Botanical Classification has been used. Works on floras are arranged according to a special geographical schedule used in the Kew Herbarium.

221. Satya Prakash, Syed Mohammed Ali, and H. S. Sharma. *Agriculture: A Bibliography* (Gurgaon, India: Indian Documentation Service, 1977). 384p.
222. The Smithsonian Institution, Washington, D.C.. *Index to Grass Species.* Agnes Chase and Cornelia D. Niles (Boston, Mass.: G. K. Hall, 1963). 3 v.
223. Torrey Botanical Club, New York. *Index to American Botanical Literature, 1886-1966* (Boston, Mass.: G. K. Hall, 1969). 4 v.

    This is an author catalog of botanical books and papers published in the Western Hemisphere covering taxonomy, phylogeny, and floristics of the fungi, pteridophytes, boryphytes, and spermatophytes; morphology, anatomy, cytology, genetics, physiology and pathology of the same groups; plant ecology and general botany; biography and bibliography. Decennial supplements are planned.
224. Deborah Truitt. *Dolphins and Porpoises: A Comprehensive Annotated Bibliography of the Smaller Cetacea* (Detroit, Mich.: Gale Research Co., 1974). 582p.
225. U.S. Department of Agriculture. *Botany Subject Index* (Boston, Mass.: G. K. Hall, 1958). 15 v.

    This is a subject index to American and foreign botanical literature from the earliest times as published in books and serials. Textbooks, voyages, and biographies are also included.
226. U.S. Department of the Interior. *Dictionary Catalog of the Department Library* (Boston, Mass.: G. K. Hall, 1967). 37 v.

    First Supplement, 1968. 4 v.

    Second Supplement, 1971. 2 v.

    Third Supplement, 1973.

    Fourth Supplement, 1975. 8 v.

    Subjects covered: geology, mines, minerals, petroleum and coal; fish, fisheries, and wildlife including ornithology; international conservation, land management, land reclamation, public land policy; the history and development of irrigation in the western states; water and power; national parks, etc.
227. University of Washington. Fisheries-Oceanography Library. *Selected References to Literature on Marine Expeditions, 1700-1960* (Boston, Mass.: G. K. Hall, 1972). 517p.

    About 9000 references on ships and expeditions involved in fisheries and oceanographic research from the 18th to the middle of the 19th century; arranged alphabetically by ship or expedition name.
228. University of the West Indies, Imperial College of Tropical Agriculture. *Catalogue of the Imperial College of Tropical Agriculture, University of the West Indies* (Boston, Mass.: G. K. Hall, 1975). 8 v.

Selected Lists

229. LeRoy Walters. *Bibliography of Bioethics* (Detroit, Mich.: Gale Research Co., 1975-). Annual.
An ongoing project of the Center for Bioethics of the Kennedy Institute, Georgetown University.
230. Yale University. Henry S. Graves Memorial Library. *Dictionary Catalogue of the Yale Forestry Library* (Boston, Mass.: G. K. Hall, 1962). 12 v.

Abstracting and Indexing Services

231. *Abstracts of Mycology* (Philadelphia, Penn.: BioSciences Information Service of Biological Abstracts, 1967-). Monthly.
Contains all abstracts and references to fungi and lichens from *Biological Abstracts* and *Biological Abstracts/RRM*.
232. *Aerospace Medicine and Biology: A Continuing Bibliography with Indexes* (Washington, D.C.: National Aeronautics and Space Administration, 1964-).
Continues an earlier publication with the same title issued 1952-1963. Subtitle varies.
233. *Agricultural Engineering Abstracts* (Slough, England: Commonwealth Agricultural Bureaux, 1976-). Monthly.
234. *Algae Abstracts: A Guide to the Literature* (New York: Plenum Publishing Corporation). Irregular (v. 3: 1972-1974 published in 1977).
Prepared from material supplied by the Water Resources Scientific Information Center, Office of Water Research, Department of the Interior, Washington, D.C.
235. *Aquatic Sciences and Fisheries Abstracts.* Aquatic Sciences and Fisheries Information System (available from Information Retrieval, Inc., Arlington, Va., 1976-). Monthly.
236. *Bibliography of Agriculture* (Beltsville, Md.: National Agricultural Library, 1942-). Monthly.
237. *Biological Abstracts* (Philadelphia, Penn.: BioSciences Information Service of Biological Abstracts, 1926-). Semimonthly.
238. *Biological Abstracts/RRM* (Philadelphia, Penn.: BioSciences Information Service of Biological Abstracts, 1980-). Monthly.
Continues *BioResearch Index*.
239. *Biological and Agricultural Index* (New York: H. W. Wilson, 1916-).
Monthly (except August); annual and biennial cumulations.
Title from 1916-1964: *Agricultural Index*.
240. *Chemical-Biological Activities (CBAC).* An index to current literature on the biological activity of chemical substances (Columbus, Ohio: Chemical Abstracts Service, 1965-). Semimonthly.
Comprises the first five sections of the Biochemistry Group of *Chemical Abstracts*.

241. *Crop Physiology Abstracts* (Slough, England: Commonwealth Agricultural Bureaux, 1975-). Monthly.
242. *Current Contents: Agriculture, Biology and Environmental Sciences* (Philadelphia, Penn.: Institute for Scientific Information, 1970-). Weekly.
243. *Current Contents: Life Sciences* (Philadelphia, Penn.: Institute for Scientific Information, 1958-). Weekly.
244. *Forestry Abstracts* (Slough, England: Commonwealth Agricultural Bureaux, 1939-). Monthly.
245. *Genetics Abstracts* (Arlington, Va.: Information Retrieval, Inc., 1968-). Monthly.
246. *Index Medicus* (Bethesda, Md.: National Library of Medicine, 1960-). Monthly.
Formerly: *Current List of Medical Literature.*
247. *Irrigation and Drainage Abstracts* (Slough, England: Commonwealth Agricultural Bureaux, 1975-). Quarterly.
248. *Maize Quality Protein Abstracts* (Slough, England: Commonwealth Agricultural Bureaux, 1975-). Quarterly.
249. *Microbiology Abstracts* (London: Information Retrieval, Ltd.). Monthly.

    Section A: *Industrial and Applied Microbiology,* 1965-

    Section B: *Bacteriology,* 1966-

    Section C: *Algology, Micology and Protozoology,* 1972-

250. *Nutrition Abstracts and Reviews* (Slough, England: Commonwealth Agricultural Bureaux, 1931-). Monthly.
251. *Ornamental Horticulture* (Slough, England: Commonwealth Agricultural Bureaux, 1975-). Monthly.
252. *Plant Growth Regulator Abstracts* (Slough, England: Commonwealth Agricultural Bureaux, 1975-). Monthly.
253. *Potato Research Abstracts* (Slough, England: Commonwealth Agricultural Bureaux, 1975-). Monthly.
254. *Poultry Abstracts* (Slough, England: Commonwealth Agricultural Bureaux, 1975-). Monthly.
255. *Protozoological Abstracts* (Slough, England: Commonwealth Agricultural Bureaux, 1977-). Monthly.
256. *Small Animal Abstracts* (Slough, England: Commonwealth Agricultural Bureaux, 1975-). Monthly.
257. *Tobacco Abstracts* (Raleigh, N.C.: North Carolina State University, School of Agriculture and Life Sciences, Tobacco Literature Service, 1957-). Monthly.
258. *Triticale Abstracts* (Slough, England: Commonwealth Agricultural Bureaux, 1975-). Monthly.
259. *Zoological Record* (London: Zoological Society of London, 1864-). Annual.
Title during 1864: *Records of Zoological Literature.*

Selected Lists 331

260. *World Fisheries Abstracts* (Rome, Italy: Food and Agriculture Organization of the United Nations, Fisheries Division, ca. 1950-).
261. *Index to Plant Distribution Maps in North American Periodicals Through 1972* (Boston, Mass.: G. K. Hall, 1976).
    Index to plant distribution maps in 267 North American periodicals. Entries are arranged alphabetically by taxa; like taxa are arranged chronologically. Also included is a list of primarily North American books containing plant distribution maps, arranged geographically according to area covered.

Environment

Guides to Literature

262. *Available Information Materials on Solid Waste Management. Total Listing, 1966 to 1978* (Washington, D.C.: U.S. Environmental Protection Agency, 1979). 179p.
    Listing of guidelines, regulations, teaching materials, reports, and other publications on solid waste management, with subject, title, and author indexes.
263. *Directory of Environment Periodicals.* K. Subramanyam and Mary M. O'Pecko (Monticello, Ill.: Vance Bibliographies, 1979). 60p.
264. *Environment U.S.A.: A Guide to Agencies, People and Resources.* Glenn L. Paulson, advisory editor. Compiled and edited by Onyx Group, Inc. (New York: R. R. Bowker, 1974). 451p.
265. *Environmental Education: A Guide to Information Sources.* William B. Stapp and Mary D. Liston (Detroit, Mich.: Gale Research Co., 1975). 225p. (Man and the Environment Information Guide Series, v. 1).
266. *Environmental Information Sources Handbook.* Garwood R. Wolff (New York: Simon and Schuster, 1974). 568p.
    Describes bibliographic tools, information services, societies, and agencies—both private and government—concerned with environment. Libraries, films, periodicals, abstracting and indexing services, bibliographies, and other sources of information are described.
267. *Environmental Law: A Guide to Information Sources.* Mortimer D. Schwartz (Detroit, Mich.: Gale Research Co., 1977). 191p. (Man and the Environment Information Guide Series, v. 6).
268. *Environmental Planning: A Guide to Information Sources.* Michael J. Meshenberg (Detroit, Mich.: Gale Research Co., 1976). 492p. (Man and the Environment Information Guide Series, v. 3).
269. *Environmental Science and Technology Information Sources.* Sydney B. Towiner (Park Ridge, N.J.: Noyes Data Corporation, 1973). 218p.
270. *Environmental Toxicology: A Guide to Information Sources.* Robert L. Rudd (Detroit, Mich.: Gale Research Co., 1977). 266p. (Man and the Environment Information Guide Series, v. 7).

271. *Environmental Values, 1860-1972: A Guide to Information Sources.* Loren C. Owings (Detroit, Mich.: Gale Research Co., 1976). 324p. (Man and the Environment Information Guide Series, v. 4).
272. *Information Sources in the Environmental Sciences.* George S. Bonn. 18th Allerton Park Institute (Urbana, Ill.: University of Illinois, Graduate School of Library Science, 1973). 242p.
273. *Noise Pollution: A Guide to Information Sources.* Clifford R. Bragdon (Detroit, Mich.: Gale Research Co., 1979). 524p. (Man and the Environment Information Guide Series, v. 5).
274. *Pollution: Sources of Information.* Proceedings of a one-day conference held at the Library Association, 27th October 1971. Kay Henderson (London: The Library Association, 1972). 101p.
275. *Sources of Information in Water Resources: An Annotated Guide to Printed Materials.* Gerald J. Giefer (Berkeley, Calif.: Water Resources Center Archives, University of California, 1976). 290p.
276. *Waste Water Management: A Guide to Information Sources.* George Tchobanoglous, Robert G. Smith, and Ronald W. Crites (Detroit, Mich.: Gale Research Co., 1976). 202p. (Man and the Environment Information Guide Series, v. 2).
    Guide to the literature on the engineering aspects of wastewater collection, treatment, disposal, and reuse systems. Economic aspects, planning, and legislative aspects of such systems are also included.

Bibliographies

277. *Air Pollution Aspects of Emission Sources: Pulp and Paper Industry. A Bibliography with Abstracts* (Research Triangle Park, N.C.: U.S. Environmental Protection Agency, Office of Air and Water Programs, 1973). 166p. (U.S. EPA Air Pollution Control Office Publication No. AP-121.).
278. *Air Pollution Technical Publications of the U.S. Environmental Protection Agency* (Research Triangle Park, N.C.: U.S. Environmental Protection Agency, 1973-). Semiannual.
    Alphanumeric listing of reports released by EPA. Superseded by *Air Pollution Technical Publications of the U.S. Environmental Protection Agency. Quarterly Bulletin,* 1975-.
279. *Air Pollution Aspects of Emission Sources: Sulfuric Acid Manufacturing. A Bibliography with Abstracts* (Research Triangle Park, N.C.: U.S. Environmental Protection Agency, Office of Air Programs, 1971). 58p. (Publication No. AP-94).
280. *Asbestos and Air Pollution: An Annotated Bibliography* (Research Triangle Park, N.C.: U.S. Environmental Protection Agency, Air Pollution Control Office, 1971). 101p. (Publication No. AP-82).
281. *Beryllium and Air Pollution: An Annotated Bibliography* (Research Triangle Park, N.C.: U.S. Environmental Protection Agency, Air Pollution Control Office, 1971). 75p. (Publication No. AP-83).

Selected Lists

282. *Bibliography of Water Quality Research Reports* (Washington, D.C.: U.S. Environmental Protection Agency, Office of Research Monitoring, 1972). 40p. (Water Pollution Control Research Series).
283. *An Annotated Bibliography of the Literature on Livability, with an Introduction and an Analysis of the Literature.* Linda Brown (Monticello, Ill.: Council of Planning Librarians, 1975). 61p. (CPL Exchange Bibliography No. 853).
284. *Chlorine and Air Pollution: An Annotated Bibliography* (Research Triangle Park, N.C.: U.S. Environmental Protection Agency, Office of Air Programs, 1971). 113p. (Publication No. AP-99).
285. *The Ecological Impact of Solid Waste.* John D. Gunter and William Carl Jameson (Monticello, Ill.: Council of Planning Librarians, 1973). 17p. (CPL Exchange Bibliography No. 406).
286. *Ecology and the Environment: A Dissertation Bibliography* (Ann Arbor, Mich.: University Microfilms International, 1976). 87p.
287. *Bibliography of Rural and Suburban Sewage Treatment and Disposal Publications* (St. Joseph, Mich.: American Society of Agriculture Engineers, 1973). 27p. (Special Publication ASAE SP-03-73).
288. *Environmental Impact Assessment Methodologies: An Annotated Bibliography.* Richard C. Vichl, Jr. and Kenneth G. M. Mason (Monticello, Ill.: Council of Planning Librarians, 1974). 32p. (CPL Exchange Bibliography No. 691).
289. *Environmental Information Sources: Engineering and Industrial Applications. A Selected Annotated Bibliography* (New York: Special Libraries Association, 1972). 72p.
Prepared for "Environmental and Ecological Literature: Where Does it All Come From?"—A continuing education seminar held during the 63rd annual conference of the Special Libraries Association, June 4-8, 1972, Boston, Mass.
290. *Environmental Pollution and Mental Health.* John S. Williams and others (Washington, D.C.: Information Resources Press, 1973). 136p.
Abstracts of 110 journal articles and books on the influence and environmental pollution upon man's mental health and behavior. Author and subject indexes.
291. *Estuarine Pollution: A Bibliography* (Washington, D.C.: U.S. Department of the Interior, Office of Water Resources Research, Water Resources Scientific Information Center, 1973). 477p. (WRSIC 73-205).
292. *Hydrochloric Acid and Air Pollution: An Annotated Bibliography* (Research Triangle Park, N.C.: U.S. Environmental Protection Agency, Office of Air Programs, Air Pollution Technical Information Center, 1971). 107p. (Publication No. AP-100).
293. *Man and the Environment: A Bibliography of Selected Publications of the United Nations System 1946-1971.* Harry N. M. Winton (New York: R. R. Bowker, 1972). 305p.

Annotated bibliography of over 1200 publications including monographs, periodicals, bibliographies, and other reference works. Author index, series and serials index, title index, and subject index.

294. *Mercury and Air Pollution: A Bibliography with Abstracts* (Research Triangle Park, N.C.: U.S. Environmental Protection Agency, Office of Air Programs, 1972). 59p. (Publication No. AP-114).
295. *Odors and Air Pollution: A Bibliography with Abstracts* (Research Triangle Park, N.C.: U.S. Environmental Protection Agency, Office of Air Programs, 1972). 257p. (Publication No. AP-113).
296. *Oil Pollution: An Index-Catalog to the Collection of the Oil Spill Information Center* (Santa Barbara, Calif.: University of California at Santa Barbara, Oil Spill Information Center, 1972). 4 v.
297. *Solid Waste Management: Economics and Operation.* John D. Gunter and William Carl Jameson (Monticello, Ill.: Council of Planning Librarians, 1973). 25p. (CPL Exchange Bibliography No. 395).
298. United States Public Health Service. Division of Air Pollution. *Air Pollution Publications: A Selected Bibliography,* 1955–. Irregular.
299. *Urban and Environmental Resources: A Bibliography of English Language Periodicals.* Raymond W. Clugh and Ambrose Klain (Monticello, Ill.: Council of Planning Librarians, 1973). 30p. (CPL Exchange Bibliography No. 494).
300. *Use of Naturally Impaired Water: A Bibliography* (Washington, D.C.: U.S. Department of the Interior, Office of Water Resources Research, Water Resources Scientific Information Center, 1973). 364p. (WRSIC 73-217).
301. *Water Resource Development in the Great Lakes Basin: A Selected Bibliography.* Marc Jay Rogoff (Monticello, Ill.: Council of Planning Librarians, 1977). 7p. (CPL Exchange Bibliography No. 1309).
302. *Water Reuse: A Bibliography* (Washington, D.C.: U.S. Department of the Interior, Office of Water Resources Research, Water Resources Scientific Information Center, 1973–). Irregular.
303. E. A. Jeanne and William Sanders, Literature on mercury: Availability of English literature, *Journal of the Water Pollution Control Federation,* 45(9):1952-1970 (September 1973).
304. *Directory of Architecture Periodicals.* K. Subramanyam (Monticello, Ill.: Vance Bibliographies, 1980). 44p.

Abstracting and Indexing Services

305. *Air Pollution Abstracts* (Research Triangle Park, N.C.: U.S. Environmental Protection Agency, Air Pollution Technical Information Center, 1970–). Monthly; semiannual cumulative subject and author indexes
306. *Air Pollution Translations: A Bibliography with Abstracts* (Research Triangle Park, N.C.: U.S. Environmental Protection Agency, Air Pollution Technical Information Center, 1969–). Irregular.

Abstracts of documents translated from French, German, Italian, Japanese, Russian, and other languages. Author and subject indexes. The translations are available from the National Translations Center, John Crerar Library, 35 West 33rd Street, Chicago, IL 60616. Some translations available from the NTIS are also listed.

307. *Current Contents: Agriculture, Biology, and Environmental Sciences* (Philadelphia, Penn.: Institute for Scientific Information, 1970-). Weekly.

308. *Ecology and Environment* (Columbus, Ohio: Chemical Abstracts Service, American Chemical Society). Biweekly.

    Computer-readable tape service containing abstracts, bibliographic information, and keyword and other index entries for documents abstracted in the following sections of *Chemical Abstracts:*

    4. Toxicology
    17. Foods
    19. Fertilizers, Soils, and Plant Nutrition
    53. Mineralogical and Geological Chemistry
    59. Air Pollution and Industrial Hygiene
    60. Sewage and Wastes
    61. Water

309. *Environment Abstracts* (New York: Environment Information Center). Monthly.

310. *Environment Index* (New York: Environment Information Center, 1971-). Annual.

    Yearbook reporting key developments in environmental pollution control and indexes to abstracts in *Environment Abstracts:* Subject index, SIC Code index, geographic index, and author index.

311. *Environmental Periodicals Bibliography: Indexed Article Titles* (Santa Barbara, Calif.: International Academy at Santa Barbara, Environmental Studies Institute, 1972-). Monthly.

    Tables of contents of about 300 environmentally related periodicals are replicated and arranged under six main parts: (1) general human ecology, (2) air, (3) energy, (4) land resources, (5) water resources, and (6) nutrition and health. Subject and author indexes.

312. *Hydata: Water Resources Index* (Minneapolis, Minn.: American Water Resources Association, St. Anthony Falls Hydraulic Laboratory, 1965-). Monthly.

313. *Index of Human Ecology.* J. Owen Jones and Elizabeth A. Jones (London: Europa Publications, 1974). 169p.

    An index to cross-disciplinary information covering all countries and all languages.

314. *Pollution Abstracts* (Louisville, Ky.: Pollution Abstracts, Inc., 1970-). Bimonthly.

315. *Selected Water Resources Abstracts* (new series) (Washington, D.C.: U.S. Department of the Interior, Office of Water Resources Research, Water Resources Scientific Information Center, 1968-). Semimonthly. Vol. 1 (January-December 1968) published monthly as *Selected Water Resources Abstracts* by the Office of Engineering Reference, Bureau of Reclamation.

Energy

Guides to Literature

316. *Energy: A Guide to Information Sources.* Virginia Sternberg (Detroit, Mich.: Gale Research Co., 1980).
317. *Energy Guide: A Directory of Information Sources.* Virginia Bemis (New York: Garland STPM Press, 1977). 250p.
An annotated guide to sources of information on the social science aspects of energy and energy alternatives. Materials and sources described are of interest to users at all levels ranging from primary school children to postgraduate researchers.
318. *Energy Information Resources: An Inventory of Energy Research and Development Information Sources in the Continental United States, Hawaii and Alaska.* Patricia L. Brown and others (Washington, D.C.: American Society for Information Science, 1975). 207p.
319. *Guide to Literature on Nuclear Engineering.* Harold N. Wiren (Washington, D.C.: American Society for Engineering Education, Engineering School Libraries Division, 1972). 21p.
320. *Information Sources in Power Engineering: A Guide to Energy Resources and Technology.* Karen Metx (Westport, Conn.: Greenwood Press, 1975). 114p.
321. *Sources of Information on Atomic Energy.* L. J. Anthony (Oxford: Pergamon Press, 1966). 245p. (International Series of Monographs in Library and Information Science, v. 2).

Bibliographies

322. *An Annotated Bibliography of Solar Energy Research and Technology Applicable to Community Buildings and Other Non-Residential Construction.* Nan C. Burg (Monticello, Ill.: Council of Planning Librarians, 1977). 30p. (CPL Exchange Bibliography No. 1263).
323. *A Bibliography of Solar Energy: Instrumentation, Measurement, Network Design.* Sandra Braze (Tempe, Ariz.: Office of the State Climatologist, 1976). 112p. (Climatological Publications: Bibliography Series, No. 1).
324. *Energy: A Bibliography of Social Science and Related Literature.* Denton E. Morrison (New York: Garland STPM Press, 1974). 173p.

Selected Lists

325. *Energy II: A Bibliography of 1975-1976 Social Science and Related Literature.* Denton E. Morrison (New York: Garland STPM Press, 1976). 250p.
326. Energy: A Classified Listing, *Library Journal,* 105(1):73-78 (January 1, 1980).
   A list of books on all aspects of energy.
327. *Energy Conservation in Building Design.* Robert Bartholomew (Monticello, Ill.: Council of Planning Librarians, 1977). 12p. (CPL Exchange Bibliography No. 1223).
328. *Energy-Environment Materials Guide.* Kathryn E. Marvine and Rebecca E. Cawley (Washington, D.C.: National Science Teachers Association, 1975).
   An annotated bibliography of about 300 books and articles in four parts, each part directed to a specific audience: (1) readings for teachers; (2) readings for students, grades 8-12; (3) readings for students, grades 5-9; and (4) readings for students, grades K-6. Films and other audiovisual materials, curriculum materials, and government documents are also listed.
329. *Energy Reference Sources.* Kitty Hsieh (Corvallis, Ore.: Oregon State University Press, 1975). 53p. (Bibliographic Series, OSU Press, No. 10).
330. *Energy Statistics: A Guide to Sources.* Sarojini Balachandran (Monticello, Ill.: Council of Planning Librarians, 1976). 51p. (CPL Exchange Bibliography No. 1223).
331. *Energy Statistics: An Update to Bibliography No. 1065.* Sarojini Balachandran (Monticello, Ill.: Council of Planning Librarians, 1977). 22p. (CPL Exchange Bibliography No. 1247).
332. *International Bibliography of Alternative Energy Sources.* G. Hutton and M. Rostron (New York: Nichols Pub., 1980).
333. *Mineral and Energy Resources of the U.S.S.R.: A Selected Bibliography of Sources in English.* Eugene A. Alexandrov (Falls Church, Va.: American Geological Institute, 1980). 91p.
334. *A Solar Energy Bibliography.* David L. Guthrie and Robert A. Riley (Rockville, Md.: Information Transfer, Inc., 1978). 360p.
335. United States. Army Engineer Waterways Experiment Station. *List of Publications of the U.S. Army Engineer Waterways Experiment Station. Compiled and Indexed in Special Projects Branch, Technical Information Center, U.S. Army Engineer Waterways Experiment Station.* Virginia Dale (Vicksburg, Miss.: The Branch, 1976). 293p.
336. United States. Energy Research and Development Administration. Library. *World Energy Resources: An Annotated Bibliography of Selected Material on the Availability and Development of World Energy Resources* (Washington, D.C.: U.S. Government Printing Office, 1975). 18p.
   A bibliography of journal articles, reports, and books in English, pertaining to statistical and general policy information, rather than technical

descriptions of energy development. Excludes references on energy resources in the United States. Author index.
337. United States. General Accounting Office. *GAO Energy Digest: A Bibliography*. Issued by the Comptroller General of the United States (Washington, D.C.: U.S. General Accounting Office, 1977). 216p.
For sale by the Superintendent of Documents, U.S. Government Printing Office, Washington, D.C.
338. United States. National Aeronautics and Space Administration. Scientific and Technical Information Office. *Energy: A Special Bibliography with Indexes* (Springfield, Va.: National Technical Information Service, 1974-). Quarterly supplements.
Annotated references to documents that are announced in *Scientific and Technical Aerospace Reports* and *International Aerospace Abstracts*.

Abstracting and Indexing Services

339. *Energy Abstracts for Policy Analysis* (Oak Ridge, Tenn.: Oak Ridge National Laboratory, 1974-). Monthly.
Supersedes *NSF-RANN Energy Abstracts.*
340. *Energy Information Abstracts* (New York: Environment Information Center). Monthly.
341. *Energy Index* (New York: Environment Information Center). Annual.
Yearbook reporting key developments in energy and indexes to abstracts in *Energy Information Abstracts*.
342. *Energy Research Abstracts* (Springfield, Va.: National Technical Information Service, 1976-). Monthly.
Vol. 1, Nos. 1 & 2 (January and February 1976) issued as *ERDA Research Abstracts* by the Energy Research and Development Administration.
343. *Atomindex* (Vienna, Austria: International Nuclear Information System, 1970-). Bimonthly.
An abstracting service generated from a centralized bibliographic database created by INIS in cooperation with IAEA member countries.

References

1. Derek J. de Solla Price, The scientific foundations of science policy, *Nature*, 206:234 (April 17, 1965).
2. J. C. R. Licklider, A crux in scientific and technical communication, *American Psychologist*, 12:1044-1051 (1966).
3. D. W. King and others, *Statistical Indicators of Scientific and Technical Communication, 1960-1980*. PB 254 060 (Washington, D.C.: U.S. Government Printing Office, 1976).
4. Edward M. Crane, The U.S. Scientific Bibliography—A big effort for a big purpose, *Publishers Weekly*, 165(15):1621-1623 (October 11, 1952).

References

5. Francis J. Witty, The beginnings of indexing and abstracting: Some notes towards a history of indexing and abstracting in antiquity and the Middle ages, *The Indexer*, 8(4):193-198 (October 1973).
6. Harold Borko and Charles L. Bernier, *Abstracting Concepts and Methods* (New York: Academic Press, 1975).
7. Bruce M. Manzer, *The Abstract Journal, 1790-1920: Origin, Development, and Diffusion* (Metuchen, N.J.: Scarecrow Press, 1977).
8. Data for this table were obtained from *CAS Today: Facts and Figures About Chemical Abstracts Service* (Columbus, Ohio: Chemical Abstracts Service, 1980).
9. E. M. R. Ditmas, Coordination of information. A survey of schemes put forward in the last fifty years, *Journal of Documentation*, 3(4):209-221 (March 1948).
10. S. C. Bradford, Extent to which scientific and technical literature is covered by present abstracting and indexing periodicals, *ASLIB Report of Proceedings*, 14:59-71 (1937).
11. John Martyn and Margaret Slater, Tests on abstract journals, *Journal of Documentation*, 20(4):212-235 (December 1964).
12. John Martyn, Tests on abstracts journals: Coverage overlap and indexing, *Journal of Documentation*, 23(1):45-70 (March 1967).
13. Irving Amron, History of Indexing and Abstracting by the ASCE. ASCE Annual Convention and Exposition, Philadelphia, September 27-October 1, 1976. Preprint 2736.
14. J. L. Wood, C. Flanagan, and H. E. Kennedy, Overlap in the lists of journals monitored by BIOSIS, CAS and EI, *Journal of the American Society for Information Science*, 23(1):36-38 (January-February 1972).
15. J. L. Wood, C. Flanagan, and H. E. Kennedy, Overlap among the journal articles selected for coverage by BIOSIS, CAS and EI, *Journal of the American Society for Information Science*, 24(1):25-28 (January-February 1973).
16. *A Study of Coverage Overlap Among Major Science and Technology Abstracting and Indexing Services*. NFAIS 77/1 (Philadelphia, Penn.: National Federation of Abstracting and Indexing Services, 1977).
17. CAS completes 9th Collective Index in record time, *Chemical and Engineering News*, 56(39):54 (September 25, 1978).
18. Martyn (Ref. 12), pp. 48-49.
19. Jeffrey J. Gardner, Pathfinders, Library, *Encyclopedia of Library and Information Science* (New York: Marcel Dekker, 1977), v. 21, pp. 468-473.
20. David Garvin, The IAC and the library, *Special Libraries*, 62:17-23 (January 1971).
21. *Directory of USAEC Information Analysis Centers* (Office of Information Services, U.S. Atomic Energy Commission, October 1974), pp. 26-27.

22. AEC and DOD information analysis centers, *Special Libraries,* 57(1):21-34 (January 1966).
23. United States. Library of Congress. National Referral Center for Science and Technology. *A Directory of Information Resources in the United States: Physical Sciences, Engineering* (Washington, D.C.: Library of Congress, 1971).
24. United States. Library of Congress. National Referral Center for Science and Technology. *A Directory of Information Resources in the United States: Federal Government* (Washington, D.C.: Library of Congress, 1974).
25. United States. Library of Congress. National Referral Center for Science and Technology. *A Directory of Information Resources in the United States: Biological Sciences* (Washington, D.C.: Library of Congress, 1972).
26. Rao Aluri and Phillip A. Yannarella, Publications of selected information analysis centers, *Special Libraries,* 65(10/11):455-461 (October/November 1974).
27. Lockheed Missiles and Space Company, Inc. DIALOG Reduced Rates, *NFAIS Newsletter,* 19(1):14 (February 1977).
28. Irving M. Klempner, *Diffusion of Abstracting and Indexing Services for Government-Sponsored Research* (Metuchen, N.J.: Scarecrow Press, 1968), pp. 187-188.
29. *Nuclear Science Abstracts* was discontinued in June 1976. The present title of USGRDR is *Government Reports Announcements and Index.*

19

CURRENT TRENDS AND PROSPECTS

19.1 Introduction

The exploitation of scientific and technical information for advancements in economic and industrial sectors, national defense, transportation, pollution control, education, and related socioeconomic issues has become a major national concern in most countries. Significant advances have been made in the technological and organizational aspects of the generation, bibliographic control, dissemination, and utilization of scientific and technical information. So numerous and diverse are the plans, projects, and systems that are being designed and implemented at international, national, and local levels that it is futile to attempt a comprehensive and coherent account of these developments. Undoubtedly, the most important single force that has touched almost every aspect of scientific and technical communication is the electronic computer. The computer has not only added new dimensions to the very process of research, it has also enhanced by several orders of magnitude both the speed of data processing and the volume of data that can be manipulated in the communication of scientific and technical information. The computer has made possible alternative modes of primary publishing, as well as speeding up the traditional modes of publishing research results. The editorial processing center and the "electronic journal" have been discussed in an earlier chapter. Three other major trends that are readily discernible are: (1) integration of primary and secondary publishing and the development of integrated information systems, (2) structural and procedural changes in the bibliographic control of scientific and technical literature, and (3) increasing international cooperation in scientific and technical communication, more particularly in the bibliographic control and dissemination of scientific and technical information. It is easy to see that these are interrelated phenomena rendered possible by the recent developments in the technology of information storage and dissemination. Optical character recognition (OCR), video-display terminals for online composition and editing, computer-output micrographics (COM), and online interactive information retrieval systems are only some of the developments that have added new dimensions to the total process of scientific and technical communication in recent years.

19.2 Integrated Primary and Secondary Publishing

For some years now, both the ACS and the AIP have been attempting the integration of primary and secondary publishing processes. Traditionally, the production of secondary journals has been a separate process, sequentially following the production of primary journals. The separate production processes have involved (1) considerable duplication of clerical and intellectual effort and (2) a long delay between the publication of primary literature and that of its surrogates in secondary journals. The integrated production of primary and secondary journals is based on two concepts: (1) secondary information (e.g., index entries, abstracts) about research articles can be generated as a by-product of and concurrently with the primary journal publication process; and (2) converting the edited and indexed manuscript into machine-readable format early in the production process facilitates manipulation of the material for generating, simultaneously or in rapid succession, a variety of products including the primary journal, separates or offprints of articles, current awareness services, and retrospective search tools. A bibliographic database containing abstracts and/or indexes may also be generated and marketed to libraries, information centers, and database vendors. Such an integrated production of a variety of primary and secondary products from a single input process holds promise of considerable savings in time and intellectual and clerical effort. The AIP, which publishes a large number of primary journals and some secondary services including the SPIN (Searchable Physics Information Notices) database, has been rigorously pursuing experimental and developmental work in this direction [1].

The ACS has been experimenting with the integrated production of primary and secondary journals through an interlinked manuscript processing system in which editing of a manuscript for publication in a primary journal (the *Journal of Organic Chemistry*) and assignment of index terms including registry numbers for new chemical compounds proceed simultaneously. Early indications are that such an approach can significantly lower the costs of production and eliminate much redundant editorial processing effort [2,3].

A similar integrated approach is also being tried by a team of researchers at the University of California, Los Angeles, to expedite the production of books with indexes by treating index preparation as an integral part of a computerized text-processing system [4].

19.3 Computer-Based Bibliographic Control Systems

19.3.1 Historical Overview

Martha Williams has identified four distinct phases in the history and development of computerized bibliographic databases [5]:

Phase 1: Computer-readable databases began to be generated as a by-product when abstracting and indexing services converted to photocomposition for producing their hard copy publications. These tapes were not designed for search purposes; but distributable tapes could be obtained by removing printing instructions from them and adding data-element tags.

Phase 2: Searchable databases are produced as a direct product from the computerized production process of abstracting and indexing services. Each record is input only once and a master database is created from which a number of products and services, including the hard copy indexing or abstracting journal, are generated.

Phase 3: Computer-readable databases will be produced and distributed with no hard copy counterparts.

Phase 4: Databases will be produced but not distributed in the form of computer-readable tapes; instead electronic online access will be provided to databases containing texts of primary journals and abstracting and indexing services.

According to Williams, we are now in the middle of this four-phase history of computerized information retrieval and dissemination. The advent of the electronic digital computer has had a very profound effect on secondary bibliographic services. The use of computers has not only greatly speeded up the production of abstracting and indexing services; it has also made possible the creation of large, integrated information systems that can generate a variety of products and services from a central database. Among the best known examples of large-scale computerized bibliographic control systems are MEDLARS (Medical Literature Analysis and Retrieval System) of the National Library of Medicine, and the Chemical Abstracts Services of the American Chemical Society. A large body of published literature exists on these and other similar systems.

Index Medicus, which has been in existence since 1879, was computerized in 1964, when it was ripe for mechanization with some 13,000 citations per month. This has now become one of the largest computerized systems with a variety of spinoff products such as the *Bibliography of Medical Reviews* (monthly), *Toxicity Bibliography* (quarterly), *TOXLINE* (an online version of

Toxicity Bibliography), and *MEDLINE* (*MEDLARS ONLINE*). At the present time, the computer is used to automate and expedite predominantly clerical tasks such as sorting, alphabetizing, collating, composing for printing, and generating permuted title word indexes. The intellectual tasks of subject analysis and assignment of index terms are still very much a human endeavor, although a great deal of experimental work on automatic indexing has been going on.

19.3.2 Chemical Abstracts Service

Computer-based methods of chemical information handling at Chemical Astracts Service began in the late 1950s when the postwar spurt of R & D activity resulted in a rising tide of scientific and technical literature. The traditional methods of producing and publishing abstracting and indexing services became very time-consuming and uneconomical, and mechanization became imperative to minimize the delay and expense involved in the production of secondary services. In the traditional, manual methods of production, human effort is involved in (1) the intellectual tasks of analyzing the primary documents, preparing abstracts, and assigning or deriving index terms, and (2) the clerical tasks of organizing and arranging the material for publication, duplicating and keeping track of the individual records in several processing streams, each stream leading to the production of one publication or service. With financial aid from the National Science Foundation, the National Institutes of Health, and the Department of Health, Chemical Abstracts Service has been developing a fully integrated secondary information system for the production of a variety of secondary services. In the computer-based system that is now emerging, the products of subject analysis and surrogation—bibliographic citations, index entries, and abstracts—are all recorded on a machine-readable database early in the production process. A variety of bibliographic packages are then produced by extracting appropriate records from the master file, repackaging the data elements in the desired format, and photocomposing the product for conversion into offset printing plates. All these clerical tasks and technical processes are accomplished largely through the computer. In the Chemical Abstracts Service, the database called CHEMICAL CONDENSATES is available for online searching. Some of the many derivative products generated from the master file are: *Chemical-Biological Activities* (CBAC) which parallels eight sections of *Chemical Abstracts* covering literature on the interaction of chemical substances with biological systems; *Polymer Science and Technology—Patents* (POST-P), and *Polymer Science and Technology—Journals* (POST-J) which parallel the macromolecular chemistry sections of *Chemical Abstracts*. Besides the generation of these and other spinoff products, the production of various indexes to *Chemical Abstracts* is also computerized. General subject index

19.3 Computer-Based Bibliographic Control Systems

entries, molecular formulas, and registry numbers of chemical compounds are ordered and merged by computer programs to produce the semiannual and five-year collective general subject index, chemical substance index, and molecular formula index.

The chief advantages of computerizing the operations at the Chemical Abstracts Service are claimed to be: (1) savings in costs of personnel directly involved in producing *Chemical Abstracts* and its indexes, (2) decrease in the average cost of processing a document, (3) more economical and faster production of indexes, and (4) additional revenues from the sale of machine-readable databases to database vendors.

The semiannual volume indexes can now be issued within six months or less after the completion of a volume of abstracts, this time lag was as much as 20 months in the late 1950s. The eighth collective index, which was the first to be produced after the system was computerized, was completed within 22 months of the end of the collective period (1967-1971); nearly four years had been required to produce the much smaller seventh collective index manually.

The ninth collective index covering the period 1972 through 1976, said to be the largest index ever produced by Chemical Abstracts Service, was also completed in record time. The entire 96,000-page index in 57 volumes was completed in less than 20 months. This collective index contains over 21 million entries for 2,024,013 papers, patents, and other documents cited in *Chemical Abstracts* during the five-year collective period. The chemical substance index alone contains 7.4 million entries and occupies more than 40,000 pages. The ninth collective index was compiled and photocomposed by computer from the computer-readable files of index entries created for the semiannual volume indexes of *Chemical Abstracts* [6].

19.3.3 Future Developments

In the last two decades, the use of computers in the production of abstracting and indexing services has steadily increased. This trend is likely to continue. Williams and Brandhorst have projected the following developments in four principal areas [7]:

1. *Hardware:* Hardware costs will continue to decline. The number of minicomputers, microprocessors, and terminals will steadily increase.
2. *Storage:* As new storage technologies such as bubble technology, charged-couple devices, electron-beam memory, and video-disk technology become available, storage capacities will dramatically increase, and storage costs will steadily decrease. For example, a video-disk with a storage capacity of one billion characters can be reproduced at a cost of about $7. Abstracting and indexing services could be distributed in this format far more speedily and far less expensively than through the print medium.

3. *Software:* Very sophisticated database management systems and operating systems will become available. A wide range of specialized application packages will become available for such applications as information storage and retrieval, text editing, and private file maintenance.
4. *Networks and Telecommunications:* The use of telecommunication networks for computer conferencing and transmission of text material will increase. Telecommunication costs will decrease from approximately $24 per hour in the early 1970s to as low as $2 per hour in the 1980s.

19.4 Online Access to Scientific and Technical Literature

19.4.1 Bibliographic Databases

A direct consequence of the application of computers to information processing has been the availability of a large number of bibliographic databases for information retrieval. Bibliographic databases are produced by national libraries, scholarly societies, mission-oriented government agencies, and commercial enterprises. In the United States, each of the three national libraries produces bibliographic databases. The MARC (Machine Readable Catalog) database of the Library of Congress is used widely for catalog card generation, literature search, and selective dissemination of information. The weekly MARC tape service is available on a subscription basis. The National Agricultural Library produces the AGRICOLA database. The *Bibliography of Agriculture* is produced from this. The MEDLARS and MEDLINE databases of the National Library of Medicine are used extensively for disseminating biomedical bibliographic information. CHEMICAL CONDENSATES of the Chemical Abstracts Service and SPIN (Searchable Physics Information Notices) of the American Institute of Physics are examples of databases produced by scholarly societies. Mission-oriented government agencies such as NASA and NTIS also produce machine-readable databases. COMPENDEX (Computerized Engineering Index) produced by Engineering Index, Inc., and SCISEARCH of the Institute for Scientific Information are two examples of databases generated by commercial concerns. About 300 bibliographic databases are now available in all branches of science and technology; not all of these, however, are searchable online.

Detailed descriptions of bibliographic databases can be obtained from the following directories:

1. *Computer Readable Databases: A Directory and Data Source Book.* Martha Williams (Washington, D.C.: American Society for Information Science, 1979). The directory is updated semiannually. The 1979 edition lists over 500 databases from all over the world. The directory is distributed by Knowledge Industry Publications, White Plains, New York.

19.4. Online Access to Scientific and Technical Literature

2. *Directory of Online Bibliographic Services: A List of Commercially Available Databases* (Rockville, Md.: Capital Systems Group, Inc., 1978). 32p.
3. *Encyclopedia of Information Systems and Services.* 3rd edition. Anthony T. Kruzas (Detroit, Mich.: Gale Research Co., 1978). 1029p.
4. *Directory of Online Databases.* Ruth N. Landau, Judith Wanger, and Mary C. Berger (Santa Monica, Calif.: Cuadra Associates, 1980). Describes about 450 bibliographic and nonbibliographic databases available for online searching. Updated quarterly.
5. *Information Market Place, 1978-1979: An International Directory of Information Products and Services* (New York: R. R. Bowker, 1978). Contains information on some 400 database producers and about 1000 machine-readable databases.
6. *Databases in Europe.* G. Pratt (London: ASLIB, 1975).

The following are the main advantages of machine-readable bibliographic databases:

1. Large volumes of bibliographic data (including abstracts) can be stored.
2. The databases can be updated frequently.
3. Bibliographic data can be manipulated at very high speeds, especially when the database is available for online searching.
4. In time-shared computer systems, a number of users can access the database simultaneously from remote locations for interactive searching and scanning.
5. Databases are useful in conducting on-demand retrospective literature searches using complex search logic. In large systems, as many as 20 to 30 searches may be conducted at once.
6. Selective dissemination of information (SDI) is also possible as a continuing service for disseminating bibliographic information on recently published literature selectively and expeditiously.

19.4.2 Database Vendors

A fairly recent development has been the advent of database vendors whose main function is to act as an intermediary between the growing number of database producers on the one side, and the numerous libraries and information centers on the other. A database vendor acquires databases from many sources, and with the help of specialized telecommunication agencies, provides "retail outlets" to small users of databases. Thus, small libraries that cannot afford to lease one or more expensive databases and support elaborate computer facilities can still access any desired database in an online mode through inexpensive terminals. The diagram in Figure 19.1 shows the intermediary role of database vendors and telecommunication networks in providing access to many databases. At present, three major database vendors—System Development Corporation

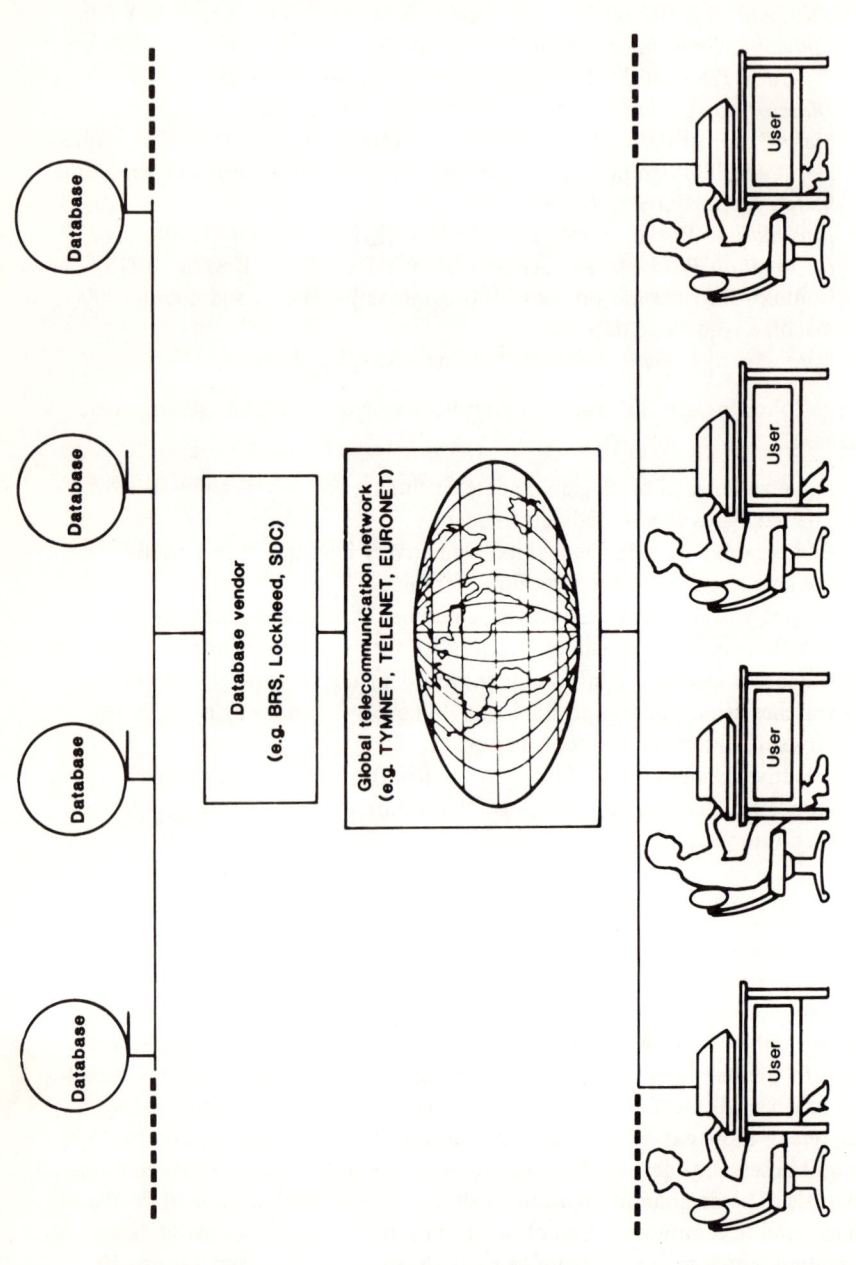

Figure 19.1 Online access to databases from remote locations.

19.4. Online Access to Scientific and Technical Literature

(ORBIT System), Lockheed Missile and Space Company (DIALOG System), and Bibliographic Retrieval Services, Inc. (BRS), are providing online access to numerous databases in the United States, Canada, and in some European countries. A selected list of bibliographic databases in science and technology accessible online is appended at the end of this chapter.

The number of databases and the extent of use made of them have been steadily increasing. In 1965, fewer than 20 databases were available to the public for information retrieval purposes. In 1977, over 3600 databases with a combined total of 71 million records were available for public use. Of these, slightly fewer than one-half (47%) of the databases were available for online access, but nearly 70% of the 71 million records were available for online search- in the United States, Canada, and some European countries.

The number of online searches has also been increasing dramatically. In 1974, about 700,000 online searches were made in the United States. This number increased to one million in 1975, and to two million in 1977. Now well over two million online searches are made annually [8-10]. These do not include online use of computers for applications other than bibliographic search- ing, such as online cataloging. The advent of database vending companies that facilitate inexpensive online access to numerous databases has been a significant factor in this dramatic increase in the volume of online bibliographic searching.

19.4.3 Standards

An important problem associated with online searching of bibliographic data- bases is the lack of standardization of record formats and access langauges used by various producers and vendors of databases. Standardization was not a crucial concern in the early days of machine-readable databases, because the databases were not numerous, and they were used almost exclusively by the database producers themselves. But with the rapidly increasing volume of online searching by remote users through vending agencies, standardization of biblio- graphic record formats and access languages has become an important concern. The American National Standards Institute has developed a standard (ANSI Z39.2-1971) for bibliographic information interchange on magnetic tape. The Technical Committee on Documentation of the International Organization for Standardization (ISO/TC 46) has produced a standard for the general form of bibliographic recording on magnetic tape [11]. A number of database producers have adopted this standard. Several associations and federations of database producers (e.g., NFAIS, ASIDIC, EUSIDIC, and ICSU/AB) are concerned with the development of standards for bibliographic databases. NFAIS has recently produced a document entitled *Data Element Definitions for Secondary Services*; this document is useful in specifying bibliographic descriptions in machine- readable databases. A subcommittee of the American National Standards

Committee Z39 is currently developing a standard for terms and symbols used in interactive bibliographic information retrieval.

19.5 Science Information: A Global Concern

19.5.1 International Cooperation in Bibliographic Control

Large-scale computerization of bibliographic control systems has opened up the possibility of interagency cooperation at both national and international levels. At the national level, the large information systems from both the private and the public sectors are gradually moving toward a concerted effort to promote more efficient bibliographic control and dissemination of scientific literature. The Chemical Abstracts Service (CAS), for example, is participating in many joint projects with other information systems: It provides the chemical substances registry data to the U.S. Environmental Protection Agency and to the National Institute of Neurological and Communicable Diseases and Stroke; supplies data in machine-readable form to support the National Library of Medicine's TOXLINE and CHEMLINE information retrieval systems; operates the National Cancer Institute's Drug Research and Development Chemical Information system; and is exploring possibilities of cooperating with BIOSIS in covering the literature of taxonomy and chemical substance identification. At the international level, CAS has concluded agreements with a number of overseas agencies such as West Germany's International Dokumentationsgesellschaft für Chemie, the Chemical Society (London), and the Centre National de l'Information Chimique in France. These organizations pay a share of the cost of producing the CAS databases, and in return obtain access to CAS databases and computer programs as well as the exclusive right to market all CAS products in their respective nations. The Chemical Society (London) prepares abstracts and index entries for papers and patents originating from the United Kingdom, and transmits the surrogate records to CAS in machine-readable form for input to the CAS database. The national chemical information systems of West Germany, Switzerland, and France are also using the registry structure files and other computer-readable products of CAS [12].

In yet another international agreement concluded recently, CAS has assigned marketing and distribution responsibilities for CAS products and services to the Japan Society for International Chemical Information (JSICI). The agreement allows JSICI to develop publications and services of its own from CAS databases for use in Japan. The JSICI is a federation of 27 chemical oriented scientific and technical societies and over 200 Japanese corporations, specifically organized to work with CAS in joint bibliographic control activities. This international collaboration is particularly significant since Japan contains the largest concentration of users of CAS products outside the United States, and is also a major

19.5 Science Information: A Global Concern

producer of the world's chemical and chemical engineering literature. Eventually, JSICI will supply computer-readable abstracts and index entries of Japanese chemical literature for input into the CAS database. In this context, it is pertinent to note that about three-quarters of American Chemical Society's published material originates outside the United States, and two-thirds of its circulation is also overseas [13].

Similar cooperative arrangements exist between the Institute of Electrical and Electronics Engineers, New York, and the Institution of Electrical Engineers, London, in the production and distribution of the products of INSPEC (International Information Services for the Physics and Engineering Communities). In the field of physics, the American Institute of Physics and the German Fachinformationszentrum Energie, Physik, Mathematik, GmbH (National Information Center for Energy, Physics, and Mathematics) have recently concluded an agreement to jointly produce an abstract journal and a computerized database covering nearly all the physics literature published in the world. The combined abstract journal will be called *Physics Briefs/Physikalische Berichte*.

Another noteworthy example of direct international cooperation is the integrated publication of *Metals Abstracts* and *Metals Abstracts Index* jointly by the American Society for Metals and the Institute of Metals, London. Avoidance of wasteful duplication by pooling resources to create a strong, unified product was the dominant consideration that led to this international cooperative venture.

19.5.2 International Nuclear Information System

The International Nuclear Information System (INIS) is a multinational cooperative venture for the bibliographic control of scientific and technical literature on the peaceful uses of atomic energy. INIS was created by the International Atomic Energy Agency (IAEA), Vienna, in 1969 to improve and expedite the exchange of scientific and technical information among IAEA member states on the basis of multilateral cooperation, and to eliminate the overlapping and duplication in the processing of literature [14]. A master file is created by the INIS from the input received from the member states, and this file forms the basis for a number of products and services including an abstracting journal entitled *Atomindex*. The INIS database has over 400,000 bibliographic references, and is growing at the rate of about 70,000 new references annually. The organization of INIS is based on an interesting combination of centralized database management with decentralized input and dissemination of output products. Comprehensive coverage of literature and elimination of wasteful duplication of effort are claimed to be the merits of this decentralized approach, in which each participating country is responsible for preparing citations only for its own national literature. Also, the cost of data gathering is shared fairly between small and large producers and users of literature [15]. The U.S. Department of Energy also sends abstracts of atomic energy literature to the INIS database. In order to prevent unnecessary duplication, publication of the *Nuclear Science Abstracts* was discontinued as from June 1976.

19.5.3 UNESCO and UNISIST

Along the same lines as INIS, the Food and Agriculture Organization (FAO) of the United Nations has established an International Information System for Agricultural Science and Technology (AGRIS). AGRIS became operational in 1975. It has accumulated a retrospective database of some 200,000 items, of which 100,000 items were added during 1977 alone [15].

A number of international agencies (e.g., UNESCO, the International Council of Scientific Unions, and the Organization for Economic Cooperation and Development) have been actively promoting the coordination of scientific information activities at a global level. UNESCO and the International Council of Scientific Unions jointly established a central committee in January 1967 to carry out a feasibility study for a world science information system. After four years of study, the committee produced in 1970 a report entitled *UNISIST: Study Report of the Feasibility of a World Science Information System by the United Nations Educational, Scientific and Cultural Organization and the International Council of Scientific Unions*. UNISIST is a world science information system based on the principle of voluntary cooperation among existing and future national and international systems in a flexible network arrangement. According to the study report, "UNISIST is a contemporary expression of a long-standing tradition of free interchange of information among the world's scientists." A major function of UNISIST is to foster information sharing by creating opportunities for further cooperative arrangements among governments, international organizations and operating services. UNISIST is not a centralized giant structure: "There will be no international center in Paris which will unify, collate, store and disseminate information. The Utopian concept of a world center conserving all documents, which was common in the 18th and also in the 19th century, was already hopelessly out of date in the inter-war years . . ." [16]. The five main objectives of UNISIST are briefly:

1. Improvement of systems interconnections
2. Strengthening the role of institutional components of the information transfer chain
3. Development of specialized manpower
4. Development of scientific information policies and structures
5. Assistance to developing countries in the development of scientific and technical infrastructures

The activities of UNISIST are regularly reported in the *UNISIST Newsletter* and also in other periodicals including *Annals of Library Science and Documentation* and *Information Hotline*.

INIS and UNISIST are examples of world information systems, based on the principle of coordination and voluntary cooperation at the international level. However, many of the developing countries which are just beginning to lay the foundation for a national information infrastructure, cannot participate in these sophisticated worldwide systems on an equal footing with the more advanced nations. To remove this imbalance in the levels of national information systems, and to enable every country, regardless of its stage of development, to participate meaningfully in the international information systems, a new worldwide program has recently been launched by UNESCO in collaboration with other international agencies. This program, known by the acronym NATIS, was established as the outcome of an Intergovernmental Conference of the Planning of National Documentation, Library and Archives Infrastructures convened by UNESCO, the International Federation of Library Associations (IFLA), the International Federation for Documentation (FID), and the International Council of Archives (ICA). The NATIS concept is based on the following two concerns: (1) the need for systematic planning of information infrastructures to enable every participating country to utilize its national information resources fully, and to benefit from existing and future world information systems; and (2) the need for coordinated planning of information resources at both national and international levels [17,18]. The NATIS concept implies that government agencies at national, state, and local levels should maximize the availability of information through documentation, library, and archives services to all categories of users. The main objective of NATIS is to enable each country to develop its own information system, in accordance with its needs and goals, so that all countries can enjoy full access to the scientific and technical information they need for their national development.

Selected List of Bibliographic Databases Available for Online Searching

1. ABSTRACT BULLETIN OF THE INSTITUTE OF PAPER CHEMISTRY (PAPER CHEM)
    Institute of Paper Chemistry, Appleton, Wisconsin. 1969-.
    Current and retrospective information on the pulp and paper industry.

2. AGRICULTURAL ONLINE ACCESS (AGRICOLA)
    National Agricultural Library, Bethesda, Maryland. 1970-.
    Worldwide coverage of all aspects of agricultural information.

3. AQUATIC SCIENCES AND FISHERIES ABSTRACTS (ASFA)
Food and Agriculture Organization of the United Nations, Rome, and the Intergovernmental Oceanographic Commission of UNESCO, 1978-.

Oceanography, pollution, and freshwater biology; limnology; and legal, political, and social aspects of sea and inland water studies.

4. BIOSIS PREVIEWS
BioSciences Information Services of Biological Abstracts, Philadelphia, Pensylvania. 1969-.

Worldwide coverage of all aspects of biological information.

5. CA SEARCH
Chemical Abstracts Service, Columbus, Ohio. 1972-.

This expanded database incorporates the CA CONDENSATES file which contains the basic bibliographic information appearing in the printed *Chemical Abstracts,* and the CASIA file which contains CA General Subject Headings and CA Registry Numbers. All aspects of chemistry and chemical engineering and technology are covered.

6. CA PATENT CONCORDANCE
Chemical Abstracts Service, Columbus, Ohio. 1967-.

Correlates patents issued by different country for the same basic invention.

7. CAB ABSTRACTS
The Commonwealth Agricultural Bureaux, Farnham House, Farnham Royal, Slough, England. 1973-.

Agricultural and biological information contained in the 26 main journals of the Commonwealth Agricultural Bureaux.

8. CHEMICAL INDUSTRY NOTES
Predicasts, Inc., and Chemical Abstracts Service, Columbus, Ohio. 1974-.

Extracts of articles from 75 business-oriented periodicals covering chemical processing industries.

9. CHEMICAL REACTION DOCUMENTATION SERVICE (CRDS)
Derwent Publications, London, England. 1975-.

10. CHEMLINE
Chemical Abstracts Service, Columbus, Ohio and National Library of Medicine, Bethesda, Maryland.

Chemical dictionary file of Chemical Abstracts Service.

Selected List of Bibliographic Databases Available for Online Searching

11. CHEMNAME
    Chemical Abstracts Service, Columbus, Ohio, and Lockheed DIALOG Retrieval Service, Palo Alto, Calif. 1972-.
    Chemical name dictionary file of Chemical Abstracts Service.

12. CLAIMS/CHEM (Class Code, Assignee, Index, Method Search—Chemistry)
    IFI/Plenum Data Co., Arlington, Va. 1950-1970.
    U.S. chemical and chemically related patents and some foreign equivalent patents.

13. CLAIMS/CLASS (Class Code, Assignee, Index, Method Search—Classification)
    IFI/Plenum Data Co., Arlington, Va.
    Classification code and title dictionary covering classes and subclasses of the U.S. patent classification system.

14. CLAIMS/US PATENTS (Class Code, Assignee, Index, Method Search—General, Electrical, Mechanical, and Chemical)
    IFI/Plenum Data Co., Arlington, Va. 1971-.
    All patents listed in the *Official Gazette* of the U.S. Patent Office. Also available as a weekly current awareness service.

15. COLD REGIONS SCIENCE AND TECHNOLOGY BIBLIOGRAPHY
    U.S. Army Corps of Engineers. 1962-.
    Physics and mechanics of snow, ice, and frozen ground as they relate to engineering activities and the ecology of Arctic and Antarctic areas.

16. COMPREHENSIVE DISSERTATION INDEX (CDI)
    University Microfilms International, Ann Arbor, Michigan. 1861-.
    Subject, title, and author guide to dissertations accepted at (predominantly) American and Canadian universities.

17. COMPUTERIZED ENGINEERING INDEX (COMPENDEX)
    Engineering Index, Inc., New York. 1970-.
    All aspects of engineering.

18. CONFERENCE PAPERS INDEX
    Data Courier, Inc., Louisville, Kentucky. 1973-.
    Papers presented at conferences in the fields of engineering, life sciences, and physical sciences.

19. CURRENT RESEARCH INFORMATION SYSTEM (CRIS)
    U.S. Department of Agriculture Cooperative State Research Service, Washington, D.C. 1974-.
    Current research in agriculture and related areas.

20. EIS KEY TO ENVIRONMENTAL IMPACT STATEMENTS
    Information Resources Press, Washington, D.C. 1977-.
    Environmental issues including air transportation, defense programs, energy, hazardous substances, parks and forests, roads and railroads, urban and social programs, wastes and water.

21. EIS INDUSTRIAL PLANTS
    Economic Information Systems, Inc., New York.
    Data and classification of industrial plants in the continental United States (with annual sales of more than $500,000).

22. EIS NONMANUFACTURING ESTABLISHMENTS
    Economic Information Systems, Inc., New York.
    Data on nonmanufacturers in the business sector that employ 20 or more people (excluding banking and nonprofit sectors).

23. ENERGYLINE
    Environment Information Center, New York. 1971-.
    All aspects of energy resources, consumption, and conservation.

24. ENVIRONMENTAL PERIODICALS BIBLIOGRAPHY (EPB)
    Environmental Studies Institute, Santa Barbara, California. 1974-.
    Covers about 250 periodicals in the fields of general human ecology, atmospheric studies, energy, land resources, water resources, nutrition, and health.

25. ENVIROLINE
    Environment Information Center, New York. 1971-.
    Environmental literature from interdisciplinary perspectives: science, technology, politics, sociology, commerce, law.

26. FOOD SCIENCE AND TECHNOLOGY ABSTRACTS (FSTA)
    International Food Information Service, Shinfield, Reading, Berkshire, England. 1969-.
    All aspects of food science and related technologies.

27. FOODS ADLIBRA
    K&M Publications, Inc., Louisville, Kentucky. 1974-.
    Food technology, beverages, and packaging industries.

28. FOREST PRODUCTS ABSTRACTS
    Forest Products Research Society. 1947-.
    All aspects of forest products excluding forestry and production of chemical pulp and paper.

Selected List of Bibliographic Databases Available for Online Searching 357

29. GEOARCHIVE
    Geosystems, P. O. Box 1024, Westminster, London, England. 1969-.
    World coverage of geosciences literature.

30. GEOLOGICAL REFERENCE FILE (GEOREF)
    American Geological Institute, Washington, D.C. 1961-.
    All aspects of geosciences.

31. INDEX TO API ABSTRACTS OF REFINING LITERATURE (APILIT)
    American Petroleum Institute, New York. 1964-.
    All aspects of petroleum technology.

32. INDEX TO API PATENTS (APIPAT)
    American Petroleum Institute, New York. 1964-.
    Petroleum patents.

33. INFORMATION SERVICES IN MECHANICAL ENGINEERING (ISMEC)
    Data Courier, Inc., Louisville, Kentucky. 1973-.
    Mechanical engineering, production engineering, and engineering management.

34. INSPEC
    The Institution of Electrical Engineers, London. 1969-.
    World coverage of physics, electrical engineering, electronics, computer science, and control engineering literature.

35. MARITIME RESEARCH INFORMATION SERVICE ABSTRACTS (MRIS ABSTRACTS)
    Maritime Research Information Service, Washington, D.C. 1970-.
    Maritime research including cold weather regions, cargo ships, materials, navigation, safety, and harbors.

36. METALS ABSTRACTS/ALLOYS INDEX (METADEX)
    American Society for Metals, Metals Park, Ohio. 1966-.
    Worldwide metallurgical literature including *ASM Review of Metal Literature* (1966-67), *Metals Abstracts* (1968-), and *Alloys Index* (1974-).

37. METEROLOGICAL AND GEOASTROPHYSICAL ABSTRACTS (MGA)
    American Meteorological Society, Boston, Mass. 1972-.
    Meteorological and geoastrophysical literature from the United States and foreign countries.

38. NATIONAL TECHNICAL INFORMATION SERVICE (NTIS)
    National Technical Information Service, U.S. Department of Commerce, Springfield, Va. 1964-.

    Multidisciplinary database of unclassified technical reports, covering the hard sciences, engineering and technology, and social sciences.

39. OCEANIC ABSTRACTS
    Data Courier, Inc., Louisville, Kentucky. 1964-.

    Oceanography, marine biology, marine pollution, ships and shipping, geology and geophysics, meteorology, and marine resources.

40. PESTICIDAL LITERATURE DOCUMENTATION
    Derwent Publications, London. 1968-.

    Pesticides and related subjects.

41. PETROLEUM ABSTRACTS (TULSA)
    University of Tulsa, Tulsa, Okla. 1965-.

    Exploration and production of gas and oil, geology, geophysics, and geochemistry.

42. PETROLEUM/ENERGY BUSINESS NEWS INDEX (P/E NEWS)
    American Petroleum Institute, New York. 1975-.

    Business news in petroleum and energy.

43. PIRASCAN (PIRA)
    Paper and Board, Printing and Packaging Industries Research Association, Randalls Road, Leatherhead, Surrey, England. 1975-.

    Scientific, technical, marketing, and managerial aspects of the paper, printing, and packaging industries.

44. POLLUTION ABSTRACTS
    Data Courier, Inc., Louisville, Ky. 1970-.

    Sources and control of air pollution, and noise pollution; environmental quality; pesticides; radiation; solid wastes; and water pollution.

45. RAPRA ABSTRACTS
    Rubber and Plastics Research Association of Great Britain, Shawbury, Shrewsbury, Shropshire, England. 1972-.

    Worldwide technical and commercial information in the polymers and associated fields.

Selected List of Bibliographic Databases Available for Online Searching

46. SAFETY SCIENCE ABSTRACTS JOURNAL (SAFETY)
    Cambridge Scientific Abstracts, Riverdale, Maryland. 1975-.
    Safety, including industrial/occupational, transportation, aviation and aerospace, environmental/ecological, medical safety and product liability.

47. SCIENCE CITATION INDEX (SCISEARCH)
    Institute for Scientific Information, Philadelphia, Penn. 1974-.
    Multidisciplinary index to the literature of science and technology.

48. SEARCHABLE PHYSICS INFORMATION NOTICES (SPIN)
    American Institute of Physics, New York. 1975-.
    Physics and astronomy journal literature.

49. SMITHSONIAN SCIENCE INFORMATION EXCHANGE (SSIE)
    Smithsonian Science Information Exchange, The Smithsonian Institution, Washington, D.C.
    Summary of ongoing research projects in all disciplines.

50. SOCIETY OF AUTOMOTIVE ENGINEERS ABSTRACTS (SAE)
    Society of Automotive Engineers, Warrandale, Penn. 1965-.
    Automotive engineering covering self-propelled vehicles applicable to aerospace, transportation, and other related areas.

51. WELDSEARCH
    The Welding Institute, Abington, Cambridge, England. 1967-.
    All aspects of welding including welding of plastics.

52. WORLD ALUMINUM ABSTRACTS
    American Society for Metals, Metals Park, Ohio. 1968-.
    All aspects of aluminum, ranging from ore processing through end uses.

53. WORLD PATENTS INDEX (WPI)
    Derwent Publications, London. 1963-.
    Worldwide patent information.

54. WORLD TEXTILES
    Shirley Institute, Didsbury, Manchester, England. 1970-.
    Machine-readable version of *World Textile Abstracts* covering world literature of textile science and technology including technical and managerial aspects.

References

1. A. W. Kenneth Metzner, Multiple use and other benefits of computerized publishing, *IEEE Transactions on Professional Communication,* PC-18(3): 274-278 (September 1975).
2. Russell J. Rowlett, Jr., Fred A. Tate, and James L. Wood, Relationships between primary and secondary information services, *Journal of Chemical Documentation,* 10(1):32-37 (February 1970).
3. Toward a modern secondary information system for chemistry and chemical engineering, *Chemical and Engineering News,* 53(24):30-38 (June 16, 1975).
4. Paul Doebler, Computer-based indexing system telescopes editing/typesetting time, *Publishers Weekly,* 203(10):59-61 (March 5, 1973).
5. Martha E. Williams, Online problems: Research today, solutions tomorrow, *Bulletin of the American Society for Information Science,* 3(4): 14-16 (April 1977).
6. CAS completes 9th Collective Index in record time, *Chemical and Engineering News,* 56(39):54 (September 25, 1978).
7. Martha E. Williams and Ted Brandhorst, Future trends in abstracting and indexing database publication, *Bulletin of the American Society for Information Science,* 5(3):27-28 (February 1979).
8. Martha E. Williams, Databases: A history of developments and trends from 1966 through 1975, *Journal of the American Society for Information Science,* 28(2):71-78 (March 1977).
9. Martha E. Williams, 1977 Database and online statistics, *Bulletin of the American Society for Information Science,* 4(2):21-23 (December 1977).
10. Charles T. Meadow, A morning in the life of an information scientist, *Bulletin of the American Society for Information Science,* 6(1):37 (October 1979).
11. Jean Lochard, Automatic processing of documentation and standardization, *UNESCO Bulletin for Libraries,* 25(3):143-150 (May-June 1971).
12. American Chemical Society Annual Report 1975: Chemical Abstracts Service, *Chemical and Engineering News,* 54:37-38 (April 5, 1976).
13. Foreign countries help pay *Chemical Abstracts* costs, *Library Journal,* 103(8):807 (April 3, 1978).
14. D. S. R. Murty, International Nuclear Information System (INIS), *Annals of Library Science and Documentation,* 18(1):22-30 (March 1971).
15. Edward J. Brunenkant, Databases—history of development and trends, (letter) *Journal of the American Society for Information Science,* 29(3):165 (May 1978).
16. Focus on UNISIST Programmes, *Annals of Library Science and Documentation,* 29(1-4):131-135 (March-December 1973).
17. C. R. Sahaer, National information systems and UNESCO, *The Bowker Annual of Library and Book Trade Information, 1975* (New York: R. R. Bowker, 1975), pp. 336-338.
18. Intergovernmental Conference on the Planning of National Documentation, Library and Archives Infrastructures, *Annals of Library Science and Documentation,* 21(4):150-155 (December 1974).

BIBLIOGRAPHY

The following is a selected bibliography of literature on scientific and technical information. Most of the items listed in this bibliography were consulted in preparing this book. Additional bibliographic references can be obtained from the following sources:

Bibliography of Research Relating to the Communication of Scientifc and Technical Information (Rutgers University Press, New Brunswick, N.J., 1967).

Subramanyam, K. Scientific Literature. *Encyclopedia of Library and Information Science* (Marcel Dekker, New York, 1979), v. 26, pp. 376-548.

Subramanyam, K. Technical Literature. *Encyclopedia of Library and Information Science* (Marcel Dekker, New York, 1980), v. 30, pp. 144-209.

Thornton, John L. and R. I. J. Tully. *Scientific Books, Libraries and Collectors.* 3rd revised edition (The Library Association, London, 1971).

Thornton, John L. and R. I. J. Tully. *Scientific Books, Libraries and Collectors. Supplement 1969-75.* (The Library Association, London, 1978).

Bibliography

AEC and DOD Information Analysis Centers. *Special Libraries,* 57(1):21-34 (January 1966).

Albert, Ted. Information programs in the new DOE. *Bulletin of the American Society for Information Science,* 4(4):34 (April 1978).

Aluri, Rao, and Phillip A. Yannarella. Publications of selected Information Analysis Centers. *Special Libraries,* 65(10/11):455-461 (October/November 1974).

American Chemical Society annual report 1975: Chemical Abstracts Service. *Chemical and Engineering News,* 54:37-38 (April 5, 1976).

American National Standards Institute. Committee Z39. Standard for technical report numbering. *Special Libraries,* 63(11):541-542 (November 1972).

American Psychological Association. An Informal Study of the Preparation of Chapters for the *Annual Review of Psychology,* in *Project on Scientific Information Exchange in Psychology. Report No. 2* (American Psychological Association, Washington, D.C., 1963).

Amron, Irving. History of Indexing and Abstracting by the ASCE. ASCE Annual Convention and Exposition, Philadelphia, September 27-October 1, 1976. Preprint 2736.

Ash, Janet E., and Ernest Hyde. *Chemical Information Systems.* (John Wiley, New York, 1975).

Atherton, Pauline. *The Role of "Letters" Journals in Primary Contributions of Information: A Survey of Authors of Physical Letters.* Report AIP/DRP-64-1 (American Institute of Physics, New York, 1964).

Auger, C. P. *Use of Report Literature* (Shoe String Press, Hamden, Conn., 1975).

Bailey, Martha J. The laboratory notebook as an R&D record. *Special Libraries,* 65(4):189-194 (April 1972).

Bailey, Martha J. The use of abbreviations and acronyms in the physics literature. *Journal of the American Society for Information Science,* 27:81-84 (March/April 1976).

Baker, Dale B. Chemical Abstracts Service. *Encyclopedia of Library and Information Science* (Marcel Dekker, New York, 1970), v. 4, pp. 479-499.

Bamford, Jr., Harold E. The editorial processing center. *IEEE Transactions on Professional Communication,* PC-16(3):82-83 (September 1973).

Barlow, D. H. A&I services and database producers: Economic, technological and cooperative opportunities. *ASLIB Proceedings,* 28(10):325-337 (October 1976).

Barr, K. P. Estimates of the number of currently available scientific and technical periodicals. *Journal of Documentation,* 23(2):110-116 (June 1967).

Barton, H. A. The publication charge plan in physics journals. *Physics Today,* 16:45-47 (June 1963).

Baruch, Jordan J., and Nazir Bhagat. The IEEE Annals: An experiment in selective dissemination. *IEEE Transactions on Professional Communication,* PC-18(3):296-308 (September 1975).

Bates, Ralph S. *Scientific Societies in the United States.* 3rd edition (MIT Press, Cambridge, Mass., 1965).

Baum, Harry. Documentation of technical and scientific meetings. *Proceedings of the American Documentation Institute. Parameters of Information Science, Philadelphia, October 5-8, 1964* (American Documentation Institute, Washington, D.C., 1964). pp. 243-246.

Becker, Joseph. *A National Approach to Scientific and Technical Information in the United States* (Joseph Becker, Los Angeles, Calif., 1976).

Bemer, Robert W., and A. Richard Shriver. Integrating computer text processing with photocomposition. *IEEE Transactions on Professional Communication*, PC-16(3):92-96 (September 1973).

Benton, John L. Patents in the chemical industry. *Chemistry and Industry*, No. 7:298-301 (April 5, 1975).

Bernal, J. D. Preliminary analysis of pilot questionnaire on the use of scientific literature. *Royal Society Scientific Information Conference, 21 June-2 July 1948. Report and Papers Submitted* (The Royal Society, London, 1948). pp. 589-637. (Paper No. 46).

Bernal, J. D. Provisional scheme for central distribution of scientific publications. *Ibid.,* pp. 253-258.

Bernal, J. D. The transmission of scientific information: A user's analysis. *Proceedings of the International Conference on Scientific Information, Washington, DC, November 16-21, 1958* (National Academy of Sciences-National Research Council, Washington, D.C., 1959). v. 1, pp. 77-95.

Bernier, Charles L. Condensed technical literatures. *Journal of Chemical Documentation*, 8(4):195-197 (November 1968).

Bernier, Charles L. Terse literatures: 1. Terse conclusions. *Journal of the American Society for Information Science*, 21(5):316-319 (September-October 1970).

Bernier, Charles L. Terse literature viewpoint of wordage problems. *Journal of Chemical Documentation*, 12(2):81-84 (May 1972).

Bernier, Charles L. Terse literatures: 2. Ultraterse literatures. *Journal of Chemical Information and Computer Sciences*, 15(3):189-192 (August 1975).

Bibliography of Research Relating to the Communication of Scientific and Technical Information (Rutgers University Press, New Brunswick, N.J., 1967).

Borko, Harold, and Charles L. Bernier. *Abstracting Concepts and Methods* (Academic Press, New York, 1975).

Borko, Harold, and Charles L. Bernier. *Indexing Concepts and Methods* (Academic Press, New York, 1978).

Bottle, Robert T., and Betty L. Emery. Information transfer by reader service cards. A response time analysis. *Special Libraries,* 62(11):469-474 (November 1971).

Bowman, Walker H. Importance of patents and information services to research workers. *Journal of Chemical Information and Computer Sciences,* 18(2):81-82 (May 1978).
There are several other articles on patents in this issue of the journal.

Boyer, Calvin J. *The Doctoral Disseration as an Information Source: A Study of Scientific Information Flow* (Scarecrow Press, Metuchen, N.J., 1973).

Boylan, N. T. G. A history of the dissemination of PB reports. *Journal of Library History,* 3:156-161 (April 1968).

Boylan, N. T. G. Technical reports, identification and acquisition. *RQ,* 10:18-21 (Fall 1970).

Bradford, S. C. On the scattering of scientific subjects in scientific periodicals. *Engineering,* 137:85-86 (1934).

Bradford, S. C. Extent to which scientific and technical literature is covered by present abstracting and indexing periodicals. *ASLIB Proceedings,* 14:59-71 (1937).

Bradford, S. C. *Documentation.* 2nd edition (Crosby Lockwood, London, 1953).

Branscomb, Lewis M. Support for reviews and data evaluations. *Science,* 187(4177):603 (February 21, 1975).

Brearley, Neil. The role of technical reports in scientific and technical communication. *IEEE Transactions on Professional Communication,* PC-16(3):117-119 (September 1973).

Brown, Charles H. Scientific serials: Characteristics and lists of most cited publications in mathematics, physics, chemistry, geology, physiology, botany, zoology and entomology. ACRL Monograph No. 16 (Association of College and Research Libraries, Chicago, 1956).

Brown, Norman B., and Jane Phillips. Price indexes for 1979: U.S. periodicals and serial services. *Library Journal,* 104(15):1600-1605 (September 1, 1979).

Brown, W. S., J. R. Pierce, and J. F. Traub. The future of scientific journals. *Science,* 158(3805):1153-1159 (December 1, 1967).

Brunenkant, Edward J. Databases—History of development and trends. (Letter) *Journal of the American Society for Information Science,* 29(3):165 (May 1978).

Brunning, Dennis. Review literature and the chemist. *Proceedings of the International Conference on Scientific Information, Washington, DC, November 16-21, 1958* (National Academy of Sciences-National Research Council, Washington, D.C., 1959). v. 1, pp. 545-570.

Buist, Eleanor. Soviet dissertation lists since 1934. *Library Quarterly,* 33(2): 192-207 (April 1963).

Calder, Ritchie. The nature and function of science. *ASLIB Proceedings,* 10(7):161-170 (July 1958).

Carroll, Kenneth D. Development of a national information system for physics. *Special Libraries,* 61:171-179 (April 1970).

Carter, L. J. Research Triangle Park succeeds beyond its promoters' expectations. *Science,* 200(4349):1469-70 (June 30, 1978).

CAS completes 9th collective index in record time. *Chemical and Engineering News,* 56(39):54 (September 25, 1978).

CAS Today: Facts and Figures about Chemical Abstracts Service (Chemical Abstracts Service, Columbus, Ohio, 1980).

Čermáková, Jiřina. International scientific congresses and conferences: Calendars, bibliographies of congress proceedings and conference technique handbooks. *Annals of Library Science and Documentatin,* 19(3):104-113 (September 1972).

Chakraborty, A. R., Sriram Janardan, and Meera G. Joshi. Translations and translation services. *Annals of Library Science and Documentation,* 22(2):60-72 (June 1975).

Chillag, J. P. Problems with reports, particularly microfiche reports. *ASLIB Proceedings,* 22(5):201-216 (May 1970).

Chirnside, R. C., and J. H. Hamence. *The "Practising Chemists." A History of the Society for Analytical Chemistry 1874-1974* (The Society for Analytical Chemistry, London, 1974).

Cochrane, Rexmond C. *Measures for Progress. A History of the National Bureau of Standards* (National Bureau of Standards, Washington, D.C., 1966).

Coile, Russell C. Information sources for electrical and electronics engineers. *IEEE Transactions on Engineering Writing and Speech,* EWS-12(3):71-78 (October 1969).

Colling, Patricia M. Dissertation Abstracts International. *Encyclopedia of Library and Information Science* (Marcel Dekker, New York, 1972). v. 7, pp. 238-240.

Commercial services/information retrieval and databases. *Journal of Library Automation,* 8(2):150 (June 1975).

Cottrell, William B. Evaluation and compression of scientific and technical information at the Nuclear Safety Information Center. *American Documentation,* 19(4):375-380 (October 1968).

Crane, Diana. *Invisible Colleges* (University of Chicago Press, Chicago, 1972).

Crane, Edward M. The U.S. Scientific Bibliography: A big effort for a big purpose. *Publishers Weekly,* 165(15):1621-23 (October 11, 1952).

Crawford, Susan. Informal communication among scientists in sleep research. *Journal of the American Society for Information Science,* 22:301-310 (September-October 1971).

Crawford, Susan. Information needs and uses. *Annual Review of Information Science and Technology* (Knowledge Industry Publications, Inc., White Plains, N.Y.; and American Society for Information Science, Washington, D.C., 1978). v. 13, pp. 61-81.

Davis, C. H., and J. E. Rush. *Information Retrieval and Documentation in Chemistry* (Greenwood Press, Westport, Conn., 1974).

Davis, R. A. How engineers use literature. *Chemical Engeineering Progress,* 61(3):30-34 (March 1965).

Davis, W. Development in auxiliary publication. *American Documentation,* 2(1):7-11 (January 1951).

A debate on preprint exchange. *Physics Today,* 19:60-73 (June 1966).

DeGennaro, Richard. Escalating journal prices: Time to fight back. *American Libraries,* 8(2):69-74 (February 1977).

Demise of scientific journals. *Nature,* 228:1025-26 (December 12, 1970).

Dible, Donald M. *What Everybody Should Know About Patents, Trademarks, and Copyrights* (The Entrepreneur Press, Fairfield, Calif., 1978).

Dirksen, Eileen F. Meetings and their publications. *Journal of Chemical Documentation,* 5(3):124-125 (August 1965).

Ditmas, E. M. R. Coordination of information. A survey of schemes put forward in the last fifty years. *Journal of Documentation,* 3(4):209-221 (March 1948).

Do technical reports belong to literature? *Nature,* 236:275 (April 7, 1972).

Doebler, Paul. Computer-based indexing system telescopes editing/typesetting time. *Publishers Weekly,* 203(10):59-62 (March 5, 1973).

Dood, Kendall J. The U.S. Patent Classification System. *IEEE Transactions on Professional Communication,* PC-22(2):95-100 (June 1979).

Drott, M. Carl, Toni C. Bearman, and Belver C. Griffith. The hidden literature: The scientific journals of industry. *ASLIB Proceedings*, 27(9):376-384 (September 1975).

Eakins, J. P. Integrated publication system: A new concept in primary publication. *ASLIB Proceedings*, 26(11):430-434 (November 1974).

Editorial Processing Centers: Feasibility and Promise (Aspen Systems Corporation; Westat, Inc., Rockville, Md., 1976).

El-Hadidy, Bahaa. Bibliographic control among geoscience abstracting and indexing services. *Special Libraries*, 66(5/6):260-265 (May-June 1975).

Elsdon-Dew, R. The library from the point of view of the research worker. *South African Libraries*, 23:51-54 (October 1955).

English, Eileen W. Hits and misses: Securing report literature. *Special Libraries*, 66:237-240 (May-June 1975).

Evan, William. Role strain and the norm of reciprocity in research organizations. *American Journal of Sociology*, 68:346-354 (November 1962).

Fitzpatrick, William H., and Monroe E. Freeman. Science Information Exchange: The evolution of a unique information storage and retrieval system. *Libri*, 15(2):127-137 (1965).

Flowers, B. H. Survey of information needs of physicists and chemists. *Journal of Documentation*, 21:83-112 (1965).

Foreign countries help pay *Chemical Abstracts* costs. *Library Journal*, 103(8): 807 (April 3, 1978).

Freeman, Monroe E. Science Information Exchange as a source of information. *Special Libraries*, 59(2):86-90 (February 1968).

Fry, Bernard M. and Herbert S. White. Economics and interaction of the publisher-library relationship in the production and use of scholarly and research journals (National Technical Information Service, Springfield, Va., November 1975). Order No. 249108.

Gannett, Elwood K. Primary publication systems and services. *Annual Review of Information Science and Technology* (American Society for Information Science, Washington, D.C., 1973). v. 8, pp. 243-275.

Gardner, Jeffrey J. Pathfinders, Library. *Encyclopedia of Library and Information Science* (Marcel Dekker, New York, 1977). v. 21, pp. 468-473.

Garfield, Eugene. Is citation frequency a valid criterion for selecting journals? *Current Contents*, (April 5, 1972), pp. 5-6.

Garfield, Eugene. Unintelligible abbreviations and sloppy words in article titles create magic (invisible) spots for indexers. *Current Contents*, 12 (November 29, 1972), pp. 5-7.

Garvey, William D., and Belver C. Griffith. Informal channels of communication in the behavioral sciences: Their relevance in the structuring of formal or bibliographic communication. *The Foundations of Access to Knowledge: A Symposium,* Edward B. Montgomery, ed., (Syracuse University, Syracuse, N.Y., 1968). pp. 129-151.

Garvey, William D., and others. Some comparisons of communication activities in the physical and social sciences. *Communication Among Scientists and Engineers,* Carnot E. Nelson and Donald K. Pollock, eds. (D. C. Heath & Co., Lexington, Mass., 1970). pp. 61-84.

Garvey, William D., and S. D. Gottfredson. Scientific communication as an integrative social process. *International Forum on Information and Documentation,* 2:9-16 (January 1977).

Garvin, David. The IAC and the library. *Special Libraries,* 62:17-23 (January 1971).

Gautney, George E., and Ronald L. Wiginton. American Chemical Society-Chemical Abstracts Service. *Encyclopedia of Computer Science and Technology* (Marcel Dekker, New York, 1975). v. 1, pp. 386-410.

Gilmore, J. S., and others. The channels of technology acquisition in commercial firms, and the NASA Dissemination Program (Denver Research Institute, June 1967). NASA CR-790.

Godfrey, L. E., and H. F. Redman. *Dictionary of Report Series Codes.* 2nd edition (Special Libraries Association, New York, 1973).

Gordon, Irving, and Russell L. K. Carr. Utilization of terse conclusions in an industrial research environment. *Journal of Chemical Documentation,* 12(2):86-88 (May 1972).

Gottschalk, Charles M., and Winifred F. Desmond. World-wide census of scientific and technical serials. *American Documentation,* 14(3):188-194 (July 1963).

Gray, Dwight E., and Staffan Rosenborg. Do technical reports become published papers? *Physics Today,* 10(6):18-21 (June 1957).

Grogan, Denis J. *Science and Technology: An Introduction to the Literature.* 3rd edition (Clive Bingley, London, 1976).

Grosch, Audrey N. A workstudy of the review production process. *Journal of the American Society for Information Science,* 29(1):48 (January 1978).

Guha, B. *Documentation and Information* (The World Press, Calcutta, 1978).

Haberer, Isabel H. House Journals. *Progress in Library Science* (Butterworths, London, 1967). v. 1, pp. 1-96.

Halbert, Michael H., and Russell L. Ackoff. An operations research study of the dissemination of information. *Proceedings of the International Conference on Scientific Information, Washington, DC, November 16-21, 1958* (National Academy of Sciences-National Research Council, Washington, D.C., 1959). v. 1, pp. 97-130.

Hammett, Louis P. Choice and chance in scientific communication. *Chemical and Engineering News*, 39:94-97 (April 10, 1961).

Hannay, N. B., and others. *Cost-Effectiveness of Information Subsystems.* (American Chemical Society, Committee on Corporation Associates, Subcommittee on the Economics of Chemical Information, Washington, D.C., May 1969).

Hanson, C. W., and M. Janes. Lack of indexes in reports of conferences: Report of an investigation. *Journal of Documentation*, 16:65-70 (June 1960).

Hanson, C. W., and M. Janes. Coverage by abstracting journals of conference papers. *Journal of Documentation*, 17:143-149 (September 1961).

Harris, Linda, and Robert V. Katter. Impact of the *Annual Review of Information Science and Technology. Proceedings of the ASIS Annual Meeting*, No. 5 (Greenwood Publishing Co., New York, 1968), pp. 331-333.

Hemenway, David. *Industrywide Voluntary Product Standards* (Ballinger Publishing Co., Cambridge, Mass., 1975).

Herner, S. The information gathering habits of American medical scientists. *Proceedings of the International Conference on Scientific Information, Washington, DC, November 16-21, 1958* (National Academy of Sciences-National Research Council, Washington, D.C., 1959). v. 1, pp. 277-285.

Herring, Conyers. Distill or drown: The need for reviews. *Physics Today*, 21(9):27-33 (September 1968).

Herring, Conyers. Critical reviews: The user's point of view. *Journal of Chemical Documentation*, 8(4): 232-236 (November 1968).

Hershey, David F. SSIE: A unique database. *Government Publications Review*, 1:209-212 (Winter 1973).

Herschman, Arthur. The primary journal: Past, present and future. *Journal of Chemical Documentation*, 10(1):37-42 (February 1970).

Herschman, Arthur. Keeping up with what's going on in physics. *Physics Today*, 24:23-29 (November 1971).

Hills, Jacqueline. *Review of the Literature on Primary Communication in Science and Technology* (ASLIB, London; 1972).

Houghton, Bernard. *Technical Information Sources.* 2nd edition (Shoe String, Press, Linnett Books, Hamden, Conn., 1972).

Hutchisson, Elmer. The role of international scientific organizations in improving scientific documentation. *Physics Today,* 15:24-26 (September 1962).

Hyatt-Mayor, A. Foreword. *A Guide to American Trade Catalogs, 1744-1900* (R. R. Bowker, New York, 1960; reprinted by Arno Press, New York, 1976).

Inhaber, H. Scientific cities. *Research Policy,* 3:182-200 (1974).

INIS Today: An Introduction to the International Nuclear Information System (International Atomic Energy Agency, Vienna, 1977).

Intergovernmental Conference on the Planning of National Documentation, Library and Archives Infrastructures. *Annals of Library Science and Documentation,* 21(4):150-155 (December 1974).

International Federation for Documentation. The content, influence and value of scientific conference papers and proceedings. *UNESCO Bulletin for Libraries,* 16(3):113-126 (May-June 1962).

International Federation for Documentation. Availability of scientific conference papers and proceedings. *UNESCO Bulletin for Libraries,* 16(4):165-176 (July-August 1962).

Joenk, R. J. Patents: Incentive to innovate and communicate: An introduction. *IEEE Transactions on Professional Communication,* PC-22(2):46-59 (June 1979).
This article has an extensive bibliography on inventing and patenting. The June 1979 issue of the *Transactions* is a special issue on patents, and contains 17 articles on various aspects of patents.

Johns Hopkins University. Center for Research in Scientific Communication. *Reports of Studies of the Publication Fate of Materials Presented at National Meetings (Two Years After the Meeting).* (National Technical Information Service, Springfield, Va., June 1969). PB 185 469.

Kase, Francis J. *Foreign Patents: A Guide to Official Patent Literature* (Oceana Publications, Dobbs Ferry, N.Y., 1972).

Kean, Pauline, and Jalrath Ronayne. Preliminary communications in chemistry. *Journal of Chemical Documentation,* 12(4):218-220 (November 1972).

King, Alexander. Concerning conferences. *Journal of Documentation,* 17(2): 69-76 (June 1961).

King, D. W., and others. *Statistical Indicators of Scientific and Technical Communication, 1960-1980* (U.S. Government Printing Office, Washington, D.C., 1976). PB 254 060.

Klempner, Irving. *Diffusion of Abstracting and Indexing Services for Government-Sponsored Research* (Scarecrow Press, Metuchen, N.J., 1968).

Klempner, Irving. The concept of "national security" and its effects on information transfer. *Special Libraries*, 64(7):263-269 (1973).

Knopp, O. R. H. The world's major patent systems. *Journal of the Patent Office Society*, 54(1):8-16 (1972).

Koch, H. William. Publication charges and financial solvency. *Physics Today*, 21:126-127 (December 1968).

Korbuly, Dorothy K. A new approach to coding displayed mathematics for photocomposition. *IEEE Transactions on Professional Communication*, PC-18(3):283-287 (September 1975).

Korfhage, Robert R. Informal communication of scientific information. *Journal of the American Society for Information Science*, 25:25-32 (1974).

Kovach, Eugene G. Country trends in scientific productivity. *Who Is Publishing in Science* (Institute for Scientific Information, Philadelphia, 1978), pp. 33-40.

Kronick, David. A. *A History of Scientific and Technical Periodicals. The Origins and Development of the Scientific and Technical Press, 1665-1790.* 2nd edition (Scarecrow Press, Metuchen, N.J., 1976).

Kuiper, Barteld E. Toward a world catalog of standards. *UNESCO Bulletin for Libraries*, 27:155-159+ (May 1973).

Kuney, Joseph H. American Chemical Society information program. *Encyclopedia of Library and Information Science* (Marcel Dekker, New York, 1968). v. 1, pp. 247-264.

Kuney, Joseph H. New developments in primary journal publication. *Journal of Chemical Documentation*, 10:42-46 (February 1970).

Kuney, Joseph H., and William H. Weissgerber. System requirements for primary journal systems: Utilization of the *Journal of Organic Chemistry*. *Journal of Chemical Documentation*, 10:150-157 (August 1970).

Kuntz, Werner, and others. *Methods of Analysis and Evaluation of Information Needs: A Critical Review* (Verlag Dokumentation, Munich, 1977).

Ladendorf, Janice M. Information flow in science, technology, and commerce. *Special Libraries*, 61:215-222 (May/June 1970).

Lerner, Rita G. American Institute of Physics. *Encyclopedia of Computer Science and Technology* (Marcel Dekker, New York, 1975). v. 1, pp. 435-443.

Licklider, J. C. R. A crux in scientific and technical communication. *American Psychologist*, 12:1044-51 (1966).

Liebesny, Felix. Lost information: Unpublished conference papers. *Proceedings of the International Conference on Scientific Information, Washington, DC, November 16-21, 1958* (National Academy of Sciences-National Research Council, 1959). v. 1, pp. 475-479.

Liebesny, Felix and others. The scientific and technical information contained in patent specifications. The extent of time factors of its publications in other forms of literature. *The Information Scientist,* 8(4):165-177 (December 1974).

Lippert, Walter. Gmelin's Handbook for Inorganic Chemistry. *Journal of Chemical Documentation,* 10(3):174-180 (August 1970).

Lochard, Jean. Automatic processing of documentation and standardization. *UNESCO Bulletin for Libraries,* 25(3):143-150 (May-June 1971).

Locke, William N. Translation by machine. *Scientific American,* 194(1):29-33 (January 1956).

Lockheed Missiles and Space Company, Inc. DIALOG Reduced Rates. *NFAIS Newsletter,* 19(1):14 (February 1977).

Loosjes, Th. P. *On Documentation of Scientific Literature* (Butterworths, London, 1973).

Lublin, Joann S. Underground papers in corporations tell it like it is—or perhaps like it isn't. *Wall Street Journal* (November 3, 1971), p. 16.

Luckenbach, Reiner. Der Beilstein. *Chemtech,* 9:612-621 (October 1979).

Machlup, Fritz, Kenneth Leeson, and associates. *Information Through the Printed Word. The Dissemination of Scholarly, Scientific, and Intellectual Knowledge* (Praeger Publishers, New York, 1978).
    v. 1: *Book Publishing*
    v. 2: *Journals*
    v. 3: *Libraries*

Mangla, P. B. Scientifc literature and documentation. *Herald of Library Science,* 3(4):286-292 (October 1964).

Manten, A. A. Scientific review literature. *Scholarly Publishing,* 5:75-89 (October, 1973).

Manzer, Bruce M. *The Abstract Journal, 1790-1920: Origin, Development, and Diffusion* (Scarecrow Press, Metuchen, N.J., 1977).

Marquis, Donald G., and Thomas J. Allen. Communication patterns in applied technology. *American Psychologist,* 21:1052-1060 (1966).

Martin, D. C. The Royal Society's interest in scientific publications and the dissemination of information. *ASLIB Proceedings,* 9(5):127-141 (May 1957).

Bibliography

Martin, Jean K., and Ronald G. Parsons. Evaluation of current awareness services for physics and astronomy literature. *Journal of the American Society for Information Science,* 25(3):156-161 (May-June 1974).

Martyn, John. Tests on abstracts journals: Coverage overlap and indexing. *Journal of Documentation,* 23(1):45-70 (March 1967).

Martyn, John, and Margaret Slater. Tests on abstracts journals. *Journal of Documentation,* 20(4):212-235 (December 1964).

Martyn, John, and A. Gilchrist. *Evaluation of British Scientific Journals* (ASLIB, London, 1968).

Matarazzo, James M. Scientific journals: Page or price explosion? *Special Libraries,* 63:53-58 (February 1972).

Mathews, Eleanor. The *Bibliography of Agriculture. Reference Services Review,* 7(2):57-62 (June 1979).

McElderry, S. Toward a national information system in the U.S. *Libri,* 25:199-212 (September 1975).

Meadow, Charles T. A morning in the life of an information scientist. *Bulletin of the American Society for Information Science,* 6(1):37 (October 1979).

Medvedev, Zhores A. *The Rise and Fall of T. D. Lysenko.* Translated by I. Michael Lerner (Doubleday, Garden City, N.Y., 1971).

Mellon, M. G. Beilstein's Handbuch, *Encyclopedia of Library and Information Science* (Marcel Dekker, New York, 1969). v. 2, pp. 283-291.

Metzner, A. W. Kenneth. Multiple use and other benefits of computerized publishing. *IEEE Transactions on Professional Communication,* PC-18(3): 274-278 (September 1975).

Middleton, M. INIS: International Nuclear Information System. *Australian Library Journal,* 23(4):136-140 (May 1974).

Miller, Elizabeth, and Eugenia Truesdell. Citation indexing history and applications. *Drexel Library Quarterly,* 8:159-172 (April 1972).

Moore, C. Alan. Preprints: An old information device with new outlooks. *Journal of Chemical Documentation,* 5(3):126-128 (August 1965).

Moore, James A. An inquiry on new forms of primary publications. *Journal of Chemical Documentation,* 12:75-78 (May 1972).

Moore, Julie L. Bibliographic control of American doctoral dissertations. *Special Libraries,* 63(5/6):227-230 (May/June 1972).

Moore, Julie L. Bibliographic control of American doctoral dissertations. *Special Libraries,* 63(7):285-291 (July 1972).

Murra, Kathrine O. Futures in international meetings. *College and Research Libraries,* 19(6):445-450 (November 1958).

Murty, D. S. R. International Nuclear Information System (INIS). *Annals of Library Science and Documentation,* 18(1):22-30 (March 1971).

National Agricultural Library: CRIS-Online. *NFAIS Newsletter,* 19(2):6 (April 1977).

Nelson, Carnot E., and Donald K. Pollock. *Communication Among Scientists and Engineers* (D. C. Heath & Co., Lexington, Mass., 1970).

Oppenheim, C. The patent coverage of *Chemical Abstracts. The Information Scientist,* 8(3):133-137 (September 1974).

Parisi, Paul A. Composition innovations at the ASCE. *IEEE Transactions on Professional Communication,* PC-18(3):244-273 (September 1975).

Parks, W. George. Gordon Research Conferences: Program for 1964. *Science,* 143(3611):1203-1205 (March 13, 1964).

Pasternack, Simon. Is journal publication obsolescent? *Physics Today,* 19:38-43 (May 1966).

Pelzer, C. W. The International Nuclear Information System. *ASLIB Proceedings,* 24:38-55 (January 1972).

Periodicals. *Illinois Monthly Magazine,* 1:302-303 (1831), quoted in Gottschalk and Desmond, *American Documentation,* 14(3):193 (July 1963).

Phelps, Ralph H. Alternatives to the scientific periodical: A report and bibliography. *UNESCO Bulletin for Libraries,* 14(2):61-75 (March-April 1960).

Phipps, T. E. Scientific communication. *Science,* 129(3342):118 (January 16, 1959).

Pirie, N. W. Note on the simultaneous publication of papers at two different levels of completeness. *The Royal Society Scientific Information Conference, 21 June-2 July 1948. Report and Papers Submitted* (The Royal Society, London, 1948). pp. 419-422.

Porter, J. R. The scientific journal—300th anniversary. *Bacteriological Reviews,* 28(3):211-230 (September 1964).

Pownall, J. F. *Organized Publication* (Elliott Stock, London, 1926), quoted in Phelps, *UNESCO Bulletin for Libraries,* 14(2):63-64 (March-April 1960).

Preface: A major development in Chemical Society Publications. *Annual Reports on the Progress of Chemistry* (The Chemical Society, London, 1967). v. 64, pp. vii-ix.

Price, Derek J. de Solla. *Science Since Babylon* (Yale University Press, New Haven, Conn., 1961).

Price, Derek J. de Solla. *Little Science Big Science* (Columbia University Press, New York, 1963).

Price, Derek J. de Solla. The scientific foundations of science policy. *Nature*, 206:234 (April 17, 1965).

Price, Derek J. de Solla. Measuring the size of science. *Proceedings of the Israel Academy of Sciences and Humanities*, 4:98-111 (1969).

Price, Derek J. de Solla. Some remarks on elitism in information and the invisible college phenomenon in science. *Journal of the American Society for Information Science*, 22:74-75 (March-April 1971).

Price, Derek J. de Solla, and S. Gursey. Some statistical results for the number of authors in the states of the United States and the nations of the world. *Who is Publishing in Science* (Institute for Scientific Information, Philadelphia, Penn., 1977). pp. 26-34.

Proceedings of the International Conference on Scientific Information, Washington, DC November 16-21, 1958 (National Academy of Sciences-National Research Council, 1959). 2 v.

Publications of the National Bureau of Standards. 1976 Catalog. SP 305, Supplement 8 (U.S. Department of Commerce, National Bureau of Standards, Washington, D.C., 1977).

Purpose in publication. *Nature*, 191(4788):527-530 (August 5, 1961).

Rawls, Rebecca L. ACS conducting dual basic journal experiment. *Chemical & Engineering News*, 54(24):28-29 (June 7, 1976).

Rea, Robert H. Defense Documentation Center. *Drexel Library Quarterly*, 10(1-2):21-38 (January-April 1974).

Redman, Helen F. Report number chaos. *Special Libraries*, 53(10):574-578 (December 1962).

Reeves, S. K. Specifications, standards and allied publications for U.K. military aircraft. *ASLIB Proceedings*, 22(9):432-448 (September 1970).

Research papers in *Analytical Chemistry*. (Editor's column). *Analytical Chemistry*, 51(9):1009A-1010A (August 1979).

Rhodes, Sarah N., and Harold E. Bamford. Editorial processing centers: A progress report. *American Sociologist*, 11:153-159 (August 1976).

Romaine, Lawrence B. American trade catalogs vs. manuscript records. *Library Resources and Technical Services*, 4(1):63-65 (Winter 1960).

Romaine, Lawrence B. *A Guide to American Trade Catalogs, 1744-1900.* (R. R. Bowker, New York, 1960).

Rosen, Edward. Copernicus published as he perished. *Nature,* 241(5390):433-434 (February 16, 1973).

Rosenbloom, Richard S., and Francis W. Wolek. *Technology, Information, and Organization: A Report to the National Science Foundation* (Harvard University Graduate School of Business Administration, Boston, Mass., 1967).

Rosenbloom, Richard S., and Francis W. Wolek. *Technology and Information Transfer. A Survey of Practice in Industrial Organizations* (Harvard University, Graduate School of Business Administration, Boston, Mass., 1970).

Rossmassler, Stephen A. Scientific literature in policy decision making. *Journal of Chemical Documentation,* 10:163-167 (August 1970).

Roth, D. L. Guide to the use of Beilstein, Gmelin, and Landölt-Bornstein. *Herald of Library Science,* 11:325-333 (October 1972).

Rowlett, Jr, Russell J., Fred A. Tate, and James L. Wood. Relationship between primary and secondary information services. *Journal of Chemical Documentation,* 10(1):32-37 (February 1970).

The Royal Society Scientific Information Conference, 21 June-2 July 1948. Report and Papers Submitted (The Royal Society, London, 1948).

Sahaer, C. R. National information systems and UNESCO. *The Bowker Annual of Library and Book Trade Information, 1975* (R. R. Bowker, New York, 1975).

Schussel, George. Advent of information and inquiry services. *Journal of Data Management,* 7(9):24-31 (September 1969).

Science, Government and Information. The Responsibilities of the Technical Community and the Government in the Transfer of Information. A Report of the President's Science Advisory Committee (The White House, Washington, D.C., 1963).

Scientific Conference Papers and Proceedings (UNESCO, Paris, 1963).

Scientific and Technical Communication. A Pressing National Problem and Recommendations for Its Solution (National Academy of Sciences, Washington, D.C., 1969).

Senders, J. W., C. M. B. Anderson, and C. D. Hecht. Scientific publication systems: An analysis of past, present, and future methods of scientific communication. PB 242 259 (National Technical Information Service, June 1975).

Shephard, D. A. E. Some effects of delay in publication of information in medical journals, and implications for the future. *IEEE Transactions on Professional Communication*, PC-16:143-147; 181-182 (1973).

Sherwin, C. W., and R. S. Inemson. *First Interim Report on Project Hindsight* (Summary). (Office of the Director of Defense Research and Engineering, Washington, D.C., October 13, 1966).

Short, P. J. Bibliographic tools for tracing conference proceedings. *IATUL Proceedings*, 6(2):50-53 (May 1972).

Siehl, George H. Environment Update. *Library Journal*, 104(9):1003-1007 (May 1, 1979). Annual feature.

Simonton, D. P. *Directory of Engineering Document Sources* (Global Engineering and Documentation Services, Inc., New Port Beach, Calif., 1972).

Singleton, A. K. J. Communication in physics: Review and first parameters. *Journal of Documentation*, 31(3):137-143 (September 1975).

Skolnik, Herman. Milestones in chemical information science. *Journal of Chemical Information and Computer Science*, 16(4):187-193 (November 1976).

Skolnik, Herman. Historical aspects of patent systems. *Journal of Chemical Information and Computer Sciences*, 17(3):119-121 (August 1977).

Slattery, William J. Standards Information Service (NBS). *Information Hotline*, 8(9):30-31 (October 1976).

Starker, Lee N. User experiences with primary journals on 16mm microfilm. *Journal of Chemical Documentation*, 10:5-6 (February 1970).

Steere, William C. *Biological Abstracts/BIOSIS. The First Fifty Years. Evolution of a Major Information Service* (Plenum Publishing Corporation, New York, 1976).

Sternberg, Virginia. Accountability. *Encyclopedia of Library and Information Science* (Marcel Dekker, New York, 1968), v. 1, pp. 55-60.

Storer, Norman W. Research orientations and attitudes towards team work. *IRE Transactions on Engineering Management*, EM-9(1):29-33 (March 1962).

Struglia, Erasmus J. *Standards and Specifications Information Sources* (Gale Research Co., Detroit, Mich., 1965).

A Study of Coverage Overlap Among Major Science and Technology Abstracting and Indexing Services. NFAIS-77/1 (National Federation of Abstracting and Indexing Services, Philadelphia, Penn., 1977).

Subramanyam, K. Information Exchange Groups: An experiment in science communication. *Indian Librarian,* 29(4):159-164 (March 1975).

Subramanyam, K. The scientific journal: A review of current trends and future prospects. *UNESCO Bulletin for Libraries,* 29(4):192-201 (July-August 1975).

Subramanyam, K. *A Bibliometric Investigation of Computer Science Journal Literature.* Ph.D. Dissertation, University of Pittsburgh, 1975.

Subramanyam, K. Core journals in computer science. *IEEE Transactions on Professional Communication,* PC-19(2):22-25 (December 1976).

Subramanyam, K. Acronymania. *Technical Communication.* 26(3):13-15 (Third Quarter 1979).

Subramanyam, K. Scientific Literature. *Encyclopedia of Library and Information Science* (Marcel Dekker, New York, 1979), v. 26, pp. 376-548.

Subramanyam, K., and Constance J. Schaffer. Effectiveness of "letters" journals. *New Library World,* 75:258-259 (December 1974).

Subramanyam, K., and Mary M. O'Pecko. Environmental research journals. *Environmental Science and Technology,* 13(8):927-929 (August 1979).

Swanson, Rowena W. A work study of the review production process. *Journal of the American Society for Information Science,* 27(1):70-72 (January-February 1976).

Symposium volumes. *Nature,* 214(5083):46 (April 1, 1967).

Tallman, J. E. History and importance of technical report literature. *Sci-Tech News,* 15:44-46 (Summer 1961).

Tallman, J. E. History and importance of technical report literature. *Sci-Tech News,* 16:13 (Spring 1962).

Tallman, J. E. History and importance of technical report literature. *Sci-Tech News,* 15:164-165+ (Winter 1962).

Taube, Mortimer. Memorandum for a conference on bibliographical control of government scientific and technical reports. *Special Libraries,* 39(5):54-60 (May-June 1948).

Tayal, A. S. Acquisition and updating of standards and specifications in technical libraries. *UNESCO Bulletin for Libraries,* 25:198-204 (July 1971).

Terapane, John F. A unique source of information. *Chemtech,* 8(5):272-276 (May 1978).

Thornton, John L., and R. I. J. Tully. *Scientific Books, Libraries and Collectors.* 3rd revised edition (The Library Association, London, 1971).

Too many chemistry journals. *Chemical & Engineering News*, 51:44 (December 10, 1973).

Toward a modern secondary information system for chemistry and chemical engineering. *Chemical & Engineering News*, 53(24):30-38 (June 16, 1975).

Townsend, Leroy B. Critical reviews: The author's point of view. *Journal of Chemical Documentation*, 8(4):239-241 (November 1968).

Translations in the U.K. *ASLIB Proceedings*, 25(7):264-267 (July 1973).

Trumbull, Richard. BIOSIS: 50 years of *Biological Abstracts*. *Bioscience*, 26(5):307 (May 1976).

UNESCO and International Council of Scientific Unions. *UNISIST: A Study Report on the Feasibility of a World Science Information System* (UNESCO, Paris, 1971).

U.S. Department of Defense. *Industrial Security Manual for Safeguarding Classified Information.* Attachment to DD form 441, DOD 5220.22-M (U.S. Government Printing Office, Washington, D.C., 1966).

U.S. Federal Council for Science and Technology. Committee on Scientific and Technical Information. Task Group on the Role of the Technical Report. *The Role of the Technical Report in Scientific and Technological Communication.* PB 180 944. 1968.

Van Cott, Harold P., and Albert Zavala. Extracting the basic structure of scientific literature. *American Documentation*, 19(3):247-262 (July 1968).

Vasilakis, Mary. Classified Material (Security). *Encyclopedia of Library and Information Science* (Marcel Dekker, New York, 1971), v. 5, pp. 174-185.

Vcerasnij, Rostislav P. Patent information and its problems. *UNESCO Bulletin for Libraries*, 23(5):234-239 (September-October 1969).

Virgo, Julie A. The review article: Its characteristics and problems. *Library Quarterly*, 41(4):275-291 (October 1971).

Voigt, Melvin J. *Scientists' Approaches to Information* (American Library Association, Chicago, Ill., 1961).

Wade, W. History of the American patent incentive system. *Journal of the Patent Office Society*, 54(1):67-71 (1972).

Walker, Albert. House journals. *Encyclopedia of Library and Information Science* (Marcel Dekker, New York, 1974), v. 11, pp. 61-74.

Wall, R. A. Trade literature problems. *Engineer*, 225:453-454 (March 15, 1968).

Wall, R. A. Trade literature problems. *Engineer,* 225:489-491 (March 22, 1968).

Weinstock, Melvin. Citation indexes. *Encyclopedia of Library and Information Science* (Marcel Dekker, New York, 1971)., v. 5, pp. 16-40.

Weisman, Herman M. Technical librarians and the National Standard Reference Data System. *Special Libraries,* 63(2):69-76 (February 1972).

Weissbach, Oskar. *The Beilstein Guide: A Manual for the Use of Beilstein's Handbuch der Organischen Chemie* (Springer-Verlag, Berlin, 1976).

West, S. S. The ideology of academic scientists. *IRE Transactions on Engineering Management,* EM-7(2):54-62 (June 1960).

Williams, Martha E. 1977 database and online statistics. *Bulletin of the American Society for Information Science,* 4(2):21-23 (December 1977).

Williams, Martha E. Databases: A history of developments and trends from 1966 through 1975. *Journal of the American Society for Information Science,* 28(2):71-78 (March 1977).

Williams, Martha E. Online problems: Research today, solutions tomorrow. *Bulletin of the American Society for Information Science,* 3(4):14-16 (April 1977).

Williams, Martha E., and Ted Brandhorst. Future trends in abstracting and indexing database publication. *Bulletin of the American Society for Information Science,* 5(3):27-28 (February 1979).

Williams, Martha E., and Laurence Lannom. Database online in 1979. *Bulletin of the American Society for Information Science,* 6(2):22-31 (December 1979).

Witty, Francis J. The beginnings of indexing and abstracting: Some notes towards a history of indexing and abstracting in antiquity and the Middle Ages. *The Indexer,* 8(4):193-198 (October 1973).

Wolfle, Dael. Gordon Research Conferences. *Science,* 148(3670):583 (April 30, 1965).

Wolfle, Dael, and Charles V. Kidd. The future market for Ph.D.'s. *Science,* 173(3999):784-793 (August 27, 1971).

Wood, D. N. The foreign language problem facing scientists and technologists in the U.K. Report of a recent survey. *Journal of Documentation,* 23(2): 117-130 (June 1967).

Wood, D. N., and D. R. L. Hamilton. *The Information Requirements of Mechanical Engineers* (The Library Association, London, 1967).

Wood, J. L., C. Flanagan, and H. E. Kennedy. Overlap in the lists of journals monitored by BIOSIS, CAS and EI. *Journal of the American Society for Information Science,* 23(1):36-38 (January-February 1972).

Wood, J. L., C. Flanagan, and H. E. Kennedy. Overlap among the journal articles selected for coverage by BIOSIS, CAS and EI. *Journal of the American Society for Information Science,* 24(1):25-28 (January-February 1973).

Woodward, A. M. Review literature: Characteristics, sources, and output in 1972. *ASLIB Proceedings,* 26(9):367-376 (September 1974).

Woodward, A. M. The role of reviews in information transfer. *Journal of the American Society for Information Science,* 28(3):175-180 (May 1977).

Wyatt, H. V. Research newsletters in the biological sciences: A neglected literature service. *Journal of Documentation,* 23(4):321-327 (December 1967).

Yagello, Virginia. Indexes to conference proceedings. *Reference Services Review,* 7(2):37-45 (June 1979).

Zaltman, Gerald. A note on an international invisible college for information exchange. *Journal of the American Society for Information Science,* 25:113-117 (March-April 1974).

Ziman, J. M. Information, communication, knowledge. *Nature,* 224:76-84 (October 25, 1969).

APPENDIX

ABBREVIATIONS

AAAS	American Association for the Advancement of Science
ACS	American Chemical Society
AD	ASTIA Document (or, Accession Document)
AGARD	Advisory Group for Aerospace Research and Development (NATO)
AGRIS	Agricultural Research Information System
AIAA	American Institute of Aeronautics and Astronautics
A.I.Ch.E.	American Institute of Chemical Engineers
AIP	American Institute of Physics
ANSI	American National Standards Institute
API	American Petroleum Institute
ARIST	*Annual Review of Information Science and Technology*
ASA	American Standards Association (now, ANSI)
ASCE	American Society of Civil Engineers
ASIDIC	Association of Information Dissemination Centers
ASIS	American Society for Information Science
ASME	American Society of Mechanical Engineers
ASTIA	Armed Services Technical Information Agency (now, DTIC)
ASTM	American Society for Testing and Materials
BIOSIS	BioSciences Information Service of Biological Abstracts
BLLD	British Library Lending Division
BP	*British Pharmacopoeia*
BSI	British Standards Institution

BUCOP	*British Union Catalogue of Periodicals*
CA	*Chemical Abstracts*
CAS	Chemical Abstracts Service
CASSI	*Chemical Abstracts Service Source Index*
CBD	*Current Bibliographical Directory*
CDI	*Comprehensive Dissertation Index*
CFSTI	Clearinghouse for Federal Scientific and Technical Information (now, NTIS)
CGD	Captured German Documents
COM	Computer Output Micrographics
COSATI	Committee on Scientific and Technical Information
DDC	Defense Documentation Center
DOD	Department of Defense
DODISS	*Department of Defense Index of Specifications and Standards*
DODSSP	Department of Defense Single Stock Point
DOE	Department of Energy
EDP	Electronic Data Processing
EI	*Engineering Index*
EPC	Editorial Processing Center
ERDA	Energy Research and Development Administration
ESA	European Space Agency
ETC	European Translations Center (now, ITC)
EURATOM	European Atomic Energy Community
FAO	Food and Agriculture Organization of the United Nations
FID	International Federation for Documentation
GRAI	*Government Reports Announcements and Index*
HMSO	Her Majesty's Stationary Office
IAA	*International Aerospace Abstracts*
IAEA	International Atomic Energy Agency
ICRS	International Research Communication System
IEEE	Institute of Electrical and Electronics Engineers
IEC	International Electrotechnical Commission

Abbreviations

IEG	Information Exchange Group
IFLA	International Federation of Library Associations
INIS	International Nuclear Information System
INSPEC	International Information Services for the Physics and Engineering Communities
ISI	Institute for Scientific Information
ISO	International Organization for Standardization
ISSN	International Standard Serial Number
ITC	International Translations Center
IUPAC	International Union of Pure and Applied Chemistry
JPRS	Joint Publications Research Service
KWIC	Keyword in context (index)
LC	Library of Congress
MARC	Machine-Readable Catalog
NACA	National Advisory Committee for Aeronautics (now, NASA)
NASA	National Aeronautics and Space Administration
NBS	National Bureau of Standards
NBS-SIS	National Bureau of Standards-Standards Information Service
NDA	National Distributing Authority
NOISH	National Institute for Occupational Safety and Health
NPFC	Naval Publications and Forms Center
NSA	*Nuclear Science Abstracts*
NSRDS	National Standards Reference Data System
NTC	National Translations Center
NTIS	National Technical Information Service
OECD	Organization for Economic Cooperation and Development
OSRD	Office of Scientific Research and Development
OSTI	Office of Scientific and Technical Information
OTS	Office of Technical Services
PB	Publications Board
PTO	Patents and Trademarks Office
R & D	Research and Development

RSC	Readers service card
SAE	Society of Automotive Engineers
SCAN	Selected Current Aerospace Notices (NASA)
SDI	Selective Dissemination of Information
SPIN	Selected Physics Information Notices
SRIM	Selected Research in Microfiche (NTIS)
SSIE	Smithsonian Science Information Exchange
STAR	*Scientific and Technical Aerospace Reports*
TAPPI	Technical Association of the Pulp and Paper Industry
TODARS	Terminal-Oriented Data Analysis and Retrieval System (NBS)
UDC	Universal Decimal Classification
U.K.	United Kingdom
UKAEA	United Kingdom Atomic Energy Authority
UNESCO	United Nations Educational, Scientific, and Cultural Organization
UNISIST	Universal System for Information in Science and Technology (UNESCO)
USAEC	United States Atomic Energy Commission
USASI	United States of America Standards Institute (now, ANSI)
VSMF	Visual Search Microfilm Files

AUTHOR INDEX

Roman numbers indicate the page in text where the author's work is mentioned. Italic numbers give the page on which the full bibliographic description of the work appears. Numbers in brackets are reference numbers.

Abercrombie, Michael, *191* [98]
Abramowitz, Milton, *215* [18]
Achee, Nicholas, *318* [102]
Achinger, William C., *228* [193]
Ackoff, Russell L., 16, *20* [33], *369*
Adams, Edwin P., *216* [25]
Agajanian, A. H., *314* [58]
Albert, Ted, *131* [34], *361*
Alexandrov, Eugene A., *337* [333]
Ali, Syed Mohammed, *328* [221]
Allen, Thomas J., *19* [18], *372*
Allen, William H., *189* [79]
Aluri, Rao, *213*, 308, *340* [26], *361*
Amron, Irving, *339* [13], *362*
Amstutz, Gerhardt C., *189* [73]
Anderson, C. M. B., *65* [78], *376*
Anderson, I. G., *196* [1], *199* [26]
Anderson, Nancy D., *313* [37]
Anthony, L. J., *336* [321]
Antony, Arthur, *288*, *317* [83]
Arbuckle, J. Gordon, *228* [189]
Armstrong, Eva, *318* [113]
Ash, Janet E., *317* [84], *362*
Ash, Lee, *199* [24]
Asimov, Isaac, *170*, *175* [31]
Askland, Carl L., *225* [152]
Atherton, Alexine L., *200* [39]
Atherton, Pauline, *18* [5], *362*

Auger, C. P., 100, 104, 111, *129* [1], *185* [21], *362*
Avallone, Eugene A., *225* [155]
Azad, H. S., *228* [201]

Baerwald, John E., *224* [137]
Bailey, Martha, *362*
Bair, Frank E., *223* [118, 128]
Baker, B. B., *181*
Baker, Dale B., *29* [7], *362*
Baker, Robert F., *224* [133]
Balachandran, Sarojini, *288*, 306, *337* [330, 331]
Ballentyne, Denis W. G., *185* [20]
Bamford, Harold E., Jr., *65* [74, 75], *362*, *375*
Barlow, D. H., *65* [55], *362*
Barnhard, John H., *176* [38]
Barnhart, Clarence L., *174* [18]
Barr, James, *223* [124]
Barr, K. P., 34, *62* [12], *362*
Bartholomew, Robert, *337* [327]
Bartlett, Neil, *318* [108]
Barton, H. A., *63* [38], *362*
Baruch, Jordan J., *64* [65], *362*
Bates, Ralph S., 25, 26, *29* [6], *362*
Battle, G. C., *316* [73]
Bauman, Lawrence, *219* [68]

387

Baum, Harry, 67, 75 [7], *363*
Baumister, Theodore, *225* [155]
Beach, Norman E., *140*
Beam, Robert E., *190* [93]
Bearman, Toni C., *154* [4], *367*
Becker, Joseph, *363*
Beckman, Kathy, *146*
Beeching, Cyril L., *185* [19]
Beeton, Alfred, *228* [190]
Beilstein, Friedrich K., 209
Belzer, Jack, *237* [11]
Bemer, Robert W., *65* [72], *363*
Bemis, Virginia, *336* [317]
Bennett, Harry, *219* [66]
Bentley, Howard B., *323* [171]
Benton, John L., *363*
Berger, Jennifer, *185* [22]
Berger, Mary C., *347* [4]
Berlman, Isadore B., *217* [38]
Bernal, J. D., 46, 48, 49, *64* [51, 63], 249, *265* [27], *363*
Bernier, Charles L., 251, *265* [35, 36, 37, 38], 293, *339* [6], *363*
Bernstein, I. M., *221* [103]
Besancon, Robert M., *237* [15]
Besterman, Theodore, *292, 293, 311*, [15, 17]
Beyer, William H., *215* [16]
Bhagat, Nazir, *64* [65], *362*
Bielski, Benon H. J., *219* [62]
Bilboul, Roger R., *84*
Bindman, Werner, *187* [54]
Birchon, Donald, *188* [70]
Black, Dorothy M., *80*
Bland, William F., *221* [93]
Bloomfield, M., *292*
Bolton, Henry C., *291*
Bolz, Harold A., *225* [157]
Bolz, Ray E., *214* [3]
Bond, Richard G., *227* [186]
Boni, Nell, *291*
Bonn, George S., *332* [272]
Borishanskii, V. M., *237* [13]
Borko, Harold, 293, *339* [6], *363*
Born, Irene, *171*
Born, Max, *171, 172*

Boskovic, Marijan, *217* [42]
Bottle, Robert T., 161, *165* [5], *209, 305, 317* [85], *326* [202], *364*
Boublik, Thomas, *221* [100]
Bourton, Kathleen, *317* [93]
Bovey, Frank A., *221* [92]
Bowman, Norman J., *223* [120]
Bowman, Walker H., *364*
Boyer, Calvin, 79, *87* [3], *364*
Boylan, N. T. G., 104, *130* [10, 11], *364*
Bradford, S. C., 36, *62* [21, 22], 299, *339* [10], *364*
Bragdon, Clifford R. *332* [273]
Bramson, Mikael A., *217* [44]
Brandhorst, Ted, 345, *360* [7], *380*
Brandup, J., *221* [94]
Branscomb, Lewis M., 246, *264* [15], *364*
Braunstein, Y. M., *106*
Braze, Sandra, *336* [323]
Brearley, Neil, 100, *130* [4], *364*
Bretherick, L., *220* [84]
Breuer, Hans, *187* [44]
Britten, James, *191* [100]
Britt, Kenneth W., *220* [83]
Brociner, Grace E., *268, 273* [3]
Brown, Barbara, *236* [4]
Brown, Charles H., *61* [7], *364*
Brown, H. S., *291*
Brown, Linda, *333* [283]
Brown, Norman B., *63* [34], *364*
Brown, Patricia L., *336* [318]
Brown, Russell, *305, 317* [92]
Brown, Stanley B., *236* [4]
Brown, W. S., *61* [4], *364*
Brunenkant, Edward J., *360* [15], *364*
Brunning, Dennis A., 248, 249, 250, 253, *265* [25], *365*
Brush, Stephen G., *315* [65]
Buchsbaum, Walter H., *226* [162]
Buist, Eleanor, 81, *87* [6], *365*
Burchell, Robert W., *227* [185]
Burgess, Eric, *240* [46]

Author Index

Burg, Nan C., *336* [332]
Burington, Richard S., *215* [19, 20]
Burkart, A., *219* [67]
Burke, John G., *206* [104]
Burkett, J., *325* [194]
Burman, Charles R., *317* [86]
Burton, Philip E., *187* [46]
Burunova, N. M., *215* [12]
Buttress, F. A., *185* [17]

Calder, Ritchie, *282* [4], *365*
Campbell, G. A., *305, 317* [92]
Caroli, S., *221* [99]
Carroll, Kenneth D., *64* [59], *365*
Carr, Russell L. K., *265* [39], *368*
Carson, Gordon B., *225* [157]
Carter, Ciel M., *314* [54]
Carter, Ernest F., *183* [3]
Carter, L. J., *178* [4], *365*
Cawley, Rebecca E., *337* [328]
Čermáková, Jiřina, 68, 74, *75* [10], *365*
Chakraborty, A. R., *365*
Chandor, Anthony, *187* [45]
Chaney, J. F., *217* [37], *326* [196]
Chapman, D. R., *183* [5]
Chen, Ching-Chih, *306, 310* [1]
Cheremisinoff, Paul N., *225* [150], *229* [213]
Cheremisinoff, Peter P., 225 [150]
Childs, James B., *117*
Chillag, J. P., 106, 129, *130* [16], *131* [38], *365*
Chinery, Michael, *191* [108]
Chirnside, R. C., *365*
Chiu, Hong-Yee, *187* [50]
Chiu, Yi Shu, *214* [4]
Ciaccio, Leonard L., *229* [204]
Clark, George L., *237* [16]
Clason, W. E., *187* [47, 53], *188* [64], *189* [71]
Clauss, Francis J., *214* [6]
Clugh, Raymond W., *334* [299]
Coblans, Herbert, *306, 315* [66]
Cochrane, Rexmond C., *148* [1], *365*

Cohen, Henry, *237* [13]
Coile, Russell C., 108, *130* [20], *365*
Colborn, Robert, *236* [3]
Coleman, Henry E., Jr., *85*
Colgate, Craig, Jr., *199* [20]
Colling, Patricia M., *87* [8], *365*
Collins, Mike, *218* [54]
Collocott, T. C., *183* [1]
Comrie, L. J., *210* [1], *215* [14]
Condon, E. U., *217* [41]
Congrat-Butlar, Stephan, *273*
Connolly, Tom F., *315* [67], *316* [73]
Considine, Douglas M., *221* [95], *229* [208], *238* [20]
Cooke, Edward I., *165, 185* [18]
Cooke, Richard W. I., *165*
Cornwell, P. B., *190* [87]
Cors, Paul B., *325* [188]
Cote, Rosalie J., *190* [97]
Cottrell, William B., 251, *265* [34], *366*
Cowan, H. J., *189* [85]
Craig, Alexander S., *188* [62]
Crane, Diana, 20 [25], *366*
Crane, Edward M., *338* [4], *366*
Crawford, Susan, 20 [26, 35], *366*
Crawshaw, Thomas, *325* [186]
Crede, Charles E., *225* [160]
Crites, Ronald G., *332* [276]
Crowley, Ellen T., *165, 180, 184,* [8], *185* [23]
Cuadra, Carlos A., 244, *264* [9]
Cunningham, John J., *190* [97]
Cyilton, C. H., *238* [22]

Dale, Virginia, *337* [335]
Dallas, Daniel B., *225* [161]
Darcy, Harry L., *189* [83]
Darling, Richard L., 289
Davidson, A., *225* [145]
Davidson, Robert L., *221* [93]
Davis, Charles H., *317* [87], *366*
Davis, Elisabeth B., *288,* 306
Davis, R. A., 11, *19* [12], *165* [4], *366*

Davis, W., *63* [43], *366*
Debus, Allen G., *166-167, 177* [54]
DeFrancis, John, *186* [36]
DeGaliana, Thomas, *239* [41]
DeGennaro, Richard, 40, *63* [33], *366*
Desmond, Winifred F., *61* [5], *368*
Devers, Charlotte M., *61, 161, 195*
DeVries, Louis, *187* [53]
Dewar, Michael J. S., *219* [69]
Dible, Donald M., *366*
Dick, Elie M., *313* [38]
Dickinson, W. C., *229* [213]
Dirksen, Eileen F., *76* [11], *366*
Ditmas, E. M. R., *339* [9], *366*
Doebler, Paul, *360* [4], *366*
Dood, Kendall J., *99* [10], *366*
Dorian, A. F., *188* [65], *189* [80]
Dorling, A. R., *313* [39]
Dowling, H. G., *327* [210]
Doyle, James M., *310* [3]
Drott, M. Carl, 151, 153, *154* [4], *367*
Dumouchel, J. Robert, *191* [110]

Eakins, J. P., 51, *64* [67], *367*
Earney, Filmore C. F., *320* [135]
Eastin, Roy B., *117*
El-Hadidy, Bahaa, *367*
Elliott, Clark A., *176* [33]
Elsdon-Dew, R., 39, *63* [32], *367*
Elving, Philip J., *254*
Emery, Betty L., 161, *165* [5], *364*
Engelbrektson, Sune, *187* [52]
English, Eileen W., *130* [21], *367*
Enrick, Norbert L., *216* [26]
Epperson, Eugene R., *216* [24]
Ernst, Richard, *190* [89]
Ettre, Leslie S., *239* [31]
Evan, William, *19* [15], *367*
Evers, Clive, *317* [88]

Fairbridge, Rhodes W., *239* [36], *241* [56]
Falconer, Warren E., *318* [108]
Fasman, Gerald D., *219* [70]

Faulkner, L. L., *224* [142]
Fedorova, R. M., *211* [3], *215* [11]
Feinberg, Barry N., *227* [180]
Feineman, George, *226* [175]
Felber, Helmut, *183*
Finer, Ruth, *201* [49]
Fintel, Mark, *224* [131]
Fitzpatrick, William H., *87* [10], *367*
Flanagan, C., 300, *339* [14, 15], *381*
Fleming, David G., *227* [180]
Fletcher, A., *210* [1], 211, *215* [14]
Flowers, B. H., *265* [29], *367*
Flugge, S., *238* [18]
Ford, Charles A., *183* [2]
Forsythe, William E., *218* [48]
Foster, William S., *224* [134]
Fowler, Maureen J., *57* [6]
Franko, Lawrence, *317* [94]
Freeman, Monroe E., *87* [10], *367*
Freiberger, W. F., *186* [38]
Fried, Vojtech, *221* [100]
Fristrom, R. M., *192* [115]
Fry, Bernard M., *63* [36], *322* [158], *367*
Fry, D. G., *215* [12]
Fulton, J. H., *291*
Funkhouser, Richard L., *325* [195]
Furia, Thomas E., *209, 220* [73]

Galtsoff, Paul S., *327* [209]
Gannett, Elwood K., 43, 44, *63* [41], *367*
Gardner, Jeffrey J., *339* [19], 367
Gardner, William, *165*
Garfield, Eugene, 249, *265* [26], *367*
Garvey, William D., *20* [27], 38, 51, *62* [27], *64* [66], *368*
Garvin, David, *339* [20], *368*
Gary, Margaret, *189* [72]
Gautney, George E., *368*
Gebicki, Janusz M., *219* [62]
Gentle, Ernest J., *189* [75]
Georgano, G. N., *240* [47]
George, J., *240* [54]
George, J. D., *240* [54]
Gettys, William, *217* [47]

Author Index

Gibson, Eleanor B., *305, 319* [129]
Giefer, Gerald J., *332* [275]
Gilboa, I., *327* [210]
Gilchrist, A., 36, *62* [24], *373*
Gillispie, Charles C., *176* [41]
Gilmore, C. Stewart, *171*
Gilmore, J. S., *99* [4], *368*
Gilpin, Alan, *188* [59], *190* [90]
Glazebrook, Sir Richard, *237* [14]
Gmelin, Leopold, 209
Godel, Jules B., *323* [172]
Godfrey, Lois E., *112, 131* [29], *368*
Golze, Alfred R., *224* [132]
Gordon, Irving, *265* [39], *368*
Gottfredson, S. D., 51, *64* [66], *368*
Gottschalk, Charles M., *61* [5], *368*
Gould, Syndey H., *187* [41], *187* [42]
Grad, Frank P., *229* [203]
Graf, Rudolf F., *191* [107]
Grant, Eugene L., *224* [141]
Grasselli, Jeanette G., *219* [63]
Gray, Dorothy A., *310* [7]
Gray, Dwight E., *130* [19], *216* [32], *368*
Gray, Peter, *240* [50]
Gray, Richard A., *310* [7]
Greene, J. H., *225* [158]
Green, Sidney, *317* [95]
Greenwood, J., *211* [2], *215* [13]
Gregory, Jean, *177* [43]
Griffith, Belver C., *20* [27], *154* [4], *367, 368*
Griffith, Edward J., *228* [190]
Grimes, George H., *310* [3]
Grogan, Denis J., *62* [23], *63* [48], *76* [14], *99* [2], *148* [5], *154* [2], *265* [42], *282* [6], *310* [2], *368*
Grosch, Audrey N., *245, 264* [11], *368*
Grzegorczyk, Dean, *226* [175]
Guerra, Francisco, *289*
Guha, B., *368*
Gunter, John D., *333* [285], *334* [297]

Gurnett, J. W., *184* [10]
Gursey, S., *178* [3], *375*
Guthrie, David L., *337* [334]

Haberer, Isabel H., *154* [5], *368*
Hala, Eduard, *221* [100]
Halbert, Michael H., 16, *20* [33], *369*
Hall, C. W., *188* [58]
Hall, David M., *318* [102]
Halsey, William D., *174* [18]
Hamence, J. H., *365*
Hamilton, D. R. L., *265* [30], *380*
Hammett, Louis P., *63* [37], *369*
Hampel, Clifford A., *188* [66], *238* [25, 28]
Hanchey, Marguerite M., *296*
Handley, William, *225* [151]
Hannah, M., *99* [7]
Hannay, N. B., *99* [5], *369*
Hanson, C. W., 72, *76* [19, 21], *369*
Harper, Charles A., *220* [82], *226* [168]
Harré, R., *170*
Harris, Cyril M., *225* [160]
Harris, Linda, 248, *264* [22], *369*
Harrison, Thomas J., *216* [29]
Hart, Ivor B., *170*
Hartley, H. O., *211* [2], *215* [13]
Hartnett, J. P., *217* [40]
Hawkins, Donald T., *318* [108]
Hawkins, R. R., 289
Hawley, Gessner G., *188* [56, 66], *238* [28]
Hayman, Charles J., *189* [81]
Haywood, B. J., *318* [109]
Hecht, C. D., *65* [78], *376*
Heinisch, Kurt F., *188* [61]
Heinl, Robert D., Jr., *189* [78]
Helbing, W., *219* [67]
Hellwege, K. H., 209
Hellyer, Arthur G. L., *240* [51]
Hemenway, David, *19* [16], 136, *148* [6], *369*
Henderson, G. P., *196* [1], *199* [25]

Henderson, Kay, *332* [274]
Henderson, S. P. A., *199* [25]
Hepple, Peter, *181*
Herner, Saul, *19* [23], *201* [50], *369*
Herrick, John W., *240* [46]
Herring, Conyers, 251, *264* [13], *265* [33], *369*
Herschman, Arthur, 54, *64* [69], *65* [77], *369*
Hershey, David F., *369*
Hertzendorf, Martin S., *227* [183], *228* [188]
Herzka, A., *179*
Hewitt, J. W., *99* [7]
Hickin, Norman E., *190* [88]
Hickman, C. J., *191* [98]
Hickock, Beverly, *323* [173]
Higgins, Lindley R., *225* [154]
Hills, Jacqueline, 48, *64* [61], *369*
Hilton, C. L., *239* [31]
Himmelsbach, Carl J., *268, 273* [3]
Hippisley, R. L., *216* [25]
Ho, C. Y., *218* [52]
Hodson, H. V., *200* [38]
Holland, Robert, *191* [100]
Holmes, Ernest, *224* [143]
Holmes, S. J., *171*
Holum, John R., *241* [59]
Holzman, Albert G., *237* [11]
Hopkins, Jeanne, *187* [55]
Horblitt, H. D., *291*
Houghton, Bernard, 147, *148* [4], *304, 310* [4], *324* [184], *369*
Howard, William E., *223* [124]
Howes, Frank N., *191* [104]
Hsieh, Kitty, *337* [329]
Hunter, D., *318* [110]
Hunter, Lloyd P., *226* [176]
Hunter, P. S., *99* [7], *372*
Huschke, Ralph E., *192* [116]
Hutchison, J. W., *225* [148]
Hutchisson, Elmer, 2, *18* [3], *370*
Hutton, G., *337* [332]
Huver, Charles W., *327* [213]
Huxley, Julian, *172*
Hyatt-Mayor, A. *165* [2], *370*

Hyde, Ernest, *317* [84], *362*
Hyslop, Marjorie R., *320* [130]

Immergut, E. H., *221* [94]
Inemson, R. S., *19* [22], *377*
Inhaber, H., *178* [2], *370*
Ireland, Norma O., *172, 173* [7]
Ireson, William G., *224* [141]
Iyanaga, Shokichi, *237* [7]

Jackson, K. G., *190* [92]
Jacobson, C. A., *238* [26]
Jameson, William C., *333* [285], *334* [297]
Janardan, Sriram, *365*
Janes, M., 72, *76* [19, 21], *369*
Jeanne, E. A., *334* [303]
Joenk, R. J., *370*
Johnson, Arnold H., *239* [30]
Johnson, M. L., *191* [98]
Jones, Bessie Z., *170*
Jones, M. L., *205* [94]
Jones, Richard, *219* [69]
Jones, Thomas H., *226* [164]
Joshi, Meera G., *365*

Kaelble, Emmett F., *217* [43]
Kaiser, Frances E., *273*
Karegeannes, Carrie E., *223* [121]
Karol, Michael, *310* [6]
Kase, Francis J., *91, 370*
Katter, Robert V., 248, *264* [22], *369*
Katz, Doris B., *61, 161, 195*
Kawada, Yukiosi, *237* [7]
Kaye, D., 94
Kaye, George W. C., *218* [50]
Kean, Pauline, *19* [6], 45, *63* [47], *370*
Keehn, Pauline A., *327* [214]
Keese, M., *316* [73]
Kemp, D. A., *315* [68]
Kemper, A. M., *227* [184]
Kennedy, H. E., 300, *339* [14, 15], *381*
Kent, Allen, *237* [11]

Author Index

Kent, James A., *220* [78]
Kharasch, Norman, *256*
Kidd, Charles V., *87* [1], *380*
King, Alexander, 66, 67, 75 [2], *370*
King, Donald W., *62* [13], *338* [3], *370*
King, Reno C., *225* [156]
King, Robert C., *191* [102]
Kingsbury, Raphaella, *306*
Kirk, Thomas G., *326* [203]
Klain, Ambrose, *334* [299]
Klein, Bernard, *197* [4]
Klempner, Irving, *130* [23], 309, 310, *340* [28], *370*, *371*
Knight, David M., *311* [22]
Knopp, O. R. H., *371*
Koch, H. William, *63* [39], *371*
Koelle, Heinz H., *223* [119]
Kolthoff, I. M., *254*
Komarick, Stephan L., *220* [77]
Konarski, Michael M., *189* [84]
Kopal, Zdenek, *218* [57]
Kopkin, T. J., *287*
Kopycinski, Joseph V., *318* [103]
Korbuly, Dorothy K., *65* [73], *371*
Korfhage, Robert R., *20* [29], *371*
Korn, Granino A., *216* [22]
Korn, Theresa M., *216* [22]
Kovach, Eugene G., *178* [5], *371*
Krivatsy, Peter, 289
Krommer-Benz, Magdalena, *183*
Kronick, David A., 31, *61* [3], 243, *264* [2], *371*
Kruskal, William H., *237* [8]
Kruzas, Anthony T., *198* [17], *201* [48], *347* [3]
Kuiper, Barteld E., *148* [15], *371*
Kuney, Joseph H., *29* [8], 39, *63* [31], *64* [68], *371*
Kuntz, Werner, *371*
Kutateladze, J. S. *237* [13]
Kuvshinoff, B. W., *192* [115]
Kyed, James M., 28, *29* [10], *288*, *311* [23]
Kyte, C. H. J., *184* [10]

Laby, T. H., *218* [50]
Ladendorf, Janice M., *19* [14], *371*
Lancaster, F. W., *183*
Lancaster, H. O., *313* [43]
Landa, Henry C., *229* [212]
Landau, Ruth N., *201* [46], *347* [4]
Landsford, Edwin M., *238* [24]
Landy, Mark, *192* [114]
Lange, Norbert A., *221* [88]
Lannom, Laurence, *380*
Lastein, Allen I., *226* [179]
Lasworth, Earl J., *310* [5]
Lauche, R., *292*
Lebedev, A. V., *211* [3], *215* [11]
Lechevalier, Hubert A., *226* [179]
Lee, Sir Sidney, *174* [13]
Leeson, Kenneth, *372*
Leftwich, A. W., *191* [105]
Legenfelder, Helga, *196*, *197* [6]
Lenk, John D., *226* [169, 174]
Lerner, Rita G., *371*
Leung, A. Y., *239* [29]
Levitan, Tina, *170*
Lewis, H. L., *240* [48]
Lewis, Walter H., *192* [112]
Licklider, J. C. R., *283*, *338* [2], *371*
Liebesny, Felix, 71, 76 [18], 93, 99 [7], *372*
Linke, W. F., *221* [96]
Linsenmaier, Walter, *240* [53]
Lippert, Walter, *229* [10], *372*
Liptak, Bela G., *228* [187]
Listokin, David, *227* [185]
Liston, Mary D., *331* [265]
List, Robert J., *223* [123]
Lochard, Jean, *360* [11], *372*
Locke, William N., 266, *282* [1], *372*
Lohwater, A. J., *186* [42]
Loosjes, Th. P., 13, *19* [21], *372*
Lovett, D. R., *185* [20]
Lublin, Joann S., 150, *154* [3], *372*
Luckenbach, Reiner, *230* [2], *372*
Ludwig, Raymond H., *226* [177]
Lund, Herbert F., *228* [199, 200]

Lunsford, E. B., *287*
Lynch, Charles T., *221* [102]
Lynch, Wilfred, *220* [85]
Lyons, Jerry L., *225* [152]

Machlup, Fritz, *372*
Maizell, R. E., *317* [98]
Malhotra, Subhash C., *320* [136]
Malinowsky, Harold R., *310* [7]
Mandl, Matthew, *226* [163]
Mangla, P. B., *62* [9], *372*
Manten, A. A., *265* [31], *372*
Manu, Adrian, *183*
Manuel, Frank E., *171*
Manzer, Bruce M., *294*, *339* [7], *372*
Marckworth, M. Lois, *84*
Mark, H. F., *239* [32]
Marks, Linda, *288*
Marks, Robert W., *186* [40], *189* [82]
Markus, John, *226* [167]
Marquis, Donald G., *19* [18], *372*
Marshall, Joan K., *197* [7]
Martin, D. C., *29* [2], *372*
Martin, Jean K., *373*
Martin, S. M., *205* [94]
Martyn, John, 36, *62* [24], *206* [100], 300, *339* [11, 12], *373*
Marvine, Kathryn E., *337* [328]
Maskelyne, Nevil, 211
Mason, Kenneth G. M., *333* [288]
Matarazzo, James M., 28, *29* [10], 43, *63* [42], *288, 311* [23], *373*
Mathews, Eleanor, *373*
May, Donald C., *215* [20]
Maynard, Harold B., *225* [149]
McCaslin, John, 241 [61]
McDonald, Rita, *324* [174]
McElderry, S., *373*
McKie, Douglas, *171*
McLean, Janice, *177* [51]
Meadow, Charles T., *360* [10], *373*
Meadows, N. G., *310* [8]
Medvedev, Zhores A., *18* [4], *373*
Meek, C. L., *237* [10]
Mellan, Ibert, *221* [87]

Mellon, M. G., *230* [3], *305, 317* [89], *373*
Merriman, Arthur D., *239* [35]
Merritt, Frederick S., *224* [136]
Meshenberg, Michael J., *331* [268]
Metcalf, Kenneth N., *324* [175]
Metx, Karen, *336* [320]
Metzner, A. W. Kenneth, *65* [70], *360* [1], *373*
Middleton, M., *131* [35], *373*
Middleton, Robert G., *226* [170]
Mildren, K. W., *310* [8]
Millard, Patricia, *201* [40], *273*
Miller, Elizabeth, *373*
Miller, J. C. P., *210* [1], *215* [14]
Millington, T. Alaric, *186* [37]
Millington, William, *186* [37]
Mitchell, Dee T., *228* [190]
Mohrhardt, Foster, *322* [158]
Monkhouse, F. J., *192* [111]
Monpetit, G., *216* [28]
Moore, C. Alan, *76* [22], *373*
Moore, James A., *62* [18], *373*
Moore, Julie L., *87* [7], *373*
Moore, Patrick, *218* [55, 56]
Moorshead, H. W., *226* [178]
Morehead, Joe, *117*
Morehouse, L. G., *227* [181]
Morrill, Chester, *314* [55]
Morrison, Denton E., *336* [324], *337* [325]
Morrow, L. C., *225* [154]
Moser, Reta C., *185* [15]
Mount, Ellis, *310* [9]
Muirden, James, *216* [31]
Mulvaney, Carol, *306*
Murra, Kathrine O., 67, *75* [3], *374*
Murty, D. S. R., *131* [36], *360* [14], *374*
Myers, Buddy L., *216* [26]

Nayler, G. H. F., *190* [91]
Nayler, Joseph L., *190* [91]
Nelson, Archibald, *188* [68], *189* [86]
Nelson, Carnot E., *20* [28], *374*

Author Index 395

Nelson, K. D., *188* [68], *189* [86]
Nosova, L. N., *210*

Oberg, Eric, *225* [153]
Oberhettinger, Fritz, *210*
Obreanu, P. E., *186* [41]
O'Callaghan, T. C., *84*
Ocran, Emanuel B., *255, 324* [178]
Odishaw, Hugh, *217* [41]
Olsen, G. H., *226* [166]
O'Pecko, Mary M., *331* [263], *378*
Oppenheim, C., 94, *99* [11], *374*
Ordway, G. L., *192* [115]
Osberg, Sally R., *206* [103]
Osenton, James, *189* [80]
Owen, Dolores B., *296*
Owen, Donald B., *216* [21]
Owings, Loren C., *332* [271]

Pade, A., *220* [80]
Page, Thornton, *202* [60]
Palfrey, Thomas R., *85*
Parisi, Paul A., *65* [71], *374*
Parker, Sybil, *177* [44]
Parks, W. George, *75* [8], *374*
Parsons, Ronald G., *373*
Pasternack, Simon, *62* [19], *374*
Paulson, Glenn L., *331* [264]
Pearl, Richard M., *239* [34]
Peck, Theodore P., *317* [96, 97], *318* [106]
Peckner, Donald, *221* [103]
Pelzer, C. W., *131* [37], *374*
Pemberton, John E., *313* [40]
Perry, John H., *208*
Perry, Robert H., *238* [22]
Persons, Caroline C., *318* [102]
Peters, Jean, *215* [17]
Peterson, Martin S., *239* [30]
Phelps, Ralph H., 48, 49, 51, *64* [60], *374*
Phillips, Jane, *63* [34], *364*
Phipps, T. E., 46, *64* [52], *374*
Pierce, J. R., *61* [4], *364*
Pirie, N. W., 46, *64* [50], *374*
Plumb, P., *325* [194]

Plunkett, Edmond R., *228* [197]
Podhorsky, Richard, *217* [42]
Pollock, Donald K., *20* [28], *374*
Pollock, Vera, *318* [112]
Porter, J. R., *18* [1], *61* [1], *374*
Potter, James H., *215* [7]
Pouchert, Charles J., *219* [61]
Pownall, J. F., 48, *64* [62], *374*
Prakash, Satya, *328* [221]
Pratt, G., *347* [6]
Press, Jaques C., *175* [25, 26, 27, 28, 29]
Price, Derek J. de Solla, 10, *19* [9], *20* [30], 33, 34, 35, *62* [10, 11], *178* [1, 3], 283, 285, *338* [1], *374*, *375*
Pritchard, Alan A., 314 [56]
Przetak, Louis, *224* [135]
Pugh, Eric, *184* [11, 12]
Putnam, T. M., *217* [37], *326* [196]

Quadling, C., *205* [94]
Quinn, Karen T., *314* [57]

Rachie, Kenneth O., *327* [218]
Ralston, Anthony, *237* [10]
Ralston, Valerie H., *318* [104]
Randle, Gretchen R., *326* [197]
Ranganathan, S. R., *171*
Rangra, V. K., *268*
Rapp, George R., Jr., *239* [37]
Rawls, Rebecca L., *64* [54], *375*
Raznjevic, Kuzman, *217* [42], *225* [146]
Rea, Robert H., 120, *131* [33], *375*
Reay, David A., *229* [209]
Rechigal, Miloslav, Jr., *327* [219]
Reddig, Jill S., *206* [104]
Redman, Helen F., *112, 131,* [29, 31], *368, 375*
Reeves, S. K., 135, *148* [3], *375*
Regan, Mary M., *61, 161, 195*
Reich, Warren T., *240* [49]
Reid, Jean P., *324* [176]
Reid, Joseph B., *216* [28]

Reid, W. Malcolm, *326* [206]
Reiter, Berle, *288*
Reithmaier, Lawrence W., *189* [75]
Reuss, Jeremias D., *288*
Reynolds, Michael M., *85*
Rhodes, Sarah N., *65* [75], *375*
Ricci, Patricia, *146*
Rider, Kenneth J., *312* [25]
Ridge, John D., *320* [137]
Riley, Robert A., *337* [334]
Rittenhouse, John B., *223* [122]
Roberts, Elizabeth P., *326* [204]
Roberts, Willard L., *239* [37]
Robinson, Eric, *171*
Robinson, J. W., *220* [86]
Robinson, Sir Robert, 1, *2* [2]
Robson, John, *216* [33]
Rodgers, Gil, *207* [113]
Rodgers, Harold A., *187* [48]
Rogoff, Marc J., *334* [301]
Rohsenow, Warren H., *217* [40]
Romaine, Lawrence B., 155, *156*, *165* [1, 3], *375*
Ronayne, Jalrath, *19* [6], 45, *63* [47], *370*
Rosenbloom, Richard S., 10, 14, *19* [10, 24], *376*
Rosenborg, Staffan, *139* [19], *368*
Rosen, Edward, *130* [5], *376*
Rosenhead, L., *210* [1], *215* [14]
Rossmassler, Stephen A., 39, *63* [30], *376*
Rostron, M., *337* [332]
Roth, D. L., *230* [4], *376*
Rowlett, Russel J., *360* [2], *376*
Rudd, Robert L., *331* [270]
Ruffner, James A., *185* [22], *223* [118, 128]
Rush, James E., *317* [87], *366*

Sahaer, C. R., *360* [17], *376*
Sainsbury, D., *191* [103]
Saltman, William M., *221* [98]
Samsonov, G. V., *220* [81]
Sanders, William, *334* [303]
Sarnoff, Paul, *192* [117]

Sax, Irving, *220* [74]
Schaaf, William L., *313* [45]
Schaefer, Barbara K., *313* [41]
Schaffer, Constance J., *19* [73], *63* [48], *378*
Schandelmier, Nancy, *222* [108]
Schicke, Walter, *221* [89]
Schmeckebier, Laurence F., *117*
Schoenung, Georgia, *185* [22]
Schofield, Robert E., *171*
Schullian, Dorothy M., 289
Schussel, George, *20* [34], *376*
Schwartz, Mortimer D., *331* [267]
Schweitzer, Philip, *225* [147]
Scudder, Samuel H., *57* [3]
Seal, Robert A., *315* [69]
Seidell, A., *221* [96]
Sen, B. K., *268*
Senders, J. W., *65* [78], *376*
Sewell, Winifred, *326* [205]
Sharma, H. S., *328* [221]
Sharma, V. K., *221* [99]
Shaw, H. K. Airy, *191* [101]
Sheehy, Eugene Paul, *81, 305,* 307, *311* [11]
Sheehy, James P., *228* [193]
Shelton, John W., *241* [62]
Shepard, D. A. E., 38, *62* [28], *377*
Sherma, Joseph, *220* [72]
Sherwin, C. W., *19* [22], *377*
Shilling, Charles W., *222* [108]
Short, P. J., 74, 75 [4], *377*
Shrive, R. Norris, *238* [21]
Shriver, A. Richard, *65* [72], *363*
Shugar, Gershon J., *219* [68]
Shugar, Ronald A., *219* [68]
Siehl, George H., *377*
Signeur, Austin V., *317* [90]
Simon, Regina A., *228* [193]
Simons, Eric N., *188* [67], *222* [104]
Simons, W. W., *219* [64]
Simonton, D. P., *13* [28], *377*
Singletary, John B., *223* [122]
Singleton, A. K. J., *377*
Singleton, P., *191* [103]
Sippl, Charles J., *216* [30]

Author Index

Sittig, Marshall, 228 [192]
Skerman, V. B., 205 [94]
Skolnik, Herman, 99 [1], 377
Slater, Margaret, 339 [11], 373
Slattery, William J., 148 [11], 377
Slocum, Robert B., 172, 173 [1, 2]
Small, J., 192 [111]
Smith, Denison L., 324 [177]
Smith, Edgar F., 318 [113]
Smithells, Colin J., 222 [106]
Smith, J. R., 127
Smith, Julian F., 317 [99]
Smith, Robert G., 332 [276]
Smith, Roger C., 326 [206]
Smits, Rudolph, 57 [7]
Snell, Foster D., 239 [31]
Sommar, Helen G., 318 [105]
Spenceley, G. W., 216 [23, 24]
Spenceley, R. M., 216 [23, 24]
Spencer, Jean M., 228 [190]
Spillner, Paul, 185 [16]
Stapp, William B., 331 [265]
Starker, Lee N., 63 [45], 377
Steere, Norman V., 220 [75]
Steere, William C., 377
Stegun, Irene A., 215 [18]
Stephan, H., 221 [97]
Stephen, Sir Leslie, 174 [13]
Sternberg, Virginia A., 131 [25], 180, 336 [316], 377
Stevens, John G., 217 [47]
Stevens, Virginia, 217 [47], 336
Storer, Norman W., 19 [13], 377
Straub, Conrad P., 227 [186]
Strong, Reuben M., 291
Struglia, Erasmus J., 138, 147, 148 [13], 377
Studdard, Gloria J., 191 [109]
Subramanyam, K., 19 [7, 8], 20 [32], 62 [14], 63 [49], 64 [57], 87 [5], 331 [263], 334 [304], 361, 378
Suchard, S. N., 218 [49]
Sullivan, Linda E., 198 [17]
Swanson, Rowena W., 245, 264 [10], 378

Swift, Lloyd H., 326 [207]
Szokolay, S. V., 228 [191]

Tallman, J. E., 130 [7, 8, 9], 378
Tamir, Abraham, 221 [91]
Tanur, Judith M., 237 [8]
Tapia, Elizabeth W., 305, 319 [129]
Tarrants, William E., 318 [111]
Tate, Fred A., 360 [2], 376
Taube, Mortimer, 131 [30], 378
Tayal, A. S., 148 [8], 378
Taylor, John, 222 [113]
Tchobanoglous, George, 332 [276]
Telberg, Ina, 168, 177 [56]
Terapane, John F., 92, 93, 99 [6], 378
Thewlis, James, 187 [51], 237 [17]
Thomas, Harry E., 226 [171]
Thomas, Robert C., 57 [9]
Thorne, J. O., 174 [9]
Thornton, John L., 21, 28 [1], 361, 378
Thurman, Albert, 229 [210]
Todd, David K., 241 [60]
Tompson, LaDonna, 146
Toomey, Alice F., 292, 311 [16]
Touloukian, Y. S., 218 [51, 52, 53]
Towiner, Sydney B., 331 [269]
Townsend, Leroy B., 265 [40], 379
Traister, John E., 226 [172]
Traub, J. F., 61 [4], 364
Trevor, William I., 176 [35]
Trillo, Robert L., 239 [38]
Truesdell, Eugenia, 373
Truitt, Deborah, 328 [224]
Trumbull, Richard, 379
Trzyna, Thaddeus C., 206 [103]
Tsao, Chia Kuei, 313 [46]
Tully, R. I. J., 21, 28 [1], 361, 378
Tuma, Jan J., 214 [5], 217 [39]
Turkevich, John, 168, 177 [47]
Turkevich, Ludmilla, 177 [47]
Turner, L. W., 226 [165]
Tuve, George L., 214 [3]
Tuve, R. R., 192 [115]

Usher, George, *191* [99]
Utermark, Walther, *221* [89]
Uvarov, E. B., *183* [5]

Van Cott, Harold P., *62* [25], *379*
Vara, Albert C., *318* [107]
Vasilakis, Mary, *131* [26], *379*
Vcerasnij, Rostislav P., 93, *99* [8], *379*
Vellucci, Matthew J., *201* [50]
Verschueren, Karel, *228* [196]
Vesenyi, Paul E., *296*
Vichl, Richard C., Jr., *333* [288]
Virgo, Julie A., 242, *263* [1], *379*
Voigt, Melvin J., 13, 18, *19* [20], *379*

Waddell, Joseph J., *224* [130]
Wade, W., *379*
Walford, A. J., *81*, *305*, 306, 307, *311* [12]
Walker, Albert, 149, *154* [1], *379*
Wallenquist, Ake, *187* [52]
Wall, R. A., 163, *165* [6], *379*, *380*
Walters, LeRoy, *329* [229]
Walton Smith, Frederick G., *221* [101]
Wanger, Judith, 347 [4]
Wang, Yen, *220* [76]
Ward, Dederick C., *84*, *320* [131]
Warring, R. H., *224* [144]
Wasserman, Paul, *177* [51]
Watkins, R. J., *200* [28]
Weast, Robert C., *208*
Weber, Julius, *239* [37]
Weigel, Jack, *288*
Weigert, A., *237* [12]
Weik, Martin H., *187* [49]
Weinberg, Alvin, 105, 106, 244, 246, 247, *264* [5]
Weiner, Jack, *318* [112]
Weinstock, Melvin, *380*
Weiser, Sylvia G., *311* [13]
Weisgerber, William H., 39, *63* [31], *371*

Weisman, Herman M., 213, *230* [6], *380*
Weissbach, Oskar, *230* [5], *380*
Weller, Frank, *226* [173]
Wells, Helen T., *223* [121]
Werts, Margaret F., *222* [108]
West, S. S., *19* [17], *380*
Whalen, George J., *191* [107]
Wheeler, Marjorie W., *320* [131]
White, D., *305*, *320* [132]
White, Herbert S., *63* [36], *367*
Whiteley, Susan H., *223* [121]
Whiteside, R. M., *206* [102]
Whitford, Robert H., *316* [70]
Whittick, Arnold, *241* [57]
Wiginton, Ronald L., *368*
Wilcox, Virginia L., *320* [133]
Wilkinson, Paul H., *222* [114]
Williams, Colin H., *200* [31]
Williams, John S., *333* [290]
Williams, Martha E., *201* [43], 343, 345, *346* [1], *360* [5, 7, 8, 9], *380*
Williams, Roger J., *238* [24]
Willis, John C., *191* [101]
Winchell, Constance M., *305*
Windholz, Martha, *221* [90]
Wiren, Harold N., *336* [319]
Wisniak, Jamie, *221* [91]
Witty, Francis J., 293, 295, *339* [5], *380*
Wold, Herman O. A., *313* [47]
Wolek, Francis W., 10, 14, *19* [10, 24], *376*
Wolfe, William L., *217* [44]
Wolff, Garwood R., *331* [266]
Wolfle, Dael, *75* [9], *87* [1], *380*
Wood, D. N., *265* [30], *282* [2], *320* [134], *380*
Wood, J. L., 300, *339* [14, 15], *360* [2], *376*, *381*
Woodburn, Henry M., *96*, *317* [91]
Woodward, A. M., 250, 252, 253, 255, *264* [8], *265* [23], *381*
Wragg, David W., *189* [76]

Author Index

Wright, Arthur Williams, 85
Wukelic, George E., 222 [116]
Wyatt, H. V., 326 [202], 381
Wyllie, T. D., 227 [181]

Yagello, Virginia, 381
Yannarella, P. A., 213, 308, 340 [26], 361
Yates, B., 316 [71]
Yescombe, E. R., 305, 318 [100]
Young, Hewitt H., 225 [157]

Yska, Gerda, 206 [100]

Zaltman, Gerald, 20 [31], 381
Zalucki, Henryk, 184 [13]
Zanger, M., 219 [64]
Zaremba, Joseph, 313 [42]
Zavala, Albert, 62 [25], 379
Zilly, Robert G., 228 [195]
Ziman, J. M., 38, 63 [29], 381
Zimmermann, H., 237 [12]
Zweig, Gunter, 220 [72]

SUBJECT INDEX

This subject index includes titles of only those publications that are described or discussed in the text. Titles of publications listed at the end of chapters or merely mentioned in the text are not indexed. Entries are arranged in letter-by-letter alphabetical sequence.

Abbreviations, 180, 184-185
Abstracting and indexing services
   abstracting services, 293-295
   aviation, aerospace, and meteorology, 323
   biology and agriculture, 329-331
   characteristics, 297-304
   chemistry, chemical engineering, and technology, 318-319
   civil engineering, 324
   computerization, 343-350
   computer science, 315
   coverage, 284, 299-301
   distribution, 309-310
   earth sciences, geology, mining, and metallurgy, 321-322
   electrical engineering and electronics, 326
   energy, 338
   environment, 334-336
   foreign literature, 266-267
   functions, 297, 298-299
   general science and engineering, 312-313
   indexes to abstracts, 301-304
   indexing services, 295-297
   lists, 296-297
   mathematics and statistics, 314
   mechanical engineering, 325
   physics and astronomy, 316
   publishing, 342
   scope, 297-299
   slant in abstracting, 300
Accounts of Chemical Research, 249
Acronyms, 180, 184-185
Acronyms, Initialisms and Abbreviations Dictionary, 180
Advances in Enzymology, 256
Aerospace *see* Aviation and aerospace
Aerospace Structural Metals Handbook, 209
AGARD, 128-129
Agriculture *see* Biology and agriculture
AGRIS, 352
Alternatives to the journal, 47-51
American Chemical Society, 44, 45, 47, 342
American Doctoral Dissertations, 82

401

American Ephemeris and Nautical Almanac, 211
American Institute of Aeronautics and Astronautics, 122
American Institute of Physics, 27, 268, 342, 351
American Medical Bibliography, 289
American Men and Women of Science, 167, 175
Analytical Chemistry, 36, 37, 253
Annual Reports of the Progress of Chemistry, 253
Annual Review of Biochemistry, 248, 253, 256
Annual Review of Information Science and Technology, 244, 245, 248, 256
Annual Review of Psychology, 245
ANSI, 115, 145-146
Applied Science and Technology Index, 297-298, 301, 303
Architecture *see* Civil engineering and architecture
ASCE, 28, 59
ASEE, 305
ASLIB, 272
ASLIB Book List, 286
ASME, 11-12, 35, 60, 137-138, 211
ASME Boiler and Pressure Vessel Code, 137-138
ASME Steam Tables, 211
ASTM, 138-139
Astronomical atlases, almanacs, and catalogs, 211, 216-219
Astronomy *see* Physics and astronomy
Atomindex, 124, 127, 301, 351-352
Attitude of scientists and engineers, 11-13
Autobiographies, 171-172
Automobile engineering, 240
Auxiliary publication, 44

Aviation and aerospace
  abstracting and indexing services, 323
  bibliographies, 322
  dictionaries, 189
  directories and yearbooks, 203-204
  encyclopedias, 239-240
  guides to literature, 322
  handbooks and tables, 222-224
  reviews, 260-261
  translations, 279-280

Beilstein's *Handbook der Organischen Chemie,* 209
Bibliographic control, 283-340
  abstracting and indexing services, 293-304
  bibliographies, 285-293
  biographical literature, 172-173
  centralization, 309-310
  computerization, 341-350
  conference literature, 69-75
  decentralization, 308-310
  dictionaries and thesauri, 181-183
  directories and yearbooks, 196
  dissertations and theses, 82-85
  guides to literature, 304-307
  house journals, 153-154
  indexing services, 295-304
  international cooperation, 350-353
  journals, 55-61
  patents, 93-97
  proliferation of literature, 283-285
  review literature, 255-256
  standards, 349-350
  technical reports, 116-129
  translations, 270-274
Bibliographic Index, 293
Bibliographies and catalogs, 285-293
  aviation, aerospace, and meteorology, 322

Subject Index

biobibliographies, 291
biology and agriculture, 326-329
books, 287-291
chemistry, chemical engineering and technology, 318
civil engineering, 324
computer science, 314-315
current bibliographies, 285-288
earth sciences, geology, mining, and metallurgy, 320-321
energy, 336-338
environment, 332-334
general science and engineering, 311-312
in reviews, 248-249
mathematics and statistics, 313
mechanical engineering, 325
of reviews, 255-256
physics and astronomy, 316
retrospective bibliographies, 286-292
special subject bibliographies, 291-292
trade bibliographies, 287
Bibliographies of bibliographies, 292-293, 311, 313
Bibliography of Aeronautics, 291-292
Bibliography of Birds, 291
Biobibliographies, 291
Biographical Dictionaries and Related Works, 172, 173
Biographical Dictionaries Master Index, 173
Biographical literature, 166-178
autobiographies, 171-172
bibliographic control, 172-173
biographical monographs, 171-172
biographical serials, 168-170
collective biographies, 170-171
Festschrift volumes, 172
general biographical works, 166-167, 174-175

indexes to biographies, 173-174
list of biographical works, 173-177
need for biographies, 166
specialized biographical works, 167-168, 175-177
Biographical Memoirs of the Fellows of the Royal Society, 168-170, 176
Biographical Memoirs of the National Academy of Sciences, 168-170, 176
Biographical Notes Upon Botanists, 168, 176
Biography Index, 173
Biological Abstracts, 267, 300, 304
Biology and agriculture
abstracting and indexing services, 329-331
bibliographies and catalogs, 326-329
dictionaries, 190-191
directories and yearbooks, 205
encyclopedias, 240
guides to literature, 326
handbooks and tables, 226-227
reviews, 262
translations, 281
BLLD Announcement Bulletin, 272
Book lists, 286-288
Book of ASTM Standards, 139
Book reviews, 286-287
Books
bibliographies, 287-291
catalogs, 290-291
proliferation, 284
British journals, 36
British Library Lending Division, 126, 272
British National Bibliography, 71-72
British patents, 92

British Standards Institution, 115,
    134, 144-145, 147
British Technology Index, 297-298
Bulletin Signaletique, 298
Buyers' guides, 195

Catalogs *see* Bibliographies and
    catalogs
Catalogue of Scientific Papers, 296
Center for Research Libraries, 81
Centralized bibliographic control,
    309-310
Chemical Abstracts
    arrangement of abstracts, 302
    collective indexes, 303, 345
    computerization, 344-345
    coverage of bibliographies, 293
    coverage of patents, 94-96
    coverage of reviews, 256
    foreign literature, 266
    growth, 294-295, 298
    indexes, 302-303, 344-345
    overlap in coverage, 300
    patent concordance, 95-96
Chemical Abstracts Service, 344-345,
    350-351
*Chemical Abstracts Service Source
    Index*, 58
Chemical engineering *see* Chemistry,
    chemical engineering and tech-
    nology
Chemical Engineering Catalog, 163
Chemical Engineers' Handbook, 208
Chemical Reviews, 256
Chemical Synonyms and Trade Names,
    165
*Chemical Technology: An Encyclo-
    pedic Treatment*, 233
Chemical Titles, 298
Chemistry, chemical engineering and
    technology

abstracting and indexing services,
    318-319
bibliographies and catalogs, 318
dictionaries, 188
directories and yearbooks, 203
encyclopedias, 238-239
guides to literature, 317-318
handbooks and tables, 219-221
reviews, 258-259
translations, 277-278
Chemistry of Carbon Compounds,
    254
Chinese scientific literature, 267
Civil engineering and architecture
    abstracting and indexing services,
        324
    bibliographies, 324
    dictionaries, 189-190
    directories and yearbooks, 204
    guides to literature, 323-324
    handbooks and tables, 224
    translations, 280
Classified reports, 109-110
Climatological data, 211, 222-224
CODATA, 214
Codes of practice, 135
Collective biographies, 170-171
COM, 43
Commonwealth Agricultural Bureaux,
    300
Commonwealth Index of Unpublished
    Translations, 272
Compaction of literature, 5, 8, 250,
    251
Company standards, 137
Composite Index of CRC Handbooks,
    209
Comprehensive Dissertation Index,
    83-84
Compression ratio, 251

Subject Index

Computerized bibliographic control, 341-350
Computerized information systems, 201-202, 341-350
Computer-Readable Databases: A Directory and Data Source Book, 346
Computer science
  abstracting and indexing services, 315
  bibliographies, 314-315
  dictionaries, 187
  directories and yearbooks, 202
  encyclopedias, 237
  guides to literature, 314
  handbooks and tables, 216
  reviews, 257
  translations, 275
Conference literature, 6, 66-76
  bibliographic control, 69-75
  languages, 70-71
  post-conference literature, 69
  preconference literature, 68-69
  review articles, 253-254
Conference Papers Index, 73
Conferences, 66-68
  announcements, 68-69, 73
  functions, 67
  literature, 68-75
  major types, 67-68
  organization, 74-75
  UNESCO-FID recommendations, 74-75
Copyright, 88-89
Corresponding patents, 95-96
COSATI subject headings, 177, 270, 272
COSATI Task Group on Technical Reports, 107, 109
CRC Handbook of Chemistry and Physics, 208
CRC Handbook of Food Additives, 209
CRC Handbook of Tables for Mathematics, 210
Current awareness, 18, 248
Current Bibliographic Directory of the Arts and Sciences, 167-169, 174
Current bibliographies, 258-288
Current Biography, 170, 174
Current Biography Yearbook, 170, 174
Current Research Information System, 87

Databases
  advantages, 347
  directories and yearbooks, 201-202, 346-347
  list of databases, 353-359
  online access, 346-349
  standards, 349-350
  vendors, 347-349
Decentralized bibliographic control, 308-310
Declassified AD Documents Index, 110
Defense Technical Information Center, 119-120
Department of Defense Index of Specifications and Standards, 140
Department of Defense Single Stock Point, 140-141
Dictionaries and thesauri, 179-192
  abbreviations and acronyms, 180, 184-185
  aviation and aerospace, 189
  bibliographies, 181, 183
  biology, 190-191
  chemistry, chemical engineering and technology, 188

civil engineering, 189-190
electrical engineering and electronics, 190
encyclopedic dictionaries, 234-235
energy, 192
environment, 191-192
eponyms, 185
general science and technology, 183
geology, mining and metallurgy, 188-189
mathematics, 186-187
mechanical engineering, 190
physics and astronomy, 187
signs and symbols, 179-180, 186
tradenames, 185
Dictionaries, Encyclopedias and Other Word-Related Works, 183
Dictionary of Report Series Codes, 110, 112-114
Dictionary of Scientific Biography, 167, 176
Dimensional standards, 135
Directories and yearbooks, 193-207
aviation and aerospace, 203-204
bibliographies, 196-197
biology and agriculture, 205
chemistry, chemical engineering and technology, 203
civil engineering and architecture, 204
computerized information systems, 201-202
computer science, 202
databases, 201-202
electrical engineering and electronics, 204-205
energy, 206-207
environment, 205-206
fisheries, 205

general directories and yearbooks, 197-199
information systems, 201-202
international organizations, 194-195, 199-201
mechanical engineering, 204
mining and metallurgy, 203
physics and astronomy, 202
Directory Information Service, 196
Directory of Engineering Document Sources, 114
Directory of Federally Supported Information Analysis Centers, 308
Directory of Online Databases, 347
Directory of Physics and Astronomy Staff Members, 193
Directory of Selected Scientific Institutions of Mainland China, 194
Directory of Technical and Scientific Translators and Services, 273
Dissertation Abstracts International, 82-83
Dissertations
acquisition, 81-83
foreign dissertations, 80-82
historical overview, 77-79
impact on science, 79
Dissertations and theses, 77-87
Dissertations in Physics, 84-85
Documentation standards, 136

Earth sciences, geology, mining, and metallurgy
abstracting and indexing services, 321-322
bibliographies and catalogs, 320-321
dictionaries, 188-189
directories and yearbooks, 203

Subject Index

encyclopedias, 239
guides to literature, 319-320
handbooks and tables, 221-222
reviews, 259-260
translations, 278-279
Editerra, 249, 251-252
Editorial processing center, 52-53
Electrical engineering and electronics
  abstracting and indexing services, 326
  dictionaries, 190
  directories and yearbooks, 204-205
  guides to literature, 325-326
  reviews, 261
  translations, 280-281
Electronic journal, 53-55
Electronics *see* Electrical engineering and electronics
Elsevier's Encyclopedia of Organic Chemistry, 234
Encyclopedia of Associations, 194, 235-236
Encyclopedia of Chemical Elements, 233
Encyclopedia of Chemical Reactions, 233
Encyclopedia of Chemistry, 232
Encyclopedia of Industrial Chemical Analysis, 234
Encyclopedias, 231-241
  automobile engineering, 240
  aviation and aerospace, 239-240
  biology, 240
  chemistry, chemical engineering and technology, 238-239
  computer science, 237
  earth sciences, geology, mining, and metallurgy, 239
  encyclopedic dictionaries, 234-235
  energy, 241
  environment, 241
  general science and engineering, 236

  mathematics and statistics, 236-237
  multi-volume encyclopedias, 233-234
  physics and astronomy, 237-238
  single-volume encyclopedias, 232-233
  specialized encyclopedias, 231
  updating, 235-236
Encyclopedic dictionaries, 234-235
Encyclopedic Dictionary of Physics, 235
Endeavour, 153
Energy
  abstracting and indexing services, 338
  bibliographies, 336-338
  dictionaries, 192
  directories and yearbooks, 206
  encyclopedias, 241
  guides to literature, 336
  handbooks and tables, 229
  reviews, 263
  translations, 282
Energy Research Abstracts, 124
Engineer Index, 60
Engineering Index, 267, 300, 303
Engineers *see* Scientists and engineers
Environment
  abstracting and indexing services, 334-336
  bibliographies, 332-334
  dictionaries, 191-192
  directories and yearbooks, 205-206
  encyclopedias, 241
  guides to literature, 331-332
  handbooks and tables, 227-229
  reviews, 262-263
  translations, 282
Environmental Protection Agency, 125-126
Eponyms dictionaries, 185

European Atomic Energy Community, 128
European Space Agency, 129
Evaluation of literature, 247, 250
Everyday approach, 18
Exhaustive approach, 18
External house journals, 150-153

Festschrift volumes, 172
FID, 74-75
Fisheries yearbooks, 205
Foreign and international organizations, 194-195, 199-201
Foreign literature, 266-282
French scientific literature, 11, 266-267
Fundamental standards, 136

General science and engineering
  abstracting and indexing services, 312-313
  bibliographies and catalogs, 311-312
  bibliographies of bibliographies, 311
  dictionaries, 183
  directories and yearbooks, 197-201
  encyclopedias, 236
  guides to literature, 310-311
  handbooks and tables, 214-215
Geology *see* Earth sciences, geology, mining, and metallurgy
German scientific literature, 266-267
Gmelin's *Handbuch der Anorganischen Chemie*, 209
Government publications, 116-117
Government Reports Announcements and Index, 102-104, 117-118, 271, 294
Government standards and specifications, 139-143

Gray Herbarium Index, 290
Guides to literature, 304-307
  aviation, aerospace, and meteorology, 322
  biology and agriculture, 326
  chemistry, chemical engineering and technology, 317-318
  civil engineering, 323-324
  computer science, 314
  earth sciences, geology, mining and metallurgy, 319-320
  electrical engineering and electronics, 325-326
  energy, 336
  environment, 331-332
  general science and engineering, 310-311
  mathematics and statistics, 313
  mechanical engineering, 324
  physics and astronomy, 315-316
Guide to American Scientific and Technical Directories, 196
Guide to American Trade Catalogs, 156
Guide to Mathematical Tables, 211
Guide to Reference Books, 305
Guide to Scientific and Technical Journals in Translation, 268
Guide to Special Issues and Indexes of Periodicals, 61, 161, 195

Handbooks and tables, 11, 208-230
  astronomical data, 211, 216-219
  aviation, aerospace, and meterology, 211, 222-224
  biology and agriculture, 226-227
  chemistry, chemical engineering and technology, 219-221
  civil engineering, 224
  climatological data, 211, 222-224
  computer science, 216

Subject Index

earth sciences, geology, mining, and metallurgy, 221-222
electrical engineering and electronics, 226
energy, 229
environment, 227-229
general science and engineering, 214-215
mathematics and statistics, 210-211, 215-216
mechanical engineering, 224-226
meteorology, 211, 222-224
physics and astronomy, 216-219
spectra, 219
Harper Encyclopedia of Science, 232
House journals, 149-154
bibliographic control, 153-154
external house journals, 150-153
historical overview, 149-150
internal house journals, 150
popular house journals, 151
technical house journals, 151-153

IEEE, 10, 28, 45, 50, 351
IEEE Annals, 50
Indexes to abstracts, 301-304
Indexes to biographies, 173-174
Indexing services *see* Abstracting and indexing services
Index Medicus, 267, 343-344
Index to All Books on the Physical Sciences in English, 290
Index to Book Reviews in the Sciences, 287
Index to Mathematical Tables, 210-211
Index to Scientific and Technical Proceedings, 73
Index to Scientific Reviews, 256
Index to Scientists of the World, 172, 173-174

Industrial Research Laboratories of the United States, 194
Information analysis centers, 307-308
Information Exchange Groups, 15-16
Information needs, 10-13, 17-18
Information systems, 201-202
INIS, 124, 127, 351-352
INSPEC, 299
Integrated journal publishing, 51-52, 342
Integration of literature, 8, 247, 250
Integrative journals, 51
Internal house journals, 150
International Aerospace Abstracts, 122
International Atomic Energy Agency, 127, 351-352
International Bibliography of Directories, 196
International Bibliography of Specialized Directories, 183
International Catalogue of Scientific Literature, 291, 296
International Conference on Scientific Information, 16, 46, 49, 244, 248
International cooperation in bibliographic control, 350-353
International Critical Tables, 210
International Directory of Translators and Interpreters, 273
International Electrotechnical Commission, 147
International Nuclear Information System, 124, 127, 351-352
International organizations, 194-195, 199-201
International Research Communication System, 51-52
International standards agencies, 146-147

International Standards Organization, 147
International Translations Center, 271-272
Invisible Colleges, 13-15
Irregular Serials and Annuals, 56

James and James Mathematical Dictionary, 179-180
Japanese scientific literature, 266-267
Japan Society for International Chemical Information, 350-351
John Crerar Library, 290
Joint Publications Research Service, 271
Journal des Sçavans, 31
Journal of Organic Chemistry, 36, 39-40, 47, 342
Journal of the American Chemical Society, 45, 47, 249

Kirk-Othmer Encyclopedia of Chemical Technology, 231, 234

Lab Guide, 161-162
Landolt-Börnstein, 208
Letters journals, 5-6, 45-46
Library pathfinders, 307
Literature searching, 16-18

Magazines for Libraries, 56
Management Information Guides, 305
Manufacturers' catalogs, 162
Materials science *see* Earth sciences, geology, mining, and metallurgy
Material standards, 135
Mathematical Reviews, 267
Mathematical tables, 210-211, 215-216
Mathematics and statistics
 abstracting and indexing services, 314
 bibliographies, 313
 bibliographies of bibliographies, 313
 dictionaries, 186-187
 directories and yearbooks, 202
 encyclopedias, 236-237
 guides to literature, 313
 handbooks and tables, 210-211, 215-216
 reviews, 257
 translations, 274-275
McGraw-Hill Basic Bibliography of Science and Technology, 233, 287
McGraw-Hill Dictionary of Scientific and Technical Terms, 179
McGraw-Hill Encyclopedia of Science and Technology, 233, 235, 287
McGraw-Hill Encyclopedia of World Biography, 167, 174
McGraw-Hill Yearbook of Science and Technology, 195
Mechanical engineering
 abstracting and indexing services, 325
 bibliographies and catalogs, 325
 dictionaries, 190
 directories and yearbooks, 204
 encyclopedias, 240
 guides to literature, 324
 handbooks and tables, 224-225
 reviews, 261
 translations, 280
Mechanical Engineering: The Sources of Information, 304
Metallurgy *see* Earth sciences, geology, mining, and metallurgy
Metals Abstracts, 301, 351
Meteorological data, 211, 222-224
Meteorology *see* Aviation and aerospace
Microform journals, 44

Subject Index

MIL Specifications, 139-140
Mining *see* Earth sciences, geology, mining, and metallurgy
Modern Science and Technology, 232

NASA, 120-123, 128
NASA Tech Briefs, 121
National Agricultural Library, 87
National Bureau of Standards, 141-143
National Federation of Abstracting and Indexing Services, 297, 309
National Institutes of Health, 15-16
National Standard Reference Data System, 143, 212-214
National standards organizations, 143-146
National Translations Center, 270-271
NATIS, 353
New Serial Titles, 58
Nonformal communication, 13-15
NTIS, 102, 117-119, 271
Nuclear Science Abstracts, 123-124, 301, 310

Oceanography *see* Earth sciences, geology, mining, and metallurgy
Office of Technical Services, 102
Official Gazette, 93-94, 97, 304
Online searching, 309, 346-349

Page charges, 42-43
Pandex: Current Index to Scientific and Technical Literature, 298
Patentability of inventions, 88
Patent concordance, 95-96
Patent Office, 89-91, 97-98
Patents, 6, 88-99
  bibliographic control, 93-96
  British patents, 92
  classification system, 94
  coverage in *Chemical Abstracts*, 94-96
  depositories, 98-99
  disclosure of inventions, 92-93
  foreign patents, 91-92, 94-96
  historical overview, 89-91
  numbering system, 90-91
  proliferation, 284
Pathfinders, 307
Performance standards, 135
Philosophical Transactions, 31
Physical Review, 38
Physics Abstracts, 267, 293, 299, 302
Physics and astronomy
  abstracting and indexing services, 316
  bibliographies and catalogs, 316
  dictionaries, 187
  directories and yearbooks, 202
  encyclopedias, 237-238
  guides to literature, 315-316
  handbooks and tables, 216-219
  journals, 43
  reviews, 257-258
  translations, 275-277
Physics Briefs, 351
Popular house journals, 151
Preliminary communication, 5-6, 45-46
Preprint distribution, 48-51
Primary journals, 30-65, 253
  advance announcement of contents, 44-45
  alternatives, 47-51
  auxiliary publication, 44
  bibliographic control, 55-61
  British journals, 36
  computerized publishing, 43, 51-55
  costs of production, 40-43
  cumulative indexes, 59-61, 256

delay in publication, 38-39
directories, 55-57
editorial processing center, 52-53
electronic journal, 53-55
functions, 32-33, 39-40
future prospects, 51-55
historical overview, 30-31
integrated publishing, 51-52, 342
integrative journals, 51
letters journals, 5-6, 45-46
microform publication, 44
page charges, 42-43
proliferation, 33-36, 283-285
publishing, 43-47, 51-55, 342
scattering, 36-37
special issues, 61, 161, 195
subscription rates, 40-43, 269
supplements, 61, 161, 195
synopsis journal, 46-47
translated journals, 268-269, 274-282
union lists, 57-59
Primary literature
compaction, 5, 8, 250, 251
evaluation, 247, 250
integration, 8, 247, 250
proliferation, 247, 283-285
publishing, 342
types, 4-7
Primary sources, 4-7
Professional societies, 21-29, 137-139, 194-195, 199-201
Publications Board, 101-102
Pure and Applied Chemistry, 254

Quality standards, 136-137

Radiation Shielding Information Center, 308
RAND Corporation, 126
Random number tables, 210

Readers Advisory Service: Selected Topical Booklists, 307
Reader service cards, 159-161
Referativnyi Zhurnal, 298-299, 310
Reference books, 286-288, 306-307
Reference Services Reviews, 288
Reference Sources, 288
Reports *see* Technical reports
Reports on Progress in Physics, 256
Research Centers Directory, 194
Research in progress, 85-87
Research paper, 6-7
Retrospective bibliographies, 287-292
Retrospective Index to Theses of Great Britain and Ireland, 84
Review literature, 242-265
authorship, 244-247
aviation and aerospace, 260-261
bibliographic control, 255-256
biology, 262
characteristics, 250-252
chemistry, chemical engineering and technology, 258-259
computer science, 257
conference papers, 253-254
earth sciences, geology, mining, and metallurgy, 259-260
electrical engineering and electronics, 261
energy, 263
environment, 262-263
functions, 242-244, 247-249
guides to reviews, 255-256
historical overview, 242-244
indexes, 256
mathematics and statistics, 257
mechanical engineering, 261
monographs, 254
output of reviews, 255
physics and astronomy, 257-258
preparation, 244-247

Subject Index

primary journals, 253
serials, 252-253
sources of reviews, 251-255
technical reports, 254-255
types of reviews, 251-252
use of reviews, 249, 251
users of reviews, 249-250, 253
Royal Society, 22-23, 30-31, 295-296
Royal Society Scientific Information Conference, 46, 48-51, 243, 246, 302
Russian scientific literature, 266-268

SATCOM Report, 42-43, 104, 105, 244, 247
Scattering of literature, 36-37
Science and technology *see* General science and engineering
Science Citation Index, 256
Science Tracer Bullets, 287, 307
Scientific American, 256
Scientific and Technical Aerospace Reports, 122
Scientific and technical communication, 1-20, 284
Scientific conferences *see* Conferences
Scientific Institutions of Latin America, 194
Scientific literature
characteristics, 2-4, 10-13
primary sources, 4-7
scattering, 36-37
secondary sources, 7-8
structure, 4-10
use patterns, 10-13, 16-18
tertiary sources, 8-10
Scientific, Medical and Technical Books, 289
Scientific Research in British Universities and Colleges, 85

Scientific societies, 21-29, 137-139, 194-195, 199-201
Scientists and engineers
attitudes, 11-13
information needs, 10-13, 17-18
population, 285
specialization, 285
SDI, 119, 122, 163-164
Secondary services *see* Abstracting and indexing services
Secondary sources, 7-8
Security classification, 109-110
Select Bibliography of Chemistry, 1492-1897, 291
Serials
bibliographic control, 55-61
cumulative indexes, 59-61, 256
directories, 55-57
review serials, 252-253, 256, 257-263
union lists, 57-59
see also Primary journals
Signs and symbols
dictionaries, 179-180, 186
standards, 136
Slanted abstracts, 300
Smithsonian Science Information Exchange, 85-87
Societies *see* Scientific societies
Sources of Information on the Rubber, Plastics and Allied Industries, 305
Special issues of journals, 61, 161, 195
Specialization, 285
Specifications, 134-135
Spectra, 219
Standard Periodicals Directory, 56
Standards and specifications, 132-148
databases, 349-350
graphic symbols, 136

historical overview, 132-134
  purposes, 132-133
  quality standards, 136-137
  sources, 137-147
  terminology standards, 136
  test methods, 135
  types of standards, 135-137
  uniformity, 136-137
Standards and Specifications Information Sources, 138
STAR, 122
Statistical tables, 215-216
Statistics *see* Mathematics and statistics
Structure of literature, 4-10
Symbols *see* Signs and symbols
Synopsis journals, 46-47

Tables *see* Handbooks and tables
Talanta, 248, 253
Technical Abstracts Bulletin, 120
Technical Book Review Index, 286-287
Technical house journals, 151-153
Technical literature, 10-13, *see also* Scientific literature
Technical reports, 6, 100-131
  bibliographic control, 116-129
  characteristics, 104-109
  contents, 105-106
  COSATI Task Group, 107, 109
  definition, 100
  historical overview, 100-104
  proliferation, 284
  quality, 104-105
  report numbers, 110-115
  republication, 107-108
  reviews, 244, 254-255
  security classification, 109-110
  status, 104, 106-109
Technological gatekeepers, 13-15

Technology Utilization Reports, 121
Technology Utilization Surveys, 122
Telecommunication networks, 346
Terminology standards, 136
Tertiary literature, 8-10, *see also* Bibliographies of bibliographies, Guides to literature
Test methods, 135
Thesauri, 181, *see also* Dictionaries and thesauri
Thesaurus of Engineering and Scientific Terms, 181-182
Theses, 79-85
  bibliographic control, 82-85
  bibliographies, 79-80
  foreign theses, 80-82
  see also Dissertations and theses
Thomas Register of American Manufacturers, 163
Trade bibliographies, 287
Trade catalogs, 11, 155-165
  acquisition, 164
  advertisements and announcements, 159-161
  buyers' guides, 195
  characteristics, 157-159
  collections, 156, 164-165
  directories, 162-163
  historical overview, 155-156
  manufacturers' catalogs, 162
  purposes, 155-157
  reader service cards, 159-161
  SDI, 163-164
  special issues of journals, 161-162
  standards, 158-159
  trade catalog services, 163-164
  trade fair catalogs, 162
  types, 159-164
Trademark, 88-89

Subject Index

Tradenames, 165, 185
Trade Names Dictionary, 165
Transactions of the ASME, 35
Transdex, 271
Translation and Translators: An International Directory and Guide, 273
Translations, 266-282
  aviation and aerospace, 279
  bibliographic control, 270-274
  biology, 281
  chemistry, chemical engineering and technology, 277-278
  civil engineering, 280
  computer science and automation, 275
  directories, 272-273
  duplication, 273
  earth sciences, geology, and metallurgy, 278-279
  electrical engineering and electronics, 280-281
  energy, 282
  environment, 282
  indexes, 273-274
  International Translations Center, 271-272
  mathematics and statistics, 274-275
  National Translations Center, 270-271
  physics and astronomy, 275-277
  translated journals, 268-269
Translations Register Index, 270
Translators and Translations: Services and Sources, 273
Treaties on Analytical Chemistry, 254
Tutorial reviews, 248

U.K. Atomic Energy Authority, 126-127
Ulrich's International Periodicals Directory, 55-56
Ulrich's Quarterly, 56
UNESCO, 74-75, 352-353
Uniformity standards, 136-137
Union List of Serials in Libraries of the United States and Canada, 58
Union lists of serials, 57-59
UNISIST, 352-353
Unpublished documents, 4-5
U.S. Department of Energy, 123-124
U.S. Government Printing Office, 116-117
U.S. Patent and Trademark Office, 89-91, 97-98
U.S. Patent classification system, 94
U.S. Standard Atmosphere, 211

Van Nostrand's Scientific Encyclopedia, 231, 232, 235
Visual Search Microfilm File, 143, 163

Weather Almanac, 211
Weinberg Panel, 105-107, 244, 246, 247
Who Is Publishing in Science, 168
Who's Who in Soviet Science and Technology, 168, 177
World Aviation Directory, 162-163, 195
World Bibliography of Bibliographies, 292
World List of Scientific Periodicals, 34
World Transindex, 271-272
World Who's Who in Science, 166-167, 177
Worldwide Directory of Computer Companies, 195

Yearbook of Agriculture, 195
Yearbook of Astronomy, 195

Yearbooks, 195-207, *see also* Directories and yearbooks